ALUMINIUM ALLOYS – RECENT TRENDS IN PROCESSING, CHARACTERIZATION, MECHANICAL BEHAVIOR AND APPLICATIONS

Edited by **Subbarayan Sivasankaran**

Aluminium Alloys – Recent Trends in Processing, Characterization, Mechanical behavior and Applications
http://dx.doi.org/10.5772/68032
Edited by Subbarayan Sivasankaran

Contributors

Tomasz Arkadiusz Tański, Przemysław Snopiński, Omotayo Sanni, Patricia Popoola, Olawale Fatoba, Esther Akinlabi, Elizabeth Makhatha, Marcela Pokusova, Rajkumar Vijayakumar, Venkatesh Kannan M, Arivazhagan Natarajan, Helmut Klocker, Christopher Yukna, Williams S. Ebhota, Tien-Chien Jen, Fawzy Samuel, Herbert Doty, Salvador Valtierra, Heng Li, Chao Lei, He Yang, Chia-Wei Lin, Fei-Yi Hung, Truan-Sheng Lui, Olawale Samuel Fatoba, Ildiko Peter, Béla Varga, Nelson Daniel Vejar, Alejandra Orrego, Mamie Sancy, Maritza Paez, Ojo Sunday Fayomi, Api Popoola, Udoye N.E, Aa Abioye, Op Abioye

Published by InTech

Janeza Trdine 9, 51000 Rijeka, Croatia

Publishing Process Manager Dajana Pemac
Technical Editor SPi Global
Cover InTech Design team

First published December, 2017
Printed in Croatia
Legal deposit, Croatia: National and University Library in Zagreb

Additional hard copies can be obtained from orders@intechopen.com

Aluminium Alloys – Recent Trends in Processing, Characterization, Mechanical behavior and Applications, Edited by Subbarayan Sivasankaran
p. cm.
Print ISBN 978-953-51-3697-2
Online ISBN 978-953-51-3698-9

For EU product safety concerns:
IN TECH d.o.o., Prolaz Marije Krucifikse Kozulić 3, 51000 Rijeka, Croatia,
info@intechopen.com or visit our website at intechopen.com.

Meet the editor

Dr. S. Sivasankaran is a faculty member in the Department of Mechanical Engineering, College of Engineering, Qassim University. He has completed his PhD degree in Production Engineering from National Institute of Technology, Tiruchirappalli, India. He has passed his bachelor degree of Mechanical Engineering in First Class with Distinction under the University of Madras, India. He did his M. Tech in Manufacturing Technology and passed in First Class with Distinction from Department of Production Engineering, National Institute of Technology, Tiruchirappalli, India. After his doctoral degree, he has worked as an associate professor in KPR Institute of Engineering and Technology (KPRIET), Coimbatore, India. He has established advanced materials research laboratory and entrepreneurship development cell in KPRIET. He has received many research and seminar grants from the government of India. Further, he has served as an associate professor in Hawassa University, East Africa where he has developed BSc, Industrial Engineering; MSc, Industrial Engineering and Management; and MSc, Manufacturing Technology curricula. He has published more than 50 research articles in refereed international journals and is acting as a reviewer for more than ten refereed international journals. He has organized more than ten national- and international-level workshops. Finally, his research interests are nanocomposites, mechanical alloying, powder metallurgy, stir casting, advanced welding, 3D printing, metal matrix composites, polymer matrix composites, mechanical behaviour, machining behaviour, advanced characterization techniques, optimization and modelling.

Contents

Preface

In the twenty-first century, aluminium-based lightweight materials are being used for many commercial applications, which save both the energy and cost. The major features are lighter in weight, considerable strength, high value of corrosion resistance and considerable ductility. Most of the products made from aluminium alloys are from their wrought products that completely replace the usage of cast iron and steel as these materials would enhance the weight of the products (both cast iron and steel). Moreover, aluminium alloys can be recyclable, which are available more in the world. The present book has several sections that address the real research works and review works contributed by several researchers in the world that cover the latest processing routes of aluminium alloys including composites through casting, forming, severe plastic deformation, various grain refinement techniques especially from equal channel angular pressing, friction stir welding and corrosion behaviour, effect of alloying elements on the properties improvement, various characterization techniques and rapid solidifications. Further, detailed mechanical behaviour of aluminium alloys is also being incorporated in the present book. This book will help all academicians, the research community and industries for selection of aluminium-based materials, its processing routes, and its appropriate applications.

Dr. S. Sivasankaran
College of Engineering,
Qassim University,
Saudi Arabia

Chapter 1

Effects of Grain Refining on Columnar-to-Equiaxed Transition in Aluminum Alloys

Hicham Tahiri, Serageldin S. Mohamed,
Herbert W. Doty, Salvador Valtierra and
Fawzy H. Samuel

Additional information is available at the end of the chapter

http://dx.doi.org/10.5772/intechopen.70346

Abstract

The effects of grain refining in ultra-pure aluminum, commercially pure aluminum (1050), and Al-7%Si binary alloy were investigated, using different additions of Al-10%Ti, Al-5%Ti-1%B, and Al-4%B master alloys. Thermal analysis and metallography were used to assess the variations in microstructure resulting from these additions, at solidification rates of 0.8°C/s and ~10°C/s. The results revealed that addition of Al-4%B to ultra-pure aluminum forms AlB_{12} and AlB_2 which have no grain-refining effect. Without grain refiner addition, the pure aluminum microstructure exhibits a mixture of columnar and equiaxed grains. Addition of 30ppm Ti is sufficient to promote equiaxed grains at ~10°C/s but requires addition of 1000 ppm B to obtain similar results at 0.8°C/s. Increasing the Si content to 7% reduces the initial grain size of pure aluminum from 2800 μm to ~1850 μm, and further to 450 μm with ddition of ~500ppm B. In commercial aluminum, the B reacts with traces of Ti forming Al_3Ti and TiB_2 phases which are active grain-refiners. In Al-7%Si, Ti reacts with Si forming $(Al,Si)_2Ti$ phase, which is a poor refining agent. This phenomenon is termed poisoning. No interaction between B and Si is observed in the commercial aluminum or Al-7%Si alloy when B is added.

Keywords: aluminum, grain refining, poisoning, columnar-to-equiaxed transition, solidification rate, macrostructure

1. Introduction

Master alloys of the type Al-B are largely used in production of ultra-pure aluminum to react with transition elements such as V, Cr, and Zr [1]. Boron is not considered as effective a grain refiner when added to pure aluminum [2]. Once Si is added to Al, grain refining is activated together with a change in the α-Al dendritic structure [3]. Two types of B-compounds exist in the Al-B master alloys: AlB_2 and AlB_{12} [4]. The AlB_2 compound is stable at room temperature and contains 44.5% B [5–7]. It is inferred from the Al-B binary diagram presented in **Figure 1(a)** that there is a peritectic reaction at 975°C: L + $AlB_{12} \rightarrow AlB_2$. A eutectic reaction (liquid $\rightarrow$ α-Al + AlB_2) takes place at 660°C, making the maximum solubility of B in Al about 20 ppm where the α-Al grains would precipitate on the AlB_2 particles. The melting points of AlB_2 and AlB_{12} are 1665 ± 50°C and 2163 ± 50°C, respectively, resulting in the formation of solid dispersoid particles in the molten liquid [8]. Fundamentally speaking, the addition of a grain refiner to the molten metal would result in the nucleation of new grains or reaction with other elements in the molten

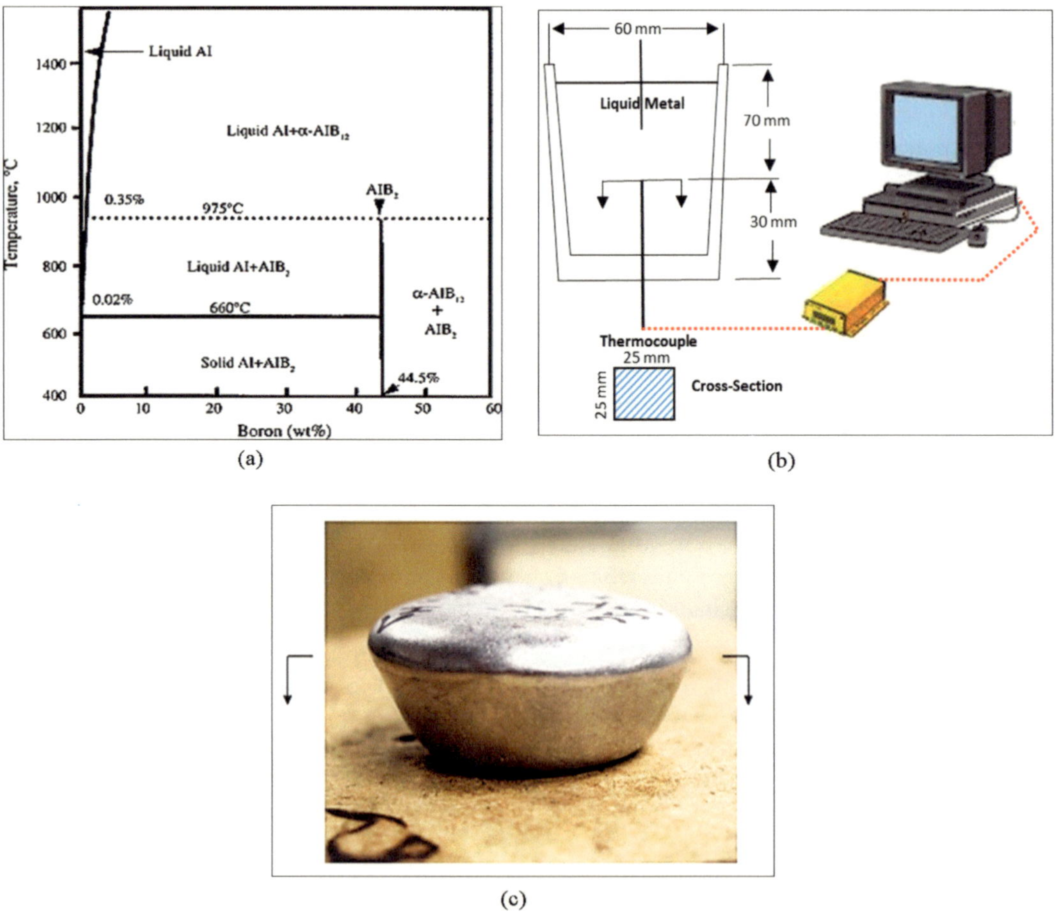

Figure 1. (a) Aluminum rich corner of Al-B phase diagram [7]. Broken arrow points to Al + AlB_2 eutectic reaction. (b) Schematic drawing showing the graphite mold used for thermal analysis. (c) Casting made in steel cups to obtain a solidification rate of 10°C/s.

metal to form nucleation sites. When Al-4%B master alloy is added to commercial aluminum containing traces of Ti, the B atoms would react with Ti forming TiB_2 phase ($Ti + B \rightarrow TiB_2(s)$). The free energy ($\Delta G$) associated with this reaction can be calculated as $\Delta G = -73381 + 38.996T$, where T is the liquid temperature [9]. The enthalpy energy associated with the formation of TiB_2 phase particles is fairly large – 326.41 kJ/mol—which allows for their stability in the molten metal.

In order to clarify the role of Si in activating the grain refiner, several castings were made using pure Al (99.999%), commercially pure aluminum (1050), and a binary Al-7%Si alloy. The study was conducted using thermal analysis, whereas another set of castings for grain refining was carried out using reduced pressure testing steel cups.

2. Experimental procedure

In this study, pure aluminum (99.999%), commercial pure aluminum (1050), and Al-7%Si were used. The compositions of the materials used are shown in **Table 1**. Each base alloy was melted in a resistance-heated clay graphite crucible and held at 750°C. The three grain refiners Al-10%Ti, Al-5%Ti-1%B, and Al-4%B were added to each melt in the amounts (Ti or B): 0.02–0.4% wt.%. Prior to casting, the liquid metal was held at 750°C for 10, 30, 60, 90, and 120 min after the grain refiner addition was made and then cast. During the holding time, the melt was continuously stirred using a graphite impeller to minimize sedimentation of Ti-containing particles and maintain uniformity. Samplings from the different melts prepared at 750°C were poured into a preheated (600°C) cylindrical graphite mold to achieve near-equilibrium solidification conditions (solidification rate ~0.8°C/s), as shown in **Figure 1(b)**. Another set of castings was made using steel cups of the reduced pressure testing machine in air. The steel cups were not preheated in order to obtain a high cooling rate of about 10°C/s, as shown in **Figure 1(c)**—solidification rate is about 10°C/s. Castings were also made using steel cups preheated at 450°C to represent an intermediate solidification rate. Additionally, the microstructure was evaluated at both cooling rates.

The temperature-time data was obtained using a K-type (chromel-alumel) thermocouple inserted through the bottom of the graphite mold along the centerline, its tip reaching up to about one-third the height of the mold from the bottom. The cooling curves were recorded using a data acquisition system attached to the thermocouple. Chemical analysis

Alloy	Al	Si	Cu	Mg	Fe	Mn	Zn	Ti	Sr
Al pure	99.998	0	0	0	0	0	0	0	0
1050	Bal.	0.25	0.05	0.05	0.4	0.04	0.07	<0.05	0
Al-7%Si	Bal.	6.78	0.02	00	0.03	0.04	0.04	<0.03	0

Table 1. Chemical composition (wt.%).

of all melts/castings was determined using a Spectrolab Jr CCD Spark Analyzer (average of three burns per sample).

The solidified castings were sectioned perpendicular to the centerline axis of the cylinder, at the level of the thermocouple tip, and polished for metallographic examination, using standard polishing procedures. The polished samples were examined using an optical microscope and an electron probe micro-analyzer (JEOL JXA-8900L operating at 20 kV), equipped with energy dispersive X-ray spectroscopic (EDS) and wavelength dispersive spectroscopic (WDS) facilities.

3. Results and discussion

Figure 2 shows the temperature-time curves of pure Al and Al-7%Si binary alloy after different melt treatments. **Figure 2(a)** shows that the addition of B to pure Al reduces the undercooling by approximately ~1°C, indicating the relative ineffectiveness of B as a grain refiner in the absence of Si. A 300 ppm level of B added to the Al-7%Si alloy eliminated the phenomenon of recalescence. Increasing the concentration of B to 800 ppm has minimal further effect on the alloy solidification behavior. As shown below, the transformation of AlB_{12} to a simple phase AlB_2 is the principal parameter in the disappearance of the observed undercooling shown in **Figure 2(a)**.

Figure 2(b) demonstrates the variation in aluminum grain size as a function of Sr-B interaction in the presence of 200 ppm Sr as well as holding time. In the absence of melt treatment, the average grain size was about 2900 μm. Once the two agents (B and Sr) were simultaneously added to the melt, the grain size dropped to 420 μm at 0.1%B. The observed reduction in the grain size may be interpreted in part as due to presence of traces of Ti either in the used aluminum or in the added Al-4%B master alloy as shown in **Table 2**. At higher B concentrations, the grain size tended to increase reaching 3300 μm at holding time of 120 min. This observation may be explained in terms of agglomeration of the nucleation sites [10]. As shown in **Table 1**, a eutectic reaction takes place at about 660°C; Liquid → $AlB_2 + \alpha$ (solid). In Si-containing aluminum alloys, i.e., 356 alloy, the melting point is less than the eutectic temperature, which facilitates the formation of AlB_2, an active nucleation agent. The melting temperature of pure Al is about 660.5°C, and hence the reported sluggishness of B as a grain refiner in pure Al [11]. **Figure 2(c)** displays the variation in the grain size of Al-7%Si alloy as a function of holding time when the alloy was grain refined using Al-4%B master alloy in the presence of 200 ppm Sr. Due to high Si content, the initial grain size dropped to approximately 1850 μm. With the addition of approximately 500 ppm B, the grain size was reduced to about 450 μm (~75%). Holding time up to 120 min seems to have a marginal effect on further decrease in the alloy grain size. These findings are in good agreement with the abovementioned discussion.

Figure 3 reveals the possibility of the coexistence of the two B-based compounds in pure aluminum. The composition of these two phases was confirmed using the WDS technique as shown in **Table 3**. **Figure 4** shows the distribution of B- (black), and Si-phases (light gray)

http://dx.doi.org/10.5772/intechopen.70346

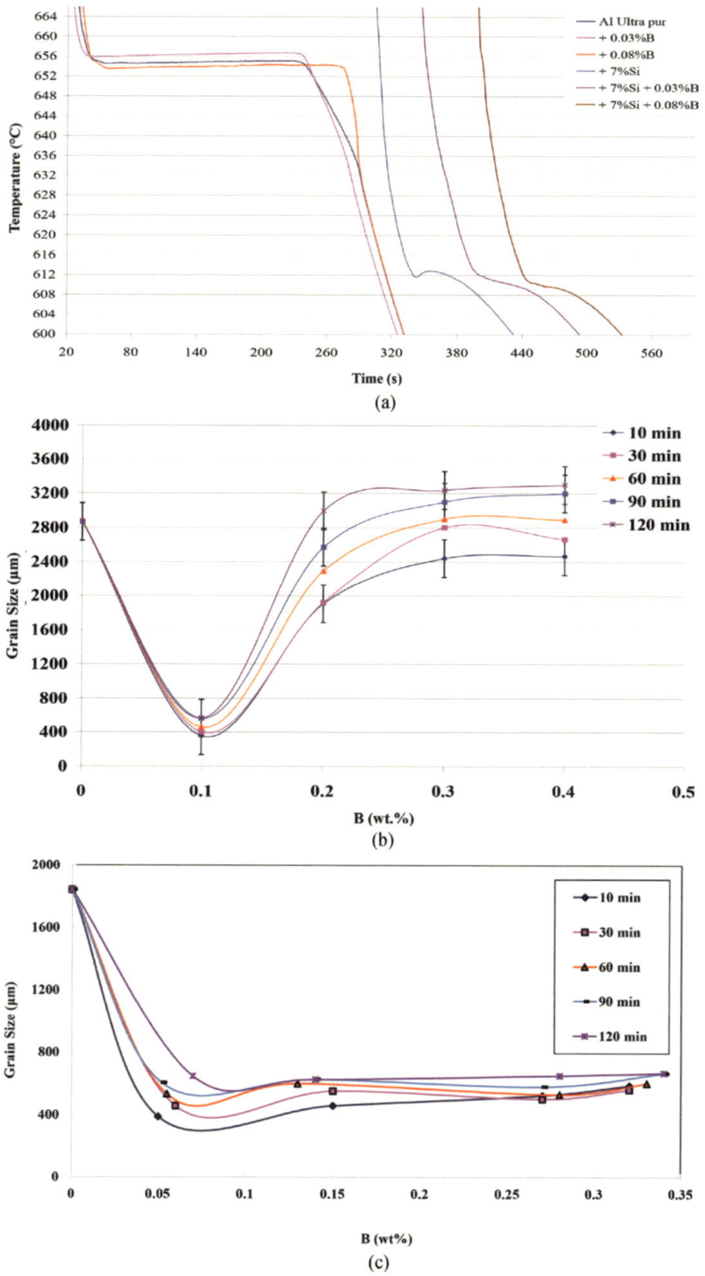

Figure 2. (a) Temperature-time solidification curves obtained from pure Al and Al-7%Si alloy following different melt treatments. (b) Effect of Sr-B interaction on the pure aluminum grain size as a function of holding time. (c) Effect of Sr-B interaction on the grain size of Al-7%Si alloy as a function of holding time.

in Al-7%Si alloy. As can be seen, these elements tend to precipitate away from each other (marked zones A and B in **Figure 4**). This observation is confirmed by the X-ray elemental distribution for each zone shown in **Figure 5**, indicating that there is no affinity for reaction

Pure Al, series Al-4%B							
Holding time (min)	**Trial #**	**wt.% B addition**	**wt.% Si**	**B (ppm)**	**% Ti**	**Added Sr (ppm)**	**Actual Sr (ppm)**
10	1	0.1	0.062	78	0.0403	200	145
	2	0.2	0.062	114	0.0104	200	195
	3	0.3	0.063	206	0.005	200	155
	4	0.4	0.068	189	0.0031	200	125
30	1	0.1	0.062	132	0.0342	200	101
	2	0.2	0.066	203	0.005	200	143
	3	0.3	0.064	147	<0.0013	200	145
	4	0.4	0.077	225	0.0019	200	95
60	1	0.1	0.054	84	0.0401	200	64
	2	0.2	0.063	93	<0.0013	200	106
	3	0.3	0.063	>360	0.0027	200	110
	4	0.4	0.061	104	0.0041	200	61
	5						
90	1	0.1	0.06	48	0.0249	200	51
	2	0.2	0.064	148	<0.0013	200	74
	3	0.3	0.07	165	<0.0013	200	80
	4	0.4	0.057	90	<0.0013	200	49
	5						
120	1	0.1	0.065	153	0.0367	200	58
	2	0.2	0.066	49	<0.0013	200	50
	3	0.3	0.0742	>360	0.0027	200	53
	4	0.4	0.059	121	0.0014	200	66

Table 2. Chemical analysis of melt treated pure aluminum samples.

between these two elements In contrast, when Ti is added with B to the 1050 commercial aluminum, a clear interaction between Ti and B takes place leading to the formation of TiB_2, as demonstrated in **Figure 6**. Thus, B has no grain-refining capacity when added as a single element to ultra-pure aluminum, whereas when the metal contains traces of Ti, adding B leads to the possible formation of different nucleation sites that coexist, such as AlB_{12}, AlB_2, and TiB_2 in Al-Si alloys.

It is well established that increasing solidification rate significantly enhances grain-refining effects [12–14]. In order to illustrate this, samples were poured into steel cups of the reduced pressure testing machine (about 200 g) in air corresponding to a solidification rate of about

Figure 3. (a) Presence of AlB_{12} in ultra-pure aluminum (possible decomposition of AlB_{12} to AlB_2), (b) X-ray image of B distribution in AlB_{12} phase particles in (a)-arrowed.

10°C/s [13, 14]. In the absence of a grain refiner, the macrostructure of pure aluminum is composed of three zones: (i) columnar near the bottom of the crucible due to the development of positive thermal gradient in the liquid metal leading to formation of grains with preferred orientation [14], (ii) fine equiaxed grained zone, and (iii) coarse grained zone where the grains are randomly oriented caused by fragmentation of the columnar grains driven by convection in the liquid metal, as illustrated in **Figure 7** for samples solidified at 0.8°C/s. The equiaxed

Al	B	Phase
34.15	63.19	AlB_2
35.40	64.47	
31.47	66.64	
33.62	65.51	
7.32	91.56	AlB_{12}
8.52	90.78	

Table 3. WDS analysis (at.%) of AlB_2 and AlB_{12} phases observed in **Figure 3(a)**.

crystals have a larger size than the chilled ones, and they are the result of volumetric solidification, which proceeds when the initially high thermal gradient is reduced as solidification progresses [15]. The refinement of the primary structure is a result of the creation of phases that act as substrates of heterogeneous nucleation for the primary aluminum phase. Therefore, active centers of aluminum heterogeneous nucleation are particles that have a high melting point and close crystallographic match with aluminum, e.g., TiC, TiN, TiB, TiB_2, AlB_2, Al_3Ti, and Zn_3Ti [16–20].

When an Al-5%Ti-1%B master alloy is added to pure aluminum, it decomposes into Al_3Ti and TiB_2 phase particles. Since the dissolution of Al_3Ti is rather fast in Al-Si alloys, it forms $Ti(Sl_{1-x},Si_x)_3$ where $x < 0.15$ [21]. The latter tends to cover the TiB_2 particle surface acting as a heterogeneous nucleation substrate for the α-Al through the peritectic reaction. The addition of 600 ppm Ti in the form of Al-10%Ti to pure aluminum (**Figure 7(b)**) resulted in the complete elimination of the columnar grains with marked reduction in the grain size of the two equiaxed zones observed in **Figure 7(a)**. Increasing the Ti concentration to 1000 ppm led

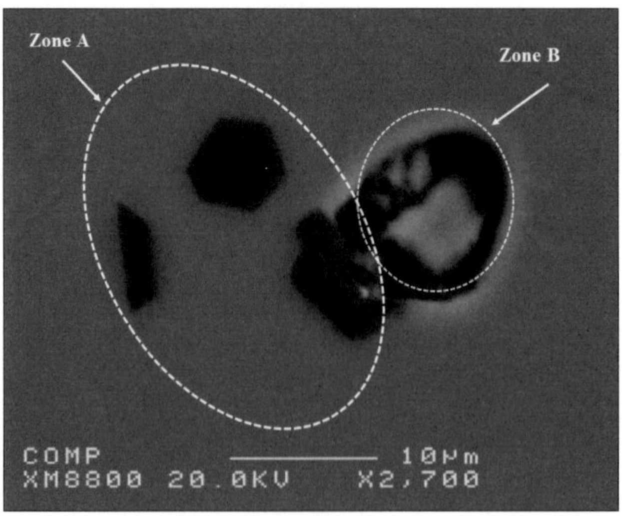

Figure 4. Presence of B and Si in Al-7%Si treated with 800 ppm B. Note the geometric shape of the particles in zone A.

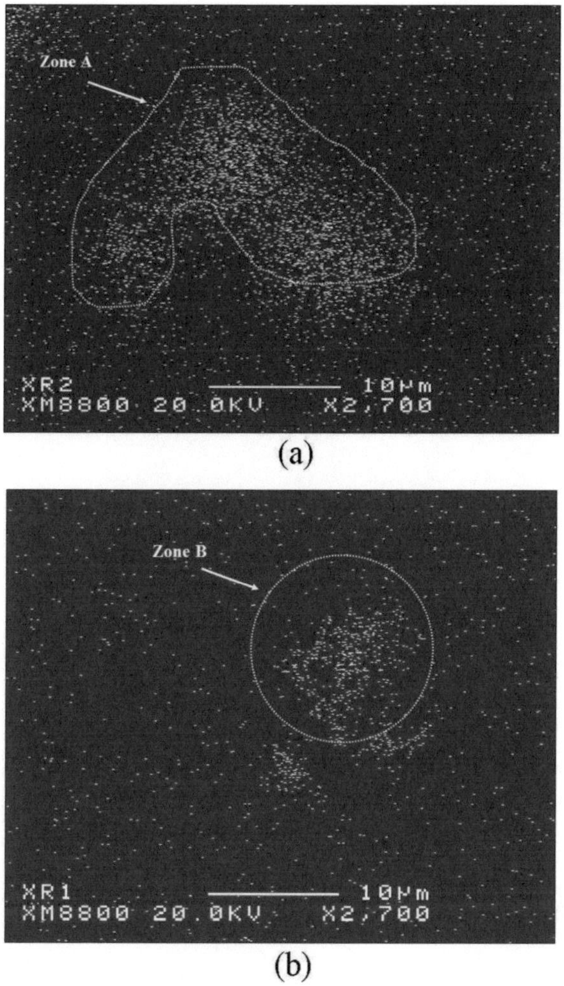

Figure 5. Selected X-ray for B and Si distribution of (a) B (b) Si in **Figure 4**—Al-7%Si alloy treated with 800 ppm B.

to formation of one zone characterized by its ultra-fine grain size (**Figure 7(c)**). Similar observations were made when the liquid aluminum was treated with Al-5%Ti-1%B master alloy (**Figure** 7(**d** and **e**)).

As previously mentioned, B has no grain-refining effect on ultra-pure aluminum. However, if its addition is associated with a relatively high cooling rate, i.e., 10°C/s as shown in **Figure 8**, it could result in some refining depending on B concentration as shown in **Figure 8(d)**, where at least 1000 ppm B is used to obtain more or less same level of grain size produced by the addition of 30 ppm TiB_2 (**Figure 7(e)**). These experiments always produced a grain size gradient from the bottom of the crucible to the top surface. This is usually attributed to the fact that the crucible surface acts as a nucleation substrate for heterogeneous nucleation of α-Al. In addition, once solidification initiates on the relatively cold bottom of the crucible, sensible and latent heat are released to the crucible, reducing the temperature gradient and effectively slowing

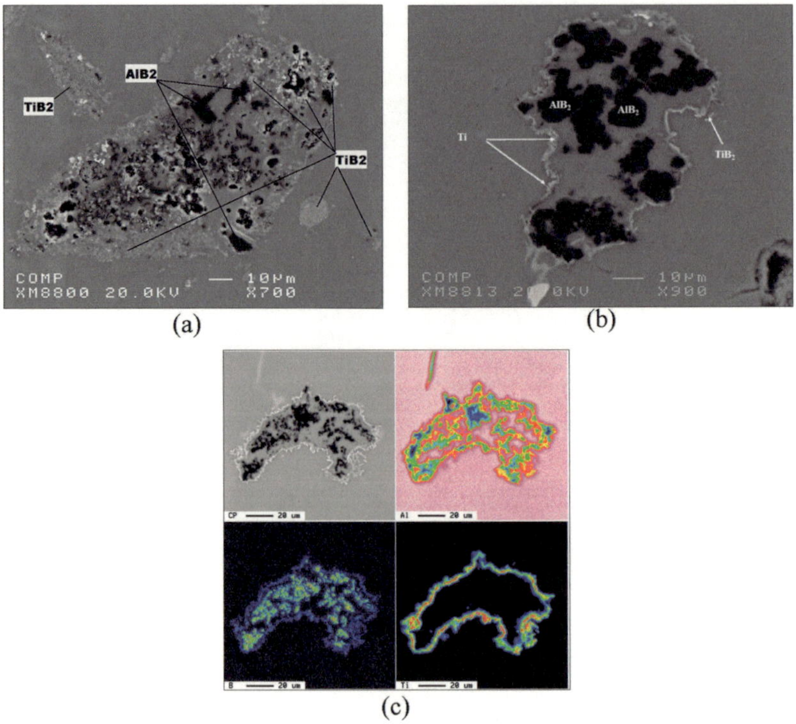

Figure 6. (a) Decomposition of Al-5%Ti-1%B master alloy, (b) backscattered electron image and (c) element distribution of Ti-B interaction in 1050 aluminum treated with 240 ppm B. Note the attraction of Ti ring toward the cluster of B-rich particles.

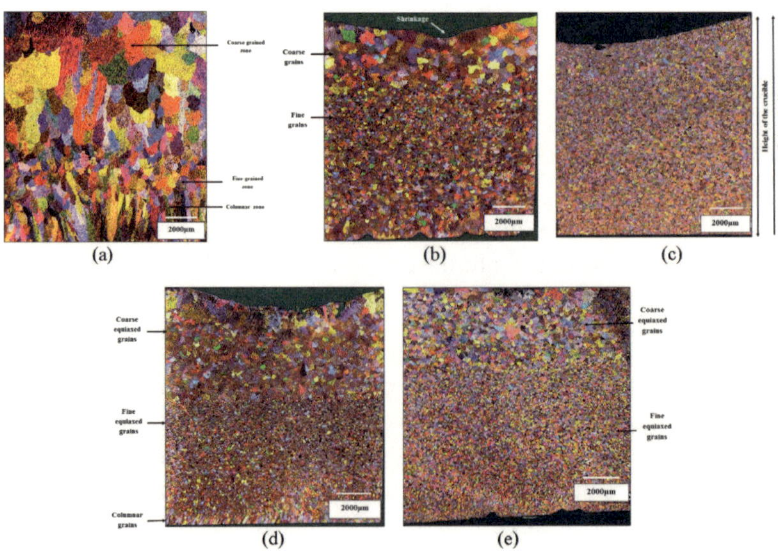

Figure 7. Macrostructures of ultra-pure aluminum treated by two types of grain refiners: (a) Al pure (no treatment), (b) Al pure + Al_3Ti (600 ppm), (c) Al pure + Al_3Ti (1000 ppm), (d) Al pure + TiB_2 (15 ppm), (e) Al pure + TiB_2 (30 ppm). All micrographs have same magnification-solidification time was ~200 s. Note the heterogeneity in grain size on going from bottom to top.

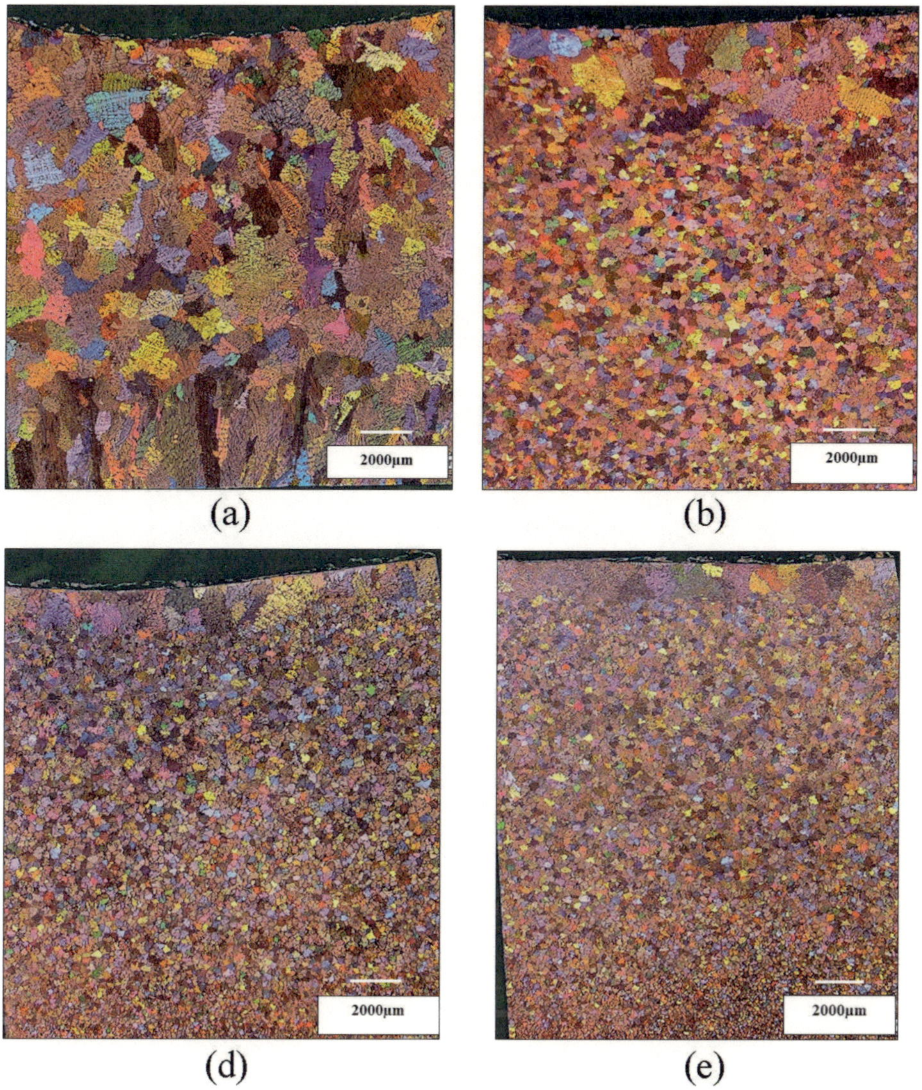

Figure 8. Macrostructures of ultra-pure aluminum treated with progressive doses of B: (a) 30 ppm B, (b) 60 ppm B, (c) 120 ppm B, (d) 1000 ppm B. All micrographs have the same magnification. Note the progressive transition from columnar-to-equiaxed (CTE) with the increase in B concentration.

the solidification rate. According to Chalmers [15], predendritic solid nuclei that formed on the mold surface during filling of the mold are driven away by circulation of the liquid metal. Depending on the degree of undercooling, some of these nuclei would survive in the liquid until the undercooling is extracted, resulting in equiaxed grains.

Considering the Al-7%Si alloy, its structure was not affected by the phenomenon mentioned above in the case of ultra-pure aluminum. **Figure 9** shows the macrostructure of the alloy with different melt treatments consisting of fine equiaxed grains caused by the addition of a suitable dose of grain refiner associated with application of high solidification rate—note

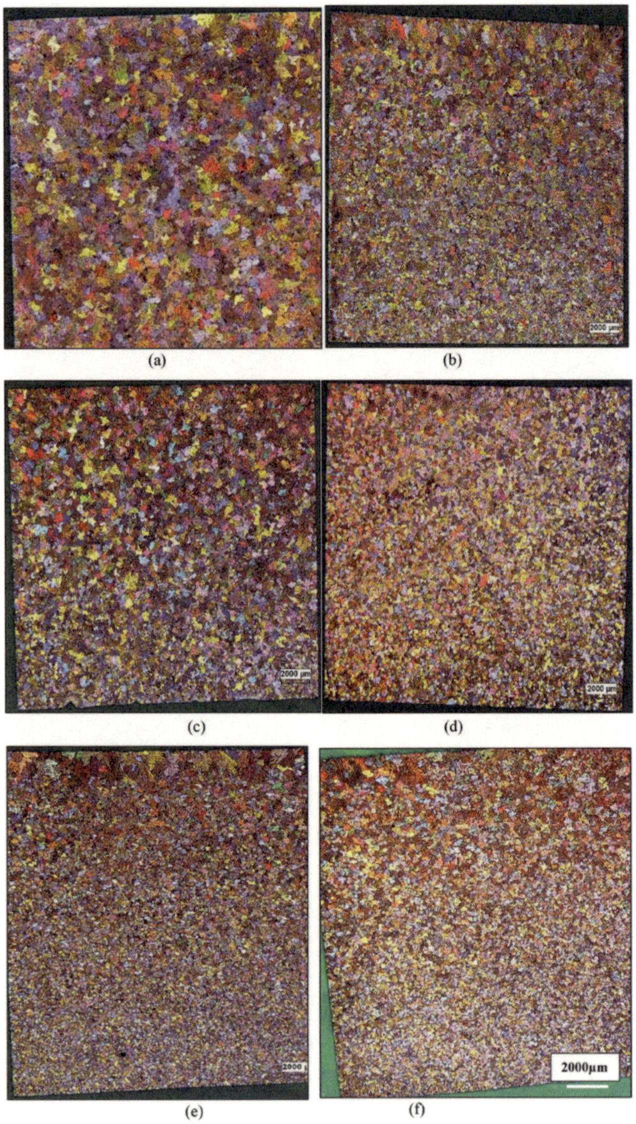

Figure 9. Variation in the grain size of Al-7%Si alloy: (a) base alloy—no addition, (b) alloy treated with Al-10%Ti (300 ppm Ti), (c) alloy treated with Al-5%Ti-1%B (150 ppm of B), (d) alloy treated with Al-5%Ti-1%B (300 ppm of B), (e) alloy treated with Al-4%B (600 ppm B), (f) alloy treated with 1000 ppm B. All micrographs have the same micron bars.

the absence of columnar grains. The transition from columnar to equiaxed grains depends on several factors: (i) mold geometry, (ii) the amount of extracted heat, (iii) composition of used alloy, (iv) density of the used grain refiner, (v) fluidity of the liquid metal, (vi) interaction between grain refiner and alloy composition and (vii) sedimentation of the added grain refiner [22]. A high solidification rate, large amount of grain refiners as well as low temperature gradient would enhance fine equiaxed grains [23]. Al-7wt.% Si alloys with and without grain refiners solidified in diffusive conditions showed columnar growth in case of nonrefined alloy, and the existence of a columnar-to-equiaxed transition (CTE) in refined alloy. A sharp CTE is

http://dx.doi.org/10.5772/intechopen.70346

observed when increasing the solidification rate and a progressive CTE when lowering the temperature gradient [24].

4. Concluding remarks

The results obtained show that when Al-4%B is added to ultra-pure aluminum, it decomposes into AlB_{12} and AlB_2 phases that have no grain-refining effect in pure aluminum. When no grain refiner is added to pure aluminum, the microstructure is a mixture of columnar and equiaxed grains. The addition of 30 ppm Ti to pure aluminum is sufficient to promote equiaxed grains when the metal is solidified at high rate (~10°C/s). A similar effect can be achieved with the addition of excessive amounts of B, in the order of 1000 ppm. Increasing the Si content to 7% enhances the reduction in initial grain size of pure aluminum from 2800 to 1850 μm in Al-7%Si alloy. In the case of Al-7%Si, only 500 ppm B is sufficient to reduce the initial grain size by about 75%. Due to a reduced temperature gradient as the solidification front travels from the bottom of the crucible to the top, the grain size varies accordingly.

In commercial aluminum, the addition of B would react with traces of Ti leading to the formation of Al_3Ti and TiB_2 phases that are effective grain-refining agents. In the case of Al-7%Si, Ti reacts with Si forming $(Al, Si)_2Ti$ phase which is a poor refining agent. This phenomenon is termed *poisoning*. No interaction was observed to take place between B and Si. In general, the grain size varied along the height of the crucible due to change in the liquid temperature.

Acknowledgements

The authors would like to thank Amal Samuel for enhancing the quality of the images and drawings presented in this article and Dr. E. Samuel for editing the manuscript.

Conflict of interest

There is no conflict of interest between the authors.

Author details

Hicham Tahiri[1], Serageldin S. Mohamed[1], Herbert W. Doty[2], Salvador Valtierra[3] and Fawzy H. Samuel[1]*

*Address all correspondence to: fhsamuel@uqac.ca

1 Département des Sciences Appliquées, Université du Québec à Chicoutimi, Québec, Canada

2 General Motors Materials Engineering, Pontiac, MI, USA

3 Nemak, S.A., Garza Garcia, N.L., Mexico

References

[1] Stiller W, Ingenlath T. Industrial boron treatment of aluminium conductor alloys and its influence on grain refinement and electrical conductivity. Aluminium. Vol. 60; 1984. pp. E577-E580

[2] Cooper PS, Kearns MA. Removal of transition metal impurities in aluminium melts by boron additives. Trans Tech Publications Ltd. Aluminium Alloys: Their Physical and Mechanical Properties. Zurich, Switzerland. Vol. 217; 1996. pp. 141-146

[3] Wang GQ, Liu SH, Li CM, Gao Q. Reaction of boron to transition metal impurities and its effect on conductivity of aluminum. Transactions of Nonferrous Metals Society of China. 2002;**12**:1112-1116

[4] Alamidari HD, Dube D, Tessier P. Behavior of boron in molten aluminum and its grain refinement mechanism. Metallurgical & Materials Transactions A. 2013;**44A**:388-394

[5] Mondolfo LF. Aluminum Alloy Structure and Properties. London: Butterworth; 1976

[6] Sigworth GK. The grain refining of aluminum and phase relationships in the aluminum-titanium-boron system. Metallurgical Transactions A: Physical Metallurgy and Materials Science. The Metallurgical Society of AIME and American Society for Metals, USA. Vol. 15A; 1984. pp. 277-282

[7] Rogl P, Schuster JC. Phase Diagrams of Ternary Boron Nitride and Silicon Nitride System. ASM International; Materials Park, OH, USA. 1992

[8] Giardini AA, Kohn JA, Toman L, Eckart DW. Boron—Synthesis, Structure and Properties. New York: Plenum Press; 1960. pp. 140-158

[9] Li H, Sritharan T, Lam YM, Leng NY. Effects of processing parameters on the performance of Al grain refinement master alloys Al-Ti and Al-B in small ingots. Journal of Materials Processing Technology. 1997;**66**:253-257

[10] Khalifa W. Rôle des inclusions dans la germination de la phase α-Aluminium et des intermétalliques contenant du fer dans le coin riche en aluminium du système ternaire Al-Si-Fe. thèse présentée à l'Université du Québec à Chicoutimi UQAC, novembre 2003. p. 340

[11] Lu HT, Wang LC, Kung SK. Grain refining in A356 alloys. Journal of Chinese Foundryman's Association. 1981;**29**:10-18

[12] Samuel AM, Samuel FH. Use of the reduced pressure test in the measurement of the hydrogen content in the foundry of composites. In: Proceedings International Symposium on "Advances in the Production of Light Metals and Metal Matrix Composites", 31st Annual Conference of Metallurgists of CIM, Edmonton, Alberta, Canada, August 23-27, 1992; Edmonton, AL, Canada. pp. 701-715

[13] Samuel AM, Samuel FH. The reduced pressure test as a measuring. Metallurgical Transactions A. 1993;**24**:1857-1868

[14] Kurz W, Bezençon C, Gäumann M. Columnar to equiaxed transition in solidification processing. Science and Technology of Advanced Materials. 2001;**2**:185-191

[15] Chalmers B. The structure of ingots. Journal of the Australian Institute of Metals. 1963;**8**:255-263

[16] Guzowski M, Sigworth G, Sentner D. The role of boron in the grain refinement of aluminum with titanium. Metallurgical Transactions A. 1987;**18**:603-619

[17] Wróbel T. Review of inoculation methods of pure aluminum primary structure. Archives of Materials Science and Engineering. 2011;**50**:110-119

[18] Greer. Structure and properties of grain-refined Al-20 wt.% Zn sand cast alloy. Metallurgy and Materials. 2009;**54**:329-334

[19] Haberl K, Krajewski WK, Schumacher P. Microstructural features of the grain-refined sand cast AlZn20 alloy. Metallurgy and Materials. 2010;**55**:837-841

[20] Wróbel T, Szajnar J. Influence of supply voltage frequency of induction coil on inoculation efficiency of pure aluminum structure. Foundry Engineering. 2010;**10**:203-208

[21] Hodaj F, Durand F. Equiaxed grains in multicomponent alloys: Effect of growth rate. Acta Materialia. 1997;**45**:2121-2127

[22] Sturz L, Zimmerman G. Investigations on columnar to equiaxed transition in binary Al alloys with and without grain refiners. Materials Science Forum. 2006;**508**:419-424

[23] Sturz L, Drevermann A, Pickmann C, Zimmermann G. Influence of grain refinement on the columnar-to-equiaxed transition in binary Al alloys. Materials Science and Engineering A. 2005;**413-414**:379-383

[24] Zimmermann G, Sturz L, Billia B, Mangelinck-Noël N, Liu DR, Nguyen Thi H, Bergeon N, Gandin Ch-A, Browne DJ, Beckermann Ch, Tourret D, Karma A. Columnar-to-equiaxed transition in solidification processing of AlSi7 alloys in microgravity-The CETSOL project. Materials Science Forum. 2014;**790**:12-21

Chapter 2

Some Considerations on the Structure Refinement in Al-Based Alloys

Ildiko Peter and Bela Varga

Additional information is available at the end of the chapter

http://dx.doi.org/10.5772/intechopen.71450

Abstract

Grain size is one of the most important characteristics that affect the processing and in turn the properties of alloys. Grain refinement determines many advantages in light alloy casting: it can be achieved using different methods, based on the available technological possibilities and on the performances that one has to obtain. By using grain refinement, important benefits can be reached, for both cast and wrought aluminium alloys: among other, the most important enhancement regards the fine distribution of the second phases, improved castability, reduction of shrinkage porosity, higher mechanical properties, as well as superior fatigue life. The present chapter is not exhaustive on this argument; however, in the first part it reports some current literature data and some perspectives about the grain refinement, while in the second part which has been mostly carried out within a current PhD Thesis focalized on the improvement of the properties of Al-based alloys by physical grain refinement methods, some experimentally obtained results have been presented and discussed.

Keywords: Al-based alloys, structure refinement, grain refinement, metallurgical methods, physical methods, morphology

1. Introduction

Due to their excellent properties (high strength-to-weight ratio compared to Fe-based alloys, good corrosion resistance, etc.), light alloys (Al- or Mg-based alloys) have a key role in various engineering applications. Generally, during traditional casting process, dendritic microstructure is developed. Dendrites can be organised in columnar or equiaxed structures in agreement with the production conditions, which involve among others, the solute concentration, the heat gradient, the solidification velocity, the presence or absence of inoculating elements, etc. The control of the shape and distribution of the grains structure is made up is of primary importance and really governs the properties of the produced alloy. In many conditions, improvement of the mechanical strength is required, especially in application where structural properties are essential. Thermal treatments and/or reducing the grain size of the constituents are commonly used methods for the improvement of some properties. Addition of grain refiners (GR) determines an increase of the mechanical properties, and in particular, it affects the strength and the fatigue resistance of the alloys, and at the same time, the quality of the cast part and the efficiency of the casting process rises [1]. According to some literature date [2–5] and some own research [6–11], one can state that the grain size is one of the most important characteristics, which affects the processing, and consequently, the properties of alloys. Presence/addition of GR determines many advantages in light alloy casting and it can be achieved using different methods, as will be presented later on. Actually at industrial level, both adding grain refiner and improving solidification conditions are two procedures for grain refinement, especially in case of Al alloys.

The reduction of the grain size has a positive effect on the microstructural homogeneity determining a fine distribution of the second phases leading to an improved castability, lower micro-porosity and shrinkage porosity as well as a reduced segregation development [4, 6].

In any case, the alloy composition has to be carefully considered in order to choose an efficient GR. Even if, Al–5Ti–B master alloy is an efficient GR for wrought Al alloys, its effect is not sufficient in case of foundry alloys: formation of Ti–Si phases occurs following the reaction between the solute Si and Ti determining the weakening of the nucleating of TiB_2 particles in the master alloy, when the Si content in the alloy is higher than 3 wt.% [12, 13].

According to [14, 15], in case of master alloys containing high amount of B and higher than 4–5.5 wt% of Si, an improvement of the grain refining can be obtained. Prior to casting, an inadequate holding time negatively influences the GR efficiency and additionally their agglomeration can take place [16–18]. For Mg alloys, Zr is the ideal GR, but its use is unsuccessful when it contains Al.

Conventional GR for Al alloys are generally produced from Al–Ti–B ternary system [19–22].

On the contrary of Fe-based alloys, in case of light alloys, the primary structure after solidification is maintained and the physical and any structural transformations induced following thermal treatment and/or plastic deformations take place within the primary structure. Any outcomes coming from such technological treatments are influenced by the characteristics of the primary structure. Additionally, light alloys have higher tendency during solidification for the development of the coarse grains with respect to iron-based alloys. From these considerations, it is evident that for light alloys, the influence of the crystallisation process with the aim of the development of fine grains and the good distribution of the different phases within a homogeneous microstructure results very important and it significantly influences the properties of the final product during their use. The schematic illustration of the relation between the structure and the properties is reported in **Figure 1**. The use of the different methods for the improvement of the structure in case of light alloys significantly influences the properties related to their use and beneficially acts on the technological properties and processing by plastic deformation as well on their behaviour during thermal treatment.

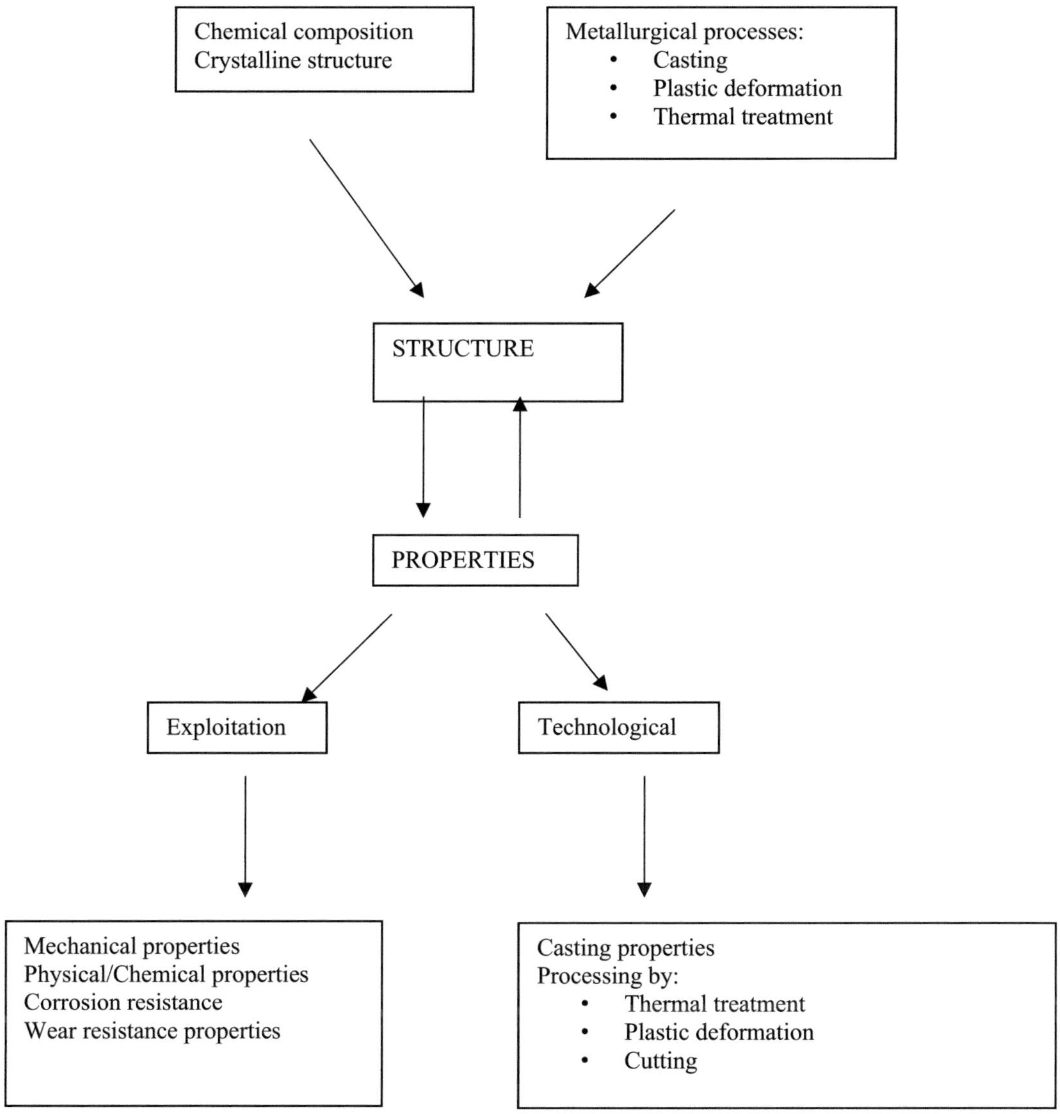

Figure 1. Schematic representation of the relation between the structure and the properties.

Solidification of the alloys is a process, which involves the passage from the liquid state into a solid state as a result of the energy loss during the release of the heat and shows at least two aspects:

1. Germination: the step when development of the nuclei or crystallisation centre has been realised. The evolution can be:
   - Homogeneous: when the centres have been developing spontaneously and has been growing during the time;

- Heterogeneous: when the development takes place close to the existent germs on the surface of the melt (on the wall of the die, near to some hold-up in the melt or during the precipitation of some phases in mini-crystal forms) and can be favoured by the introduction of some particles as modifier into the melt;
- Mechanical: such effect is induced by mechanical ultrasonic vibration and/or by using turbulent convection currents.

2. Development and growing of the crystals by the combination of the atoms from the melt on the crystal surface in form of monoatomic layers with critical size representing bi-dimensional germs [23].

During very slow cooling, a preferential direction of the heat release could be observed. During the development of dendrites, the dendrite arms are growing encountering during their expansion other neighbouring dendrites. When crystallisation finishes, the evolution continues on the perpendicular direction with respect to the former one and the whole crystal grows into a complex form.

Industrially, there is the need to obtain enhanced quality product. To realise this, one has to consider and satisfy some of the following conditions:

- To increase the cooling rate and the number of the crystalline germs, simultaneously with the reduction of the solidification time in some regions of casting;
- Managing the heat transfer process and the mass transfer, which directly influences the size of crystalline germs, determining the development of a fine equiaxed structure;
- Homogenising as much as possible the structure;
- Degassing the melt;
- Reducing the presence of internal stress avoiding in this way the crack formation [24].

The solidification process can be governed by the control of the time required for the cooling and acting directly on the alloy, on the casting profile or on the final product. The control of the solidification is significantly complex especially when the alloy shows high shrinkage in liquid state and during solidification.

Figure 2 summarises the most important methods used in order to act on the solidification. Application of a magnetic stirring field to the melt, which solidify has a grain refining effect too. There are two types of magnetic stirring:

- Electromagnetic stirring, which stimulates the development of an equiaxed crystalline zone; refinement of the solidification structure occurs with a decrease of the inclusions content and
- Permanent magnetic stirring, which has some benefits over the first one, because it consists of lower cost tools, high magnetic field intensity and with no skin effect.

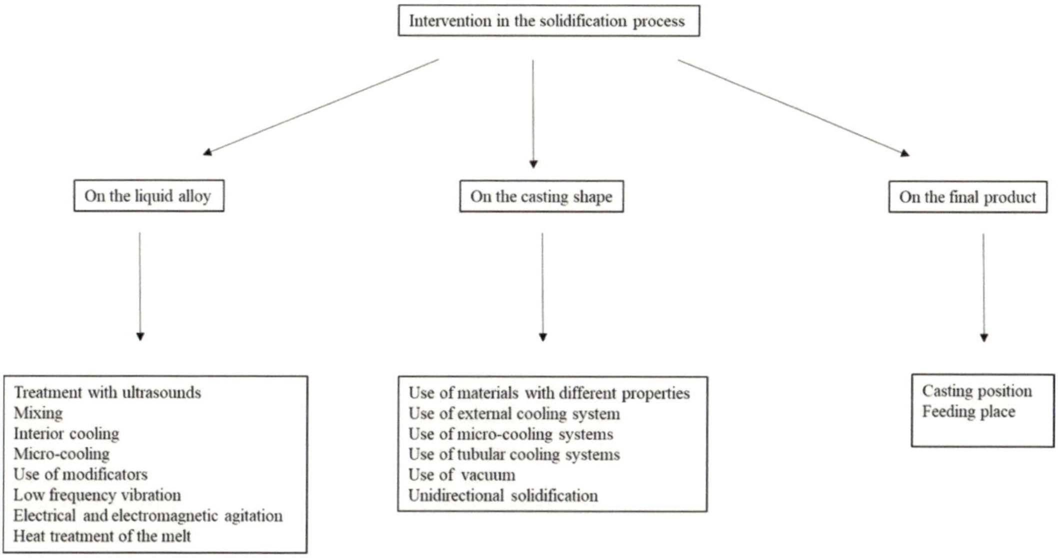

Figure 2. Schematic representation of the most important methods used for the refinement of the structure.

Moving backward the stirring direction of the electromagnetic stirring, the oxide film on the melt surface can be protected, which is important for maintaining it undamaged. Contrarily, the layer can be fragmented by the forced melt flow losing its protective character. Some studies [23, 25, 26] reveal how the effect of a magnetic field on the crystal direction and on the phase arrangement during solidification is a valid method to increase the physical properties of the alloy.

In [23, 27, 28], the authors have demonstrated that the rotating magnetic field can be effectively used to reduce the segregation appearance during solidification and to enhance the structure of metallic alloys.

In order to shortly appreciate the variations taking place during solidification, **Figure 3** reports, schematically, the enhancement of the structure using high cooling rate contemporary with the addition of some chemical element: the refinement of the structure and consequently, the improvement of the general behaviour is evident.

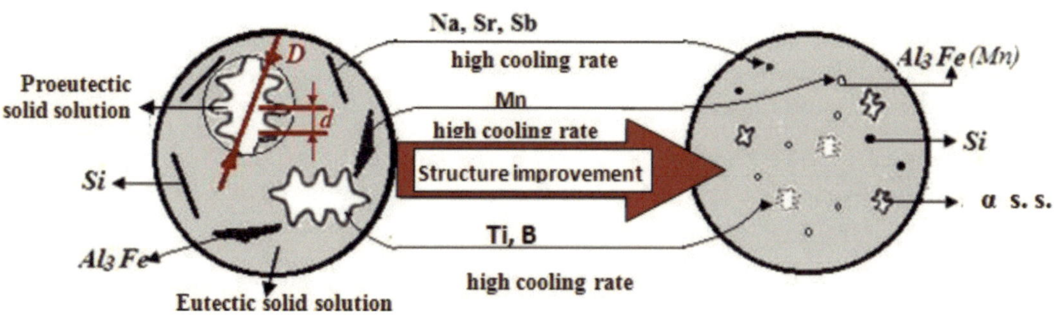

Figure 3. Schematic representation of the main modifications in hypoeutectic Al-Si alloy.

2. Experimentally used devices and methods

In the following part, some results of the current research performed within a recent PhD Thesis [10] oriented to the enhancement of the characteristics of Al-based alloys by physical grain refinement methods that have been presented and talked over, including some real case studies.

The purpose of using the dynamic methods is twofold, as follows:

A. For enhancing the structure of the alloys reaching a fine microstructure and a homogeneous product, both chemically and structurally and

B. For reducing the scrap formation during casting with a consequent time saving possibility.

Both foundry and deformable alloys have been used for such investigations. The study is realised analysing:

- The effect of the vibration during solidification on the development of the microstructure, lowering the porosity content and furthermore enhancing the mechanical properties;
- The effect of magnetic stirring applied on the melt, considered as a clean solution for the hating and for the treatment of the alloys; such approach allows, with no any interaction, the necessary thermal and mechanical energy for obtaining a homogeneous alloy with fine grains and enhancement of the mechanical properties;
- Solidification with high cooling rate, leading to positive results concerning the refinement of the structure, a method which allows obtaining high performance product with no any thermal treatment.

The alloys, with the chemical composition used for the research have been casted and details on such part have been found in some previously published paper [6–12]. The vibration system, the electromagnetic stirring system and the device used as electromagnetic induction system have been projected and realised during the present research and here is shortly presented. Results on the benefits of the physical treatment performed, with some comparison related to other similar condition for the solidification have been reported and discussed integrated with some preliminary considerations related to their eventually technological transfer. Additionally, enhancement of the structure following high cooling rate using different processing methods: melt spinning, static and dynamic casting have been recalled with some consideration about the final benefits. The analysis are carried out and the results obtained can be used as guide-line for people working in this area in an economically convenient way, saving time, energy and materials.

The research has been carried out using binary and more complex alloys, belonging to the families: Al-Si, Al-Cu, Al-Zn and Al-Mg. The chemical compositions have been reported in the previously published papers [6–12].

The mechanical vibration of the melts during solidification has been performed using a laboratory device, usually employed in a dental laboratory with the details reported in **Figure 4**. The parameters adopted for the experiment are as follow:

- Voltage: 110 V ~ 220 V/60 Hz ± 10%
- Power: 100 W
- Acceleration:0.8–0.07 m/s^2
- Velocity: 2.8–0.22 mm/s
- Amplitude: 0.07
- Dimension: 14 × 13.5 × 11.3 cm
- The device projected for the research has made possible the variation of the frequency in the range of 4÷94 Hz.

Castings have the size as reported in **Figure 5** and the K type thermocouple has been introduced in the position indicated in figure, at about ¾ height of the crucible's total length.

Castings has been obtained in a static and dynamic mode and the chemical composition has been selected to evaluate as well as the influence of the working condition on the nature of the phases developed during solidification (primary phase, solid solution and intermetallic particles, eutectic phase). Structural modification has been monitored following the dynamic solidification condition for the hypoeutectic Al-Si foundry alloy and for the deformable Al-Cu, Al-Zn, Al-Mg alloys. The effect of the dynamic condition on the areas containing columnar crystals has been assessed by observing some samples when solidified with high cooling rate, using cooling channels placed around the crucible (**Figure 6**). Directed solidification has been performed on AlSi7Mg0.3 and on AlZn10Si7 alloy to study the effect of the mechanical vibration on the properties of the columnar disposition of the dendrites.

The tool reported in **Figure 7** has been used for the study of the solidification in rotating electromagnetic field. The parameters adopted are: tri-phasic power supply, nominal power 1.5 kW, frequency 50 Hz.

Casting, industrially, has been performed using the forms reported in **Figure 8**, and they have been carried out using different cooling rates (guaranteed by the different material used for the casting).

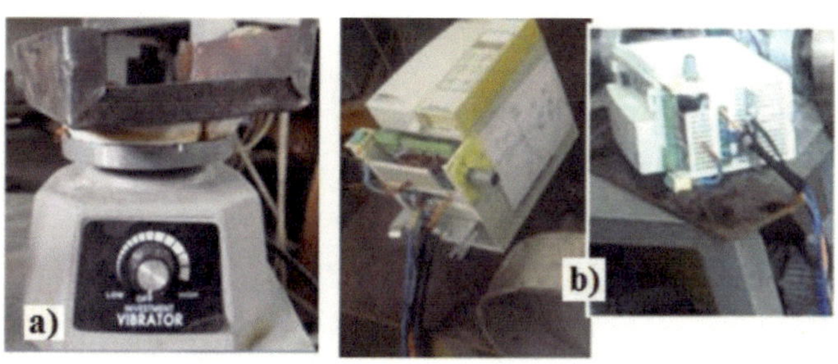

Figure 4. The set-up used experimentally for the mechanical vibration of the melt (a) and command block used for the variation and the registration of the frequency (b) [10].

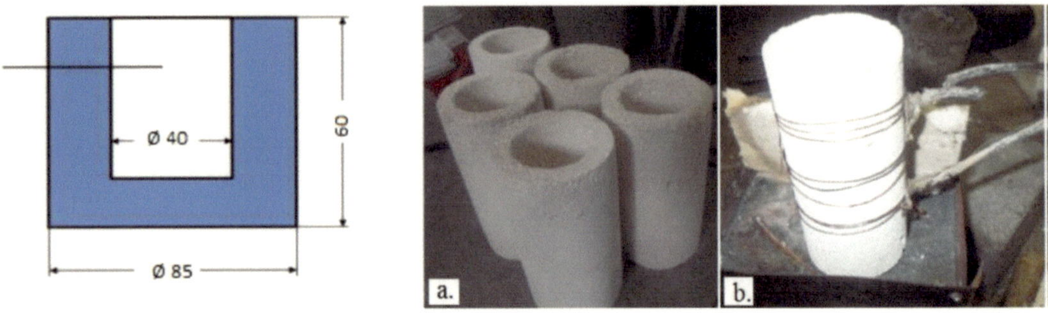

Figure 5. The shape and the size of the forms used for the casting; (a) crucible in refractory material and (b) thermocouples placed for monitoring the cooling [10].

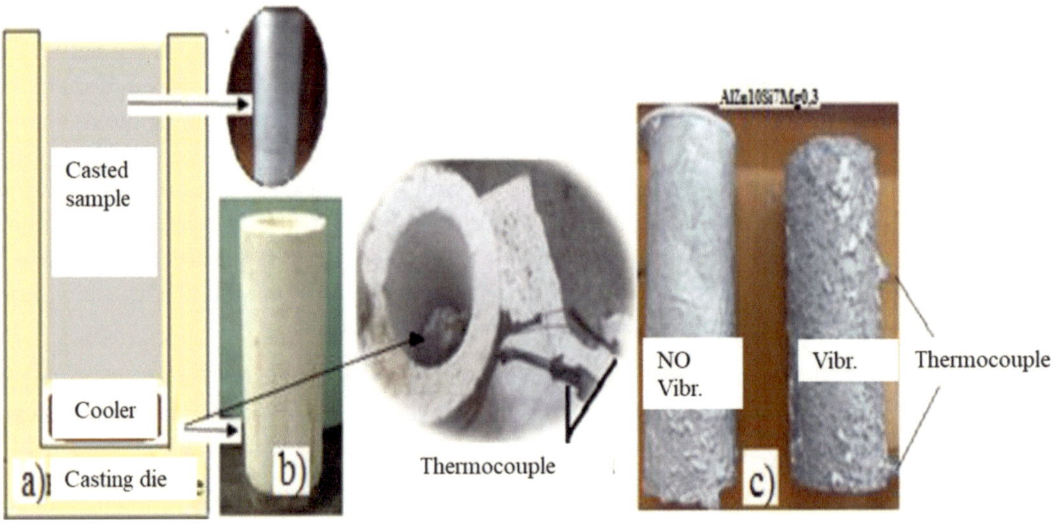

Figure 6. Scheme of the casting (a), crucible with the cooler and the thermocouples before located casting (b) and the samples with no vibration and under vibration [10].

Figure 7. Experimental set-up (a): A-inductor coils, B-casting shape in refractory material, C-Al alloy, D-termocouples; (b): Rotating electrical field generator, (c) command block and (d) system for the registration of the frequency [8, 10].

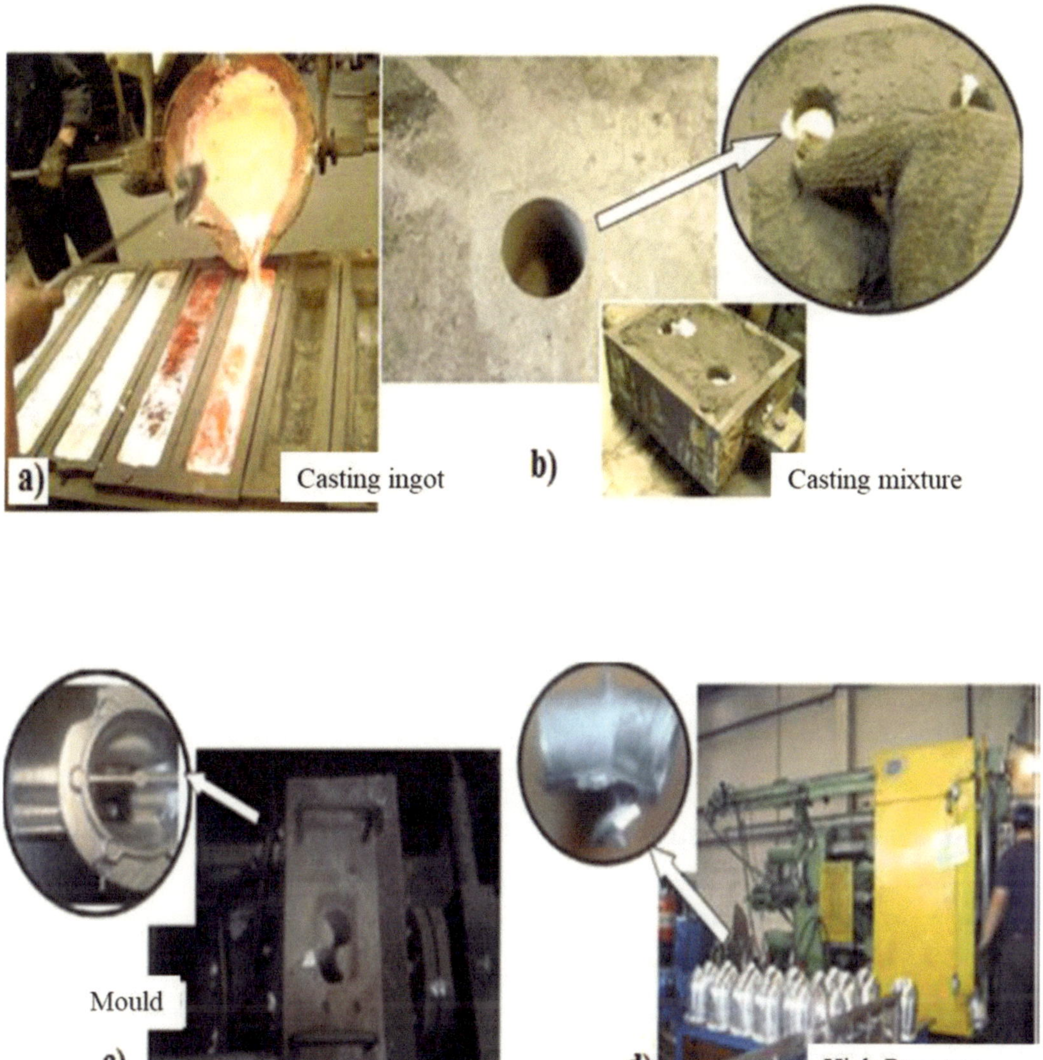

Figure 8. Casting of Al alloys in different dies [10].

The experimental device (**Figure 9**) realised for the simulation of the solidification in different cooling rates has the benefits to associate the cooling technology of the melt jet on spinning disc with the principle of the moulds feeding from low pressure casting procedure. In case of low pressure casting, the transfer of the liquid metal from the furnace to the die cavity placed on the top of the furnace, is guaranteed by an overpressure acting on the liquid alloy enclosed in the crucible, which is hermetically sealed. As a result, the liquid metal is forced to go vertically through the connecting tube. In case of the experimental device indicated in **Figure 9**,

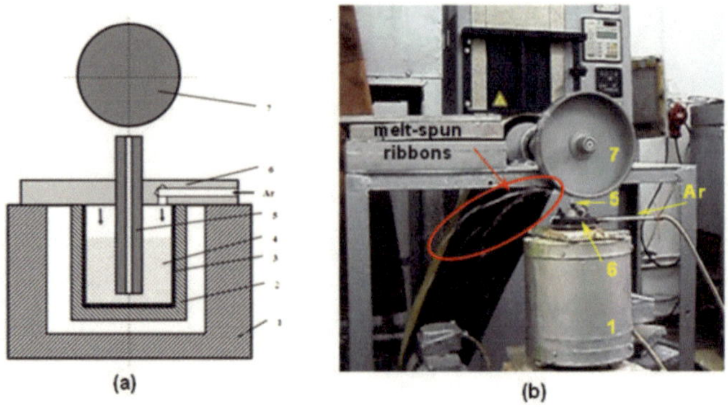

Figure 9. Experimental device for obtaining of melt-spun ribbons: (a) schematic plant, (b) melt spinning-low pressure (MS-LP) device during processing [12].

the working gas (Ar) introduced by the cover 6 acts upon the liquid metal 4 from the crucible 3 causing the fused metal raising in the feeding tube 5 followed by the melt placement on the rotating disc. The furnace and port-crucible is (1), respectively (2).

The diameter of the feeding tube (5) corresponds to 1 mm and the distance to the disc results to be 5 mm. The disc rotation speed is 2600 rot/min and the working gas pressure is 1.5 bars. The aforementioned parameters have an important role, since they establish the thickness of obtained ribbons. Major details on such devise used can be found in [10].

3. Experimental study results

3.1. Casting in steel die, refractory material die under vibration

AlCu4 and AlSi7Mg0.3 alloys have been casted in steel die, refractory material with no vibration and with vibration (f = 50 Hz, a = 15 m/s^2, v = 14 mm/s, A = 0.07 mm). The cooling curves are reported in **Figure 10**. Higher cooling rate can be observed when the alloy is subjected to vibration compared to the static cooling condition, confirmed by the microstructural analysis: the finest microstructure corresponds to the high cooling rate and in the presence of vibration for both alloys, as reported in **Figure 11**.

Directed solidification has been performed on AlSi7Mg0.3 and AlZn10Si7 alloy to investigate the influence of the mechanical vibration on the properties of the columnar zone. For this experiment, cylindrical casted samples positioned on Cu rings (Ø30 × 13 mm, weight 22 g) have been used. The thermocouples for the registration of the temperature variation as a function of time has been mounted on the top and on the bottom part of the refractory mould at a distance of 15 mm from the exterior part of the sample, as reported in **Figure 12**. The experiment has been carried out both with no vibration and under vibration conditions, using different parameters and varying the frequency in the range of 50÷90 Hz. The resonance frequency is 62 Hz and the other parameters adopted are: acceleration = 8.8 mm/s^2,

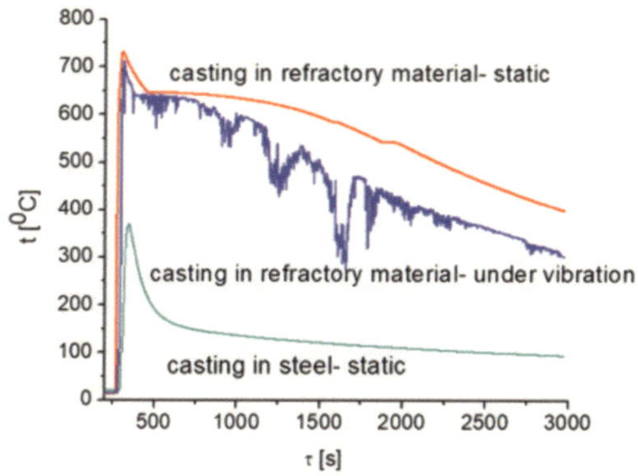

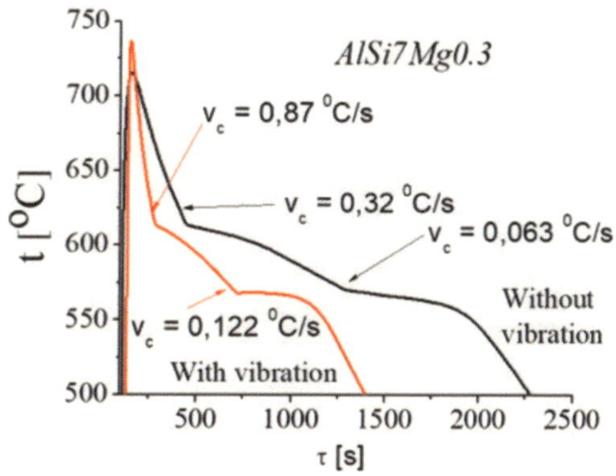

Figure 10. Cooling curves for AlCu4 casted in different mould and with no vibration and under vibration (a) and AlSi7Mg0.3 alloy (b) casted with no and under vibration [9, 10].

velocity = 99 mm/s^2 and amplitude = 0.7 mm. The distance from the cooler has been considered during the microstructural analysis and the most important areas (top, middle and bottom) have been analysed.

The finest grains have been obtained for the areas situated at the bottom part, which is in direct contact with the cooler and as the distance increases, the microstructure becomes coarser as reported in **Figure 13**. Following vibration, the roughness of the sample increases (**Figure 14**).

The results have the same tendency concerning the microstructure evolution (**Figure 15**) and for the roughness in case of AlZn10Si7 alloy (**Figure 16**) as well. There are some differences as their workability concerns: the second one is harder and seems to be more sensible to the impurities, which are on their surface. In this case, there are segregation of intermetallic particles made of Fe. With no vibration, trans-crystallisation is observed (columnar crystallisation), while in case of vibration, the equiaxed grain crystallisation has been observed.

On the upper part of the sample with vibration, some darker zones have been observed, while in the region in contact with the cooler development of finer grains have been observed. The lighter and longer appearance segregations are made of Fe, coming from the scraps re-used and this is one of the causes related to the more difficult workability. The bottom part of the sample is made of the solid solution and contains segregation of Si (darker colour than the Fe-made particles).

Industrially employed alloys have been investigated. In the case of hypoeutectic AlSi7Mg0.3 alloy is assumed that enhancement of the microstructure determines refinement of the microstructure and the modification of the eutectic phase. For the eutectic composition AlSi12CuMgNi alloy with complex composition, the vibration is of interest both for the structure of the dendrites and for the morphology of the eutectic and for the shape and distribution of the intermetallic particles segregation. For the binary AlCu4 alloy, it is important that the refinement of the dendrites of solid solution and for the development of non-equilibrium eutectic phase.

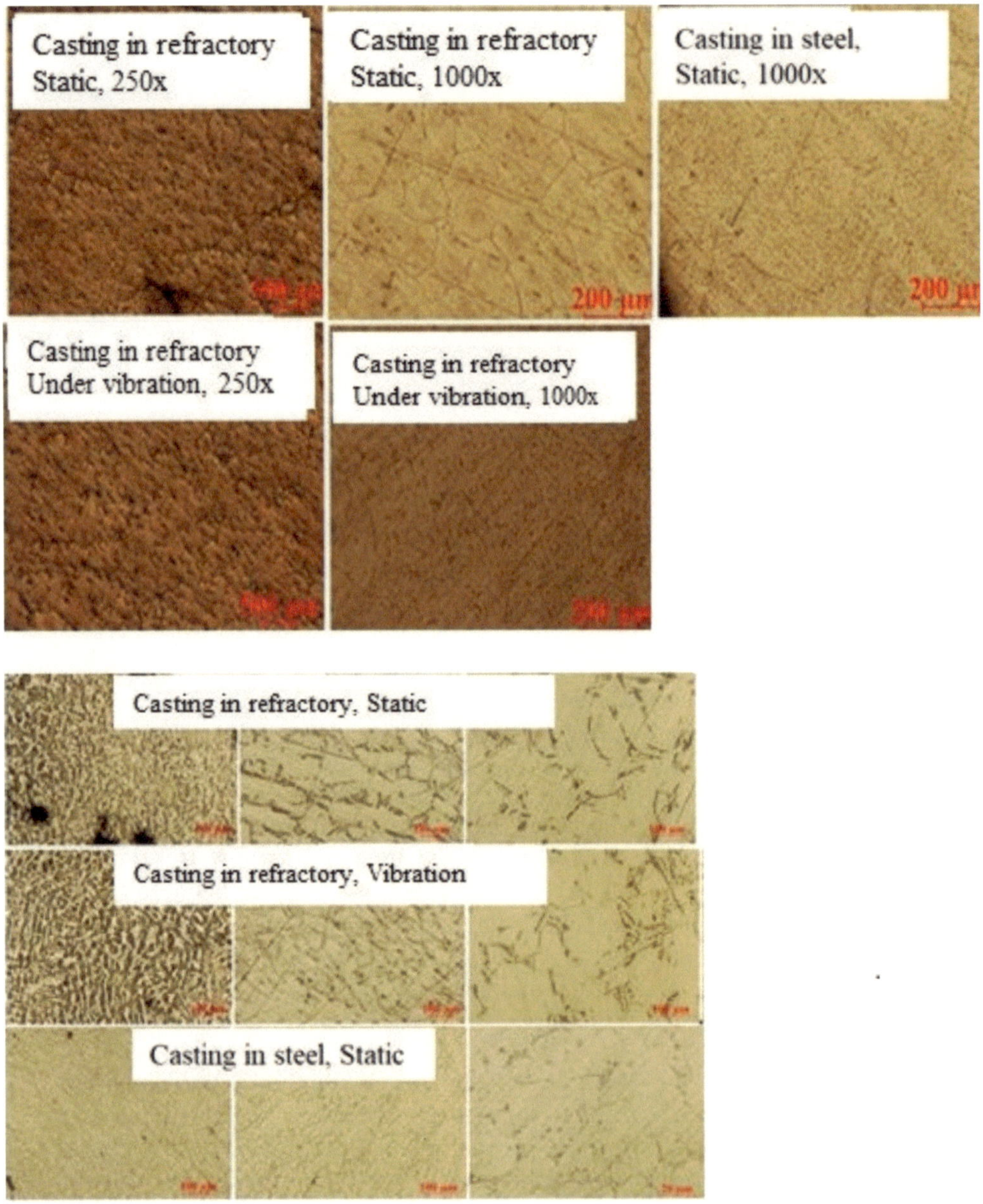

Figure 11. OM microstructure of the AlCu4 alloy (top) and AlSi7Mg0.3 (bottom) casted in different dies and with no vibration and under vibration [10].

3.2. Effect of the vibration and of the cooling velocity on the solidification condition

Comparison of the effect of the vibration and of the cooling velocity on the solidification condition has been carried out. For the thermal analysis, samples of size 14×80×160 mm have been used, labelled as C00. **Figure 17** (left) reports the results of the thermal analysis for the AlSi7Mg0.3 alloy. The analysis of the cooling curves indicates that in case of vibration, the

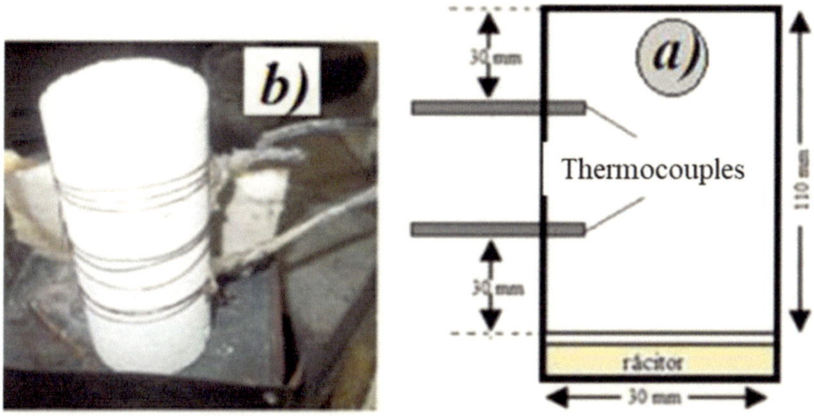

Figure 12. Set-up of the directed solidification experiment [10].

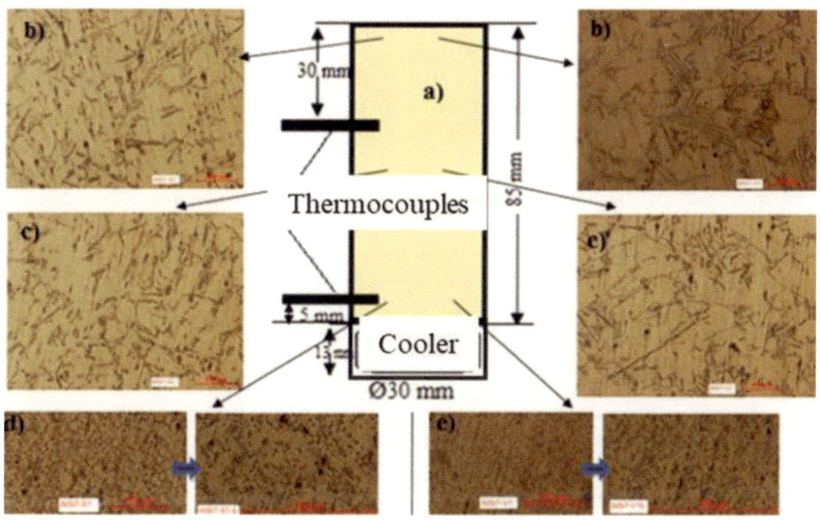

Figure 13. OM microstructures of the different areas within the AlSi7Mg0.3 alloy after directed solidification [10].

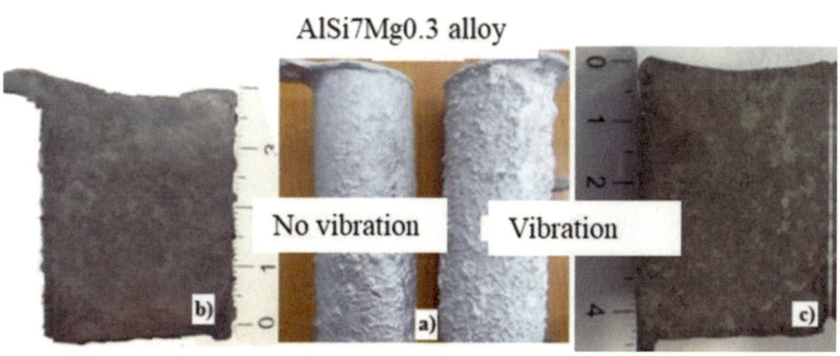

Figure 14. Photographs showing the roughness of the samples in AlSi7Mg0.3 alloy with and with no vibration (a), top part of the sample under vibration (b) and with no vibration (c) [10].

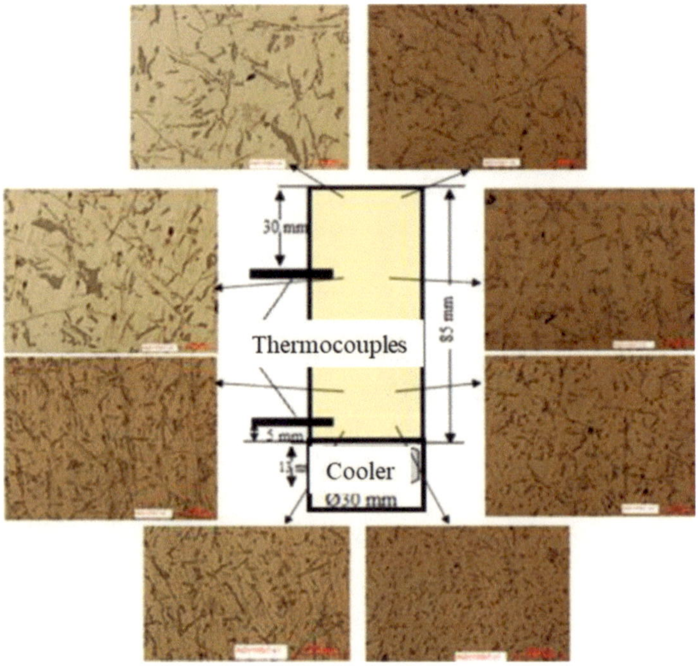

Figure 15. OM microstructures of the different areas within the AlZn10Si7 alloy after directed solidification [10].

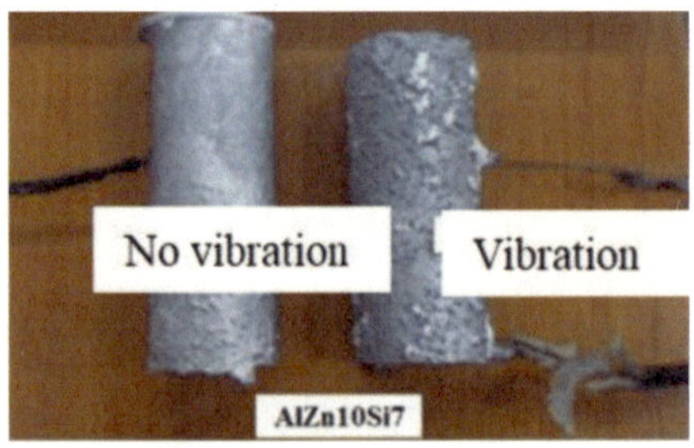

Figure 16. Photographs showing the roughness of the samples in AlZn10Si7 alloy with and with no vibration [10].

cooling rate is higher. In both cases, the liquidus temperature is 613.35°C, while the solidus temperature (eutectic) for the vibrated sample is 1.25°C lower that the sample without vibration (**Figure 17** right hand side).

It can be noticed that the eutectic transformation has a more evident appearance in the presence of vibration. The maximum cooling velocity in liquid state is 4.6°C/s and 6.7°C for the non-vibrated and vibrated sample, respectively. In case of the sample solidified under vibration, the primary solidification and the eutectic transformation take place at higher cooling velocity compared to the situation with no vibration. The period of the transformation is significantly different for the two samples: in case of vibration, a decrease of both the extent of the crystallisation and of the period of the eutectic transformation of about 50 and 31%, respectively occurs. Such reduction is determined by the increase of the cooling velocity.

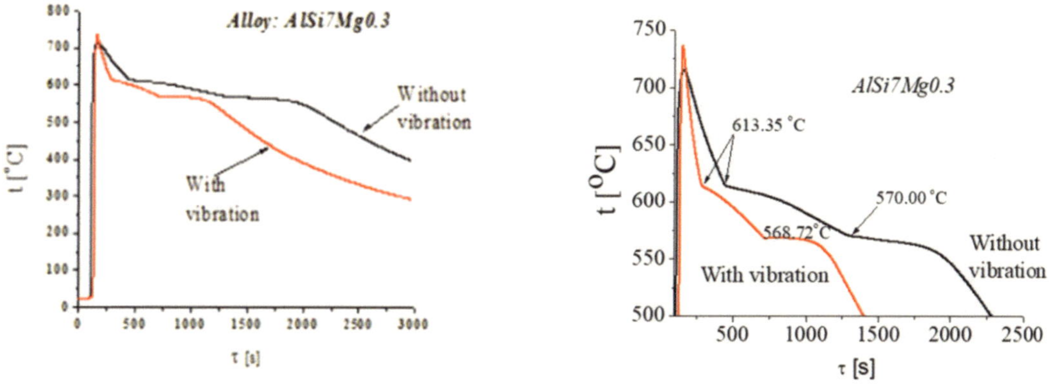

Figure 17. Cooling curves and the transformation temperatures for the AlSi7Mg0.3 alloy [9, 10].

For the AlCu4 alloy, the maximum cooling velocity in liquid state is (**Figure 18**):

- 1.15°C/s for the sample with no vibration;
- 1.6°C/s and 2.1°C/s for the samples solidified in refractory moulds;
- 17°C/s for the sample solidified in steel moulds (C00).

It can be noticed that as the vibration increases, the cooling velocity significantly increases also in the biphasic domain (liquid + solid). C0 corresponds to the liquidus temperature, the peaks labelled as C1 makes in evidence the temperature when the development of solid solution dendrites concludes, while C2 (very evident) indicates the formation of the eutectic Al-Al_2Cu phase. The region correspondent to the formation of the eutectic is preceded by the overcooling of the alloy, intensified by the vibration of the melt. The sample, solidified under vibration, shows higher roughness: intensification of the heat exchange between sample and mould occurs, appreciated by the value of the overall heat transfer coefficient, which in the presence of vibration has a favourable effect on the interaction between the alloy and the

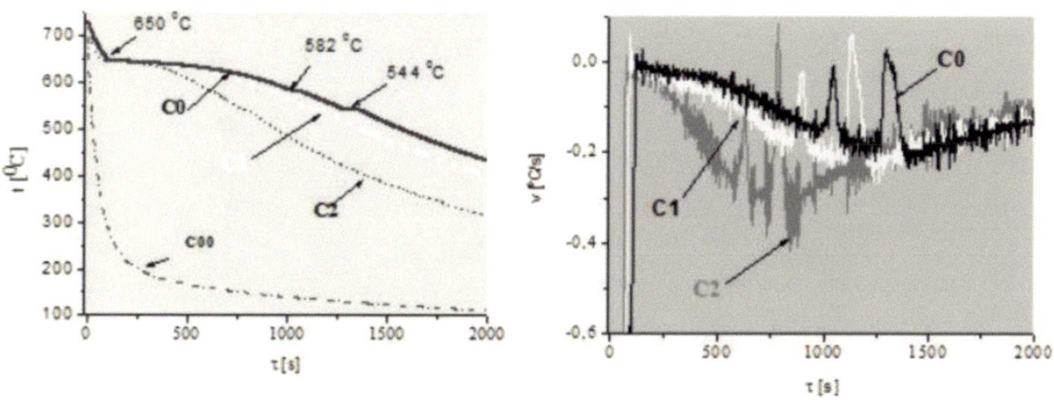

Figure 18. Cooling curves and the transformation temperatures for the AlCu4 alloy [9, 10].

material made up the mould. [7, 10] reports more details on the calculation on the overall heat transfer coefficient, based on the cooling curves.

Figure 19 reports, for comparison, the optical microstructure of the alloys investigated after solidification (no vibration and under vibration). In the case of hypoeutectic AlSi7Mg0.3 alloy (**Figure 19** top), the vibration during solidification determines, first of all, the reduction of the size of the separation of Si and of the intermetallic particles, both in length and width. The α solid solution dendrites becomes more uniform and of rounded shape, with no significant variation of their size (130 μm). For the AlSi12CuMgNi alloy (**Figure 19** middle), the secondary dendrite arm spacing under vibration is reduced (from about 70–40 μm) and the proportion of the eutectic phase is significantly increased and refinement of the Si particles occurs. In the case of the binary AlCu alloy (**Figure 19** bottom), the effect of the vibration determines the reduction (from about 245 to 200 μm) of the secondary dendrite arm spacing and more globular grains have been obtained. Additionally, the Al_2Cu particles become more fragmented.

As cooling velocity increases, the dendritic parameter decreases. Under vibration, germination of the gas bubbles is favoured and there is no promising condition for their expulsion from the melt because of the reduction of the viscosity and of the formation of the solid phase.

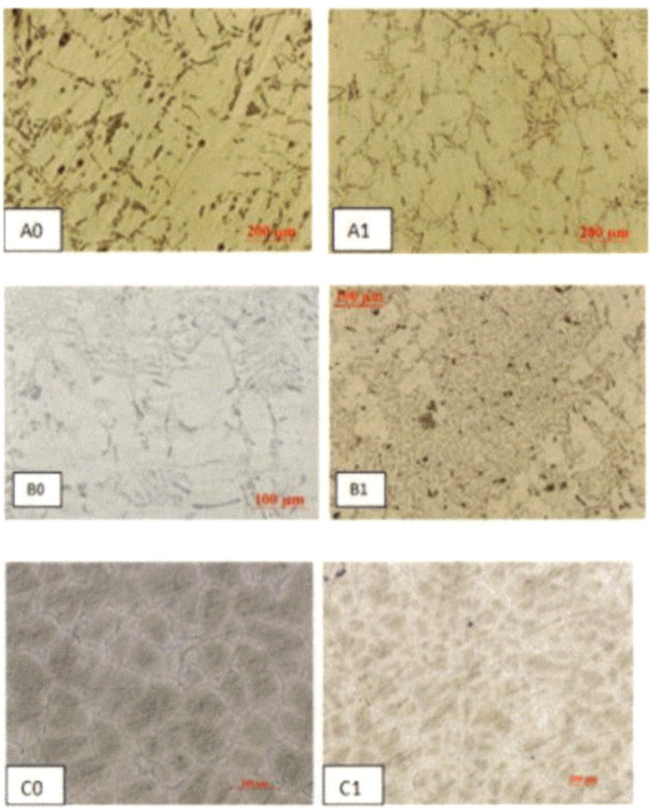

Figure 19. OM microstructure of the alloys investigated: Top: AlSi7Mg 0.3, middle AlSi12CuMgNi and bottom AlCu4; left hand side with no vibration, right hand side under vibration during solidification [9, 10].

The results obtained indicate an increase of the overall heat transfer coefficient value. However, they do not determine the enhancement of the cooling velocity in such a level to provide a significant modification of the casting structure. The schematic interdependence of such elements are summarised in **Figure 20**.

The effect of the rotator electromagnetic field on the structural evolution of light alloys has been further investigated. The casting of some cylindrical samples (Ø30 × 100 mm) in ZnAl4Cu1and AlSi7Mg0.3 alloy has been performed in refractory mould. Under the electromagnetic field action, the melt is subjected to the centrifugal force, like in the traditional centrifugal casting, driving the liquid metal against the die walls of the mould. In the refining procedure, such effect can be considered as a limiting factor and has to be minimised using a refractory material placed on the superior part of the mould when the electromagnetic field is applied. During the experiments, it has been found that the effect of the centrifugal electromagnetic field really has an effect when the frequency arrives to a certain value. At the same time, when the electromagnetic field is coupled, the rotation speed of the melt caused by the electromagnetic phenomenon in the system coil-melt decreases and do not permit sufficient grade of agitation. For this reason, after reaching the frequency, coupling-decoupling and the modification of the rotation senses have been carried out.

Such device has been used for the preliminary investigations using a low melting point alloy, in particular ZnAl4Cu1 alloy.

In case of the hypoeutectic ZnAl4Cu1 alloy, a significant modification takes place: the transformation of the columnar dendrite into a more globular shape and the appearance of a higher level (18%) and finely dispersed eutectoid phase can be observed, as pointed out in the phase diagram. During solidification, the electromagnetic field stimulate the rearrangement of the alloying elements. The evolution of the microstructure is reported in **Figure 21**.

During solidification in the case of AlSi7Mg0.3 alloy, the electromagnetic field determines a decrease of the grain size and the α solid solution globular grains development arises. The eutectic Si does not suffer any significant modification neither in shape nor in size. In the central part of the sample, refining of the α solid solution has been observed and a higher distribution of the eutectic phase has been reached. The most relevant microstructures have been reported in **Figure 22**.

In the first step of the research, the influence of the cooling velocities (encountered in the different casting technology processes) on the structure development has been analysed.

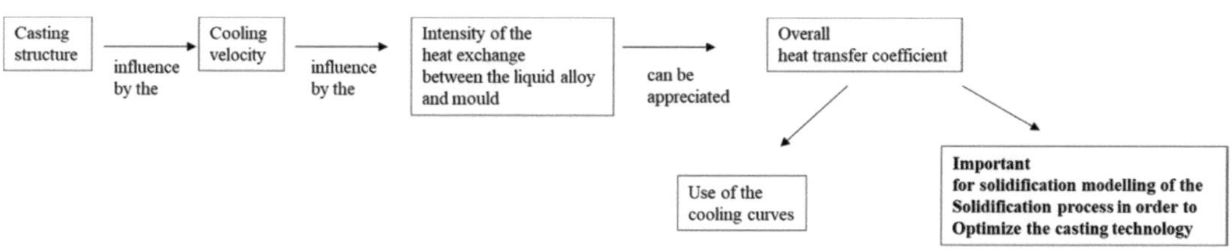

Figure 20. Schematic representation of the interdependence between some features affecting the solidification of the alloys.

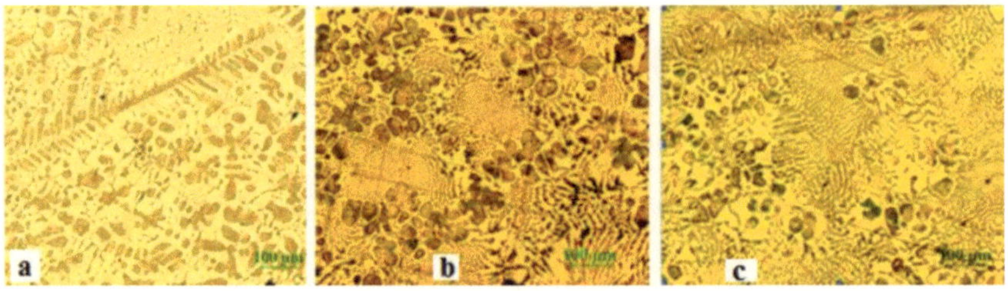

Figure 21. OM microstructure of the ZnAl4Cu1alloy: (a) with no rotator electromagnetic field and under rotator electromagnetic field, (b) top surface and (c) middle part of the sample [8, 10].

Higher the cooling rate finer becomes the structure, as reported in **Figure 23**. When the cooling rate is lower, the presence of acicular Si phase and the separation of the acicular shape particles made of Al_3Fe can be observed. In case of rapid solidification, displacement of the structure in the hypoeutectic zone can be observed with the appearance of α solid solution. At the same time, the acicular separation of small eutectic Si particles appears. Because of the high cooling rate, Fe is present as AlFeMn intermetallic particles in a polygonal shape (lighter shade than the primary Si particles). In case of casting under pressure, the morphology is close to the hypoeutectic structure, where the separation of the eutectic Si and the intermetallic particles are globular (the length/weight fraction tends to 1) and they are uniformly spread out close to the α solid solution. The chemical analysis carried out confirms that the differences between the structures observed are associated to the different solidification conditions and cooling rate.

The presence of Fe as Al9Fe2Si2 or Al12Fe3Si determines minor plasticity to the alloy, which can be balanced by high cooling rate, decreasing the negative effect especially in the case of secondary foundry alloys, where the presence of Fe is higher.

3.3. Casting using melt spinning technology

Significant structural changes, represented by the appearance of α phase, have been obtained in hypereutectic Al-Cu alloy produced by MS-LP technology. Applying a liquid quenching technique, increased solubility of Cu in the Al matrix was reached up to a value of 25.51 wt%.

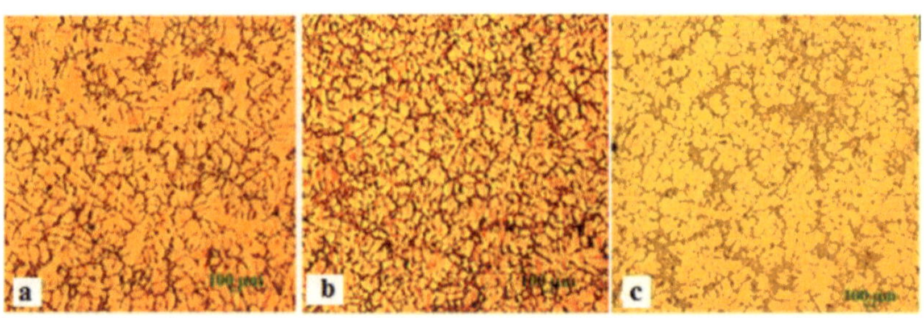

Figure 22. OM microstructure of the AlSi7Mg0.3 alloy: (a) with no rotator electromagnetic field and under rotator electromagnetic field, (b) top surface and (c) middle part of the sample [8, 10].

Figure 23. OM microstructure of the AlSi12CuMgNi alloy in different casting conditions [10].

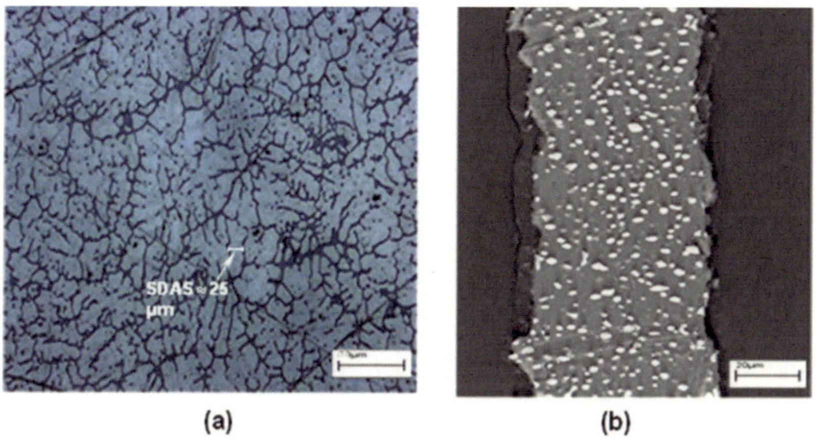

Figure 24. Microstructure of Al10Cu alloy: (a) as-cast in metallic die (OM), (b) ribbon after T_O annealing treatment (SEM) [12].

After annealing treatment applied to melt-spun ribbons, the coarse separations of the intermetallic phase (Al_2Cu) was observed, which results to be embedded into the Al matrix with high plasticity and low content of Cu. Due to a grain refined structure of melt-spun ribbons, Al_2Cu incoherent phase is uniformly distributed in the annealed ribbons, as illustrated in **Figure 24**.

4. Conclusions

Grain refinement determines many advantages in Al-based alloy casting and it can be reached by both metallurgical and physical methods. By using grain refinement method, for both cast and wrought aluminium alloys, a fine distribution of the phases, better castability, decrease of shrinkage porosity, higher mechanical properties, as well superior fatigue life were obtained.

However, the purpose of this chapter was twofold, it is not exhaustive on this topic. The first section was dedicated to present a short outlook about some possibilities to refine the structure of Al-based alloys based on some literature data and continues with the presentation and discussion of some experimental results obtained recently during the research performed within the PhD Thesis [10] targeted to the improvement of the properties of Al-based alloys by physical grain refinement methods.

The solidification process can be governed by the control of the time required for the cooling and acting directly on the alloy, on the casting profile or on the final product. Generally, the electromagnetic field and solidification under vibration are advantageous determining a decrease of the grain size; consequently in such conditions, a fine microstructure development is favoured. It comes out that significant microstructural changes can be obtained using physical methods, which based on the characteristics of the alloy composition and on the real use of the final component can be managed in such a way to produce a better quality Al-based alloy castings.

Author details

Ildiko Peter[1]* and Bela Varga[2]

*Address all correspondence to: ildiko.peter@polito.it

1 Department of Applied Science and Technology, Politecnico di Torino, Torino, Italy

2 Material Science Department, Transilvania University of Brasov, Brasov, Romania

References

[1] McCartney DG. Grain refining of aluminium and its alloys using inoculants. International Materials Review. 1989;**34**:247-260

[2] Kashyap KT, Chandrashekar T. Effects and mechanisms of grain refinement in aluminium alloys. Bulletin of Materials Science. 2001;**24**(4):345-353

[3] Sigworth GK, Kuhn TA. Grain Refinement of Aluminum Casting Alloys, AFS Transactions 2007 © Schaumburg, IL, USA: American Foundry Society; 1-12

[4] Hu S, Dai Y, Gagnoud A, Fautrelle Y, Moreau R, Ren Z, Deng K, Li C, Li X. Effect of a magnetic field on macro segregation of the primary silicon phase in hypereutectic Al-Si alloy during directional solidification. Journal of Alloys and Compounds. 2017;**722**:108-115

[5] Lu HT, Wang LC, Kung SK. Grain refining in A356 alloys. Journal of Chinese Foundrymen's Association. 1981:10-18

[6] Rosso M, Peter I, Castella C, Lombardo S. Grain Refinement of Self-Hardening Aluminum Alloys. TMS (The Minerals, Metals & Materials Society) 145th Annual Meeting and Exhibition Febr. 14-18. 2016. Nashville: Tennessee (USA).

[7] Popescu M, Bedo T, Varga B. Determination of overall heat transfer coefficient by simple thermal analyse. The Scientific Bulletin of Valahia University–Materials and Mechanics. 2016;**11**:11-15

[8] Popescu M, Varga B. Microstructure of aluminium alloys solidified by rotating electric field. The Scientific Bulletin of Valahia University – Materials and Mechanics. 2015;**10**: 44-48

[9] Popescu M, Varga B. Structures of Aluminum base alloys solidified by vibrations. Advanced Materials Research. 2015;**1128**:88-97

[10] M.Popescu, PhD Thesis. Improving of Aluminum Alloys Structure by Metallurgical and Physical Methods; 2017. Tutor: B. Varga.

[11] Peter I, Varga B, Rosso M. Eutectoid transformation in Zn-al alloys solidified by rapid cooling. Materials Science Forum. 2014;**790-791**:223-228

[12] Lichioiu I, Peter I, Varga B, Rosso M. Preparation and structural characterization of rapidly solidified Al-cu alloys. Journal of Materials Science and Technology. 2014;**30**(4): 394-400

[13] Varga B, Peter I. Structural transformations in Zn-Al22 alloy solidified by ultra-rapid cooling. Metalurgia International. 2013;**18**(6):9-12

[14] Easton MA, Qian M, Prasad A, St John DH. Recent advances in grain refinement of light metals and alloys. Current Opinion in Solid State and Materials Science. 2016;**20**(1):13-24

[15] Lee YC, Dahle AK, John DHS, Hutt JEC. The effect of grain refinement and silicon content on grain formation in hypoeutectic Al-Si alloys. Materials Science and Engineering A. 1999;**259**(1):43-52

[16] Chen Z, Wang T, Gao L, Fu H, Li T. Grain refinement and tensile properties improvement of aluminum foundry alloys by inoculation with Al-B master alloy. Materials Science and Engineering A. 2012;**553**:32-36

[17] Chen Z, Kang H, Fan G, Li J, Lu Y, Jie J, Zhang Y, Li T, Jian X, Wang T. Grain refinement of hypoeutectic Al-Si alloys with B. Acta Materialia. 2016;**120**:168-178

[18] Birol Y. AlB3 master alloy to grain refine AlSi10Mg and AlSi12Cu aluminium foundry alloys. Journal of Alloys and Compounds. 2012;**513**:150-153

[19] Limmaneevichitr C, Eidhed W. Fading mechanism of grain refinement of aluminum–silicon alloy with Al-Ti–B grain refiners. Materials Science and Engineering A. 2003; **349**:197-206

[20] Bunn AM, Schumacher P, Kearns MA, Boothroyd CB, Greer AL. Grain refinement by Al–Ti–B alloys in aluminium melts: A study of the mechanisms of poisoning by zirconium. Journal of Materials Science and Technology. 1999;**15**:1115-1123

[21] Birol Y. An improved practice to manufacture Al–Ti–B master alloys by reacting halide salts with molten aluminium. Journal of Alloys and Compounds. 2006;**420**(1-2):71-76

[22] Arnberg L, Backerud L, Klang H. Intermetallic particles in Al–Ti–B–type master alloys for grain refinement of aluminium. Metals Technology. 1982;**9**:7-13

[23] Arnberg L, Backerud L, Klang H. Evidence of metastable phase in Al-Ti-B system. Metals Technology. 1982;**9**:14-17

[24] Birol Y. Effect of the salt addition practice on the grain refining efficiency of Al–Ti–B master alloys. Journal of Alloys and Compounds. 2006;**420**:207-212

[25] Eckert S, Nikrityuk PA, Räbiger D, Eckert K, Gerbeth G. Efficient melt stirring using pulse sequences of a rotating magnetic field: Part I. Flow field in a liquid metal column. Metallurgical and Materials Transactions B: Process Metallurgy and Materials Processing Science. 2007;**38**:977-988

[26] Mikelson AE, Karklin YK. Control of crystallization processes by means of magnetic fields. Journal of Crystal Growth. 1981;**52**:524-529

[27] Morikawa H, Sassa K, Asai S. Control of precipitating phase alignment and crystal orientation by imposition of a high magnetic field. Materials Transactions, JIM. 1998;**39**(8): 814-818

[28] Li X, Du DF, Fautrelle Y, Ren ZM, Moreau R. Investigation on macrosegregation and dendrite morphology during directional solidification of Al-cu hypereutectic alloys under a strong magnetic field. Europhysics Letters. 2015;**111**:48002-48008

Chapter 3

Continuous Casting of Thin Aluminum Strips Using the Electromagnetic Levitation

Marcela Pokusová

Additional information is available at the end of the chapter

http://dx.doi.org/10.5772/intechopen.70681

Abstract

In the development of the aluminum (Al) alloys continuous casting, the significant change was involved by replacing the mechanical contact between solidified molten metal and the mould's wall with the electromagnetic force action. Here, the function of the strip's surface shaping takes the special inductor known as the electromagnetic mould (EMM). The method for calculation the horizontal EMM's parameters for continuous casting of the thin strip from the Al alloys is presented. It is based on computing the distribution of the magnetic field using the interaction between an inductor and surface of the shaped molten metal. The operating frequency of alternating current passing through the inductor, which depends on the density, the surface tension, specific electric conductivity of the liquid metal, and the pool thickness, is determined. To shape the upper and bottom peripheries of the strip of 20 × 150 mm, the system of inductors was developed, and parameters for pouring of the AlSi8.5Cu and AlSn20 alloys were computed. The proposed EMM design considers the magnetic field distribution inside the inductor and the size of air gap between the liquid metal and inductor.

Keywords: aluminum alloys, strip, continuous casting, liquid metal, surface shaping, electromagnetic force, levitation

1. Introduction

Introducing the continuous casting of metal materials has made an increase in the productivity and quality of semi-products possible, but has also shown many shortcomings and reserves of this technology. In addition to the internal negative traits of the molten metal determined by the physical phenomena related to the solidification, in many materials, the surface defects occurring have already arisen in the mould—crystallizer itself, where the liquid metal has the mechanical contact with its walls. The friction forces between the solidified molten metal and crystallizer's walls cause the formation of the skin breakings or cracks, which damage the part of a surface and make harder the next processing, for example, rolling. Therefore, the research and development have focused firstly on the issue of improving ingots' surface quality. For the continuous casting technologies of the Al alloys, development has been initially concentrated on narrowing the zone, where the shaping of liquid metal occurs. However, the results achieved did not give satisfactory improvement in the surface quality.

An important milestone was reached in the development of continuous casting for aluminum alloys, with the substitution of the mechanical contact of the solidifying molten metal with the mould's walls for shaping a free surface, with the use of an electromagnetic force instead. The function of confining and shaping was taken over by a special coil—inductor known as an electromagnetic mould (EMM) [1]. The conventional vertical EMM has been successfully used on a slab section up to 1 m^2. Insights from the practice as well as theory have concurred that the vertical EMM is not suitable for continuous casting of the Al alloys semi-products with a cross section less than 0.01 m^2, and in particular for strips with a thickness of less than 25 mm [2]. Likewise, the laboratory testing of a conductive method for generating the levitation with the additional induction confining and shaping did not provide results which would justify the continuing development in this direction.

Today, when the trend is toward reducing the slab section and producing a near-net-shape product, horizontal electromagnetic casting can offer a solution to this problem [2]. Many studies have been conducted on the vertical EMM, focusing on free surface shaping [3–6], but only a small number of papers have been published on computing the parameters of horizontal EMM and determining their boundary values [7]. The EMM development has begun with the aim of replacing the technology of continuous casting of a 150 × 30 mm strip into a horizontal graphite mould from which the desired semi-product with a section of 150 × 20 mm, and without any surface defects was achieved by milling the 5 mm layer from each surface.

2. Horizontal electromagnetic mould

The development of an electromagnetic mould arranged horizontally began when it was discovered experimentally that as the molten metal approaches the mould, the value of the electromagnetic repelling force increases simultaneously. When the air gap is reduced from 5 to 0 mm, a steep increase in the repelling force ensures a steady positioning of the molten metal's surface to the inductor. In the early stages of investigation, the calculation of the parameters for the horizontal EMM (HEMM) started with the experimentally determined dependence of the air gap a on the magnetic field intensity H. This was then transformed into a mathematical form using regression analysis [8]. In the next stage [9], increasing the pressure generated by the magnetic field when the air gap is reduced was explained as the result of interactions between the electrical current of the inductor and the conductive liquid metal. On this basis, the relations were formulated to compute the dependence of a-H in the working space of a real inductor using the finite element method [10].

2.1. Interaction between magnetic field of a conductor and its mirror image

If a conductor, carrying an alternating current I, is getting closer to an electric conductive surface, the electric current is induced in this plane. Its intensity corresponds to the local value of the magnetic field intensity, but it has an opposite sign. Here, the magnetic field and the spatial distribution of lines with the same value of the intensity $\boldsymbol{H}$ are determined by the interaction between this conductor and a fictitious one which is its mirror image below a plane which the magnetic field penetrates through, as it is shown in **Figure 1**.

Let us choose the conductor of a circular cross section with a radius R, which is passed by the electric current with a density $\boldsymbol{J}$. From Biot-Savart's law, it follows that in this single conductor, at any radius r the circular magnetic field of intensity $\boldsymbol{H}_r$ induced by electric current $\boldsymbol{I}_r = \boldsymbol{J}\pi r^2$ will act, where

$$\boldsymbol{H}_r = \frac{\boldsymbol{I}_r}{2\pi r} = \frac{\boldsymbol{J}\pi r^2}{2\pi r} = \frac{\boldsymbol{J}r}{2} \tag{1}$$

In generating the magnetic field at the radius r, the only current $\boldsymbol{I}_r$ is involved which is passing through the surface enclosed by a circular vector $\boldsymbol{H}_r$, hence the current flowing in a cylindrical space with the radius from zero to r. The electric current passing through a circular ring between R and r is not involved in producing the magnetic field $\boldsymbol{H}_r$, because of the contribution of this area into the cylinder with radius r is zero, where $\oint \boldsymbol{H}ds = 0$. According to Eq. (1), at conductor's axis, the intensity of $\boldsymbol{H}_r$ is zero, and on its surface, since $r = R$, it is the highest. When the radius increases beyond conductor's contour, that is, $r > R$, the magnetic field is already produced by the current $\boldsymbol{I} = \boldsymbol{J}\pi R^2$, and the value of $\boldsymbol{H}_r$ is $\boldsymbol{H}_r = \boldsymbol{I}/2\pi R$. At any point on the connecting line of axes of both conductors, the magnetic field intensity in scalar form will be equal to a sum of contributions H' and H'' from both conductors, and the course of total values $H = H' + H''$ is presented in **Figure 1a**. From **Figure 1a**, it follows that in the conductors at the distance l'', there are located points 1 and $1'$, at which the value of H is zero, when

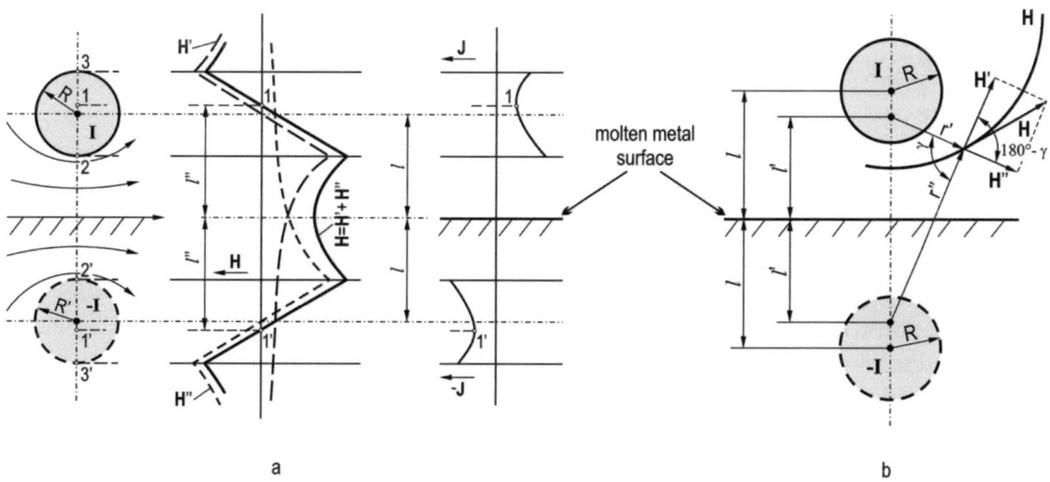

Figure 1. Interaction between the magnetic fields of a real conductor and its mirror image in a liquid metal: (a) distribution of current densities $\boldsymbol{J}$ and magnetic field intensities $\boldsymbol{H}$ on the connection line of axes of both conductors; (b) asymmetrical distribution of the magnetic field lines of force around a real conductor and its mirror image.

$$H = H' + H'' = 0 = \frac{J\left(l'' - l\right)}{2} - \frac{JR^2\pi}{2\pi\left(l'' + l\right)} = \frac{J\left(l'' - l\right)\left(l'' + l\right) - JR^2}{2\pi\left(l'' + l\right)} \tag{2}$$

The last term on the right side will be null if its numerator is null.

$$0 = J\left(l'' - l\right)\left(l'' + l\right) - JR^2 = l''^2 - l^2 - R^2 \text{ when } l''^2 = l^2 + R^2 \tag{3}$$

At the connection line of conductors' axis at the distances 1 and 1′, the density of electric current passing through the conductors will be also the lowest. At both conductors, the asymmetrical distribution of the magnetic field lines of force around inductor's conductor and its mirror image will cause an increase in the current density in the areas adjacent to a plane of the magnetic field penetration, and hence shifting the axes of symmetry of the electric current flowing on the value $l - l'$ toward each other (**Figure 1b**). From the position of points with the null intensity of *H*, using Eq. (3), we can calculate the distance of the axis of symmetry of the current flow in space [10–12].

$$l^2 = l'^2 + R^2 \tag{4}$$

At a general point of space determined by an intersection point of arcs of the circles with radii r' and r'', the localized value *H* is given by the contribution from both the conductors $H = H' + H''$. We can regard the components H' and H'', defined by Eq. (1), as straight and lying in the tangent lines of circles in which the centers are in a joining line of conductors' centers at a distance l' from a considered plane. According to **Figure 1b**, the absolute value *H* can be calculated using the law of cosines.

$$H = \sqrt{H'^2 + H''^2 - 2H'H'' \cos\gamma} = \frac{2Il'}{\pi r'^2 r''^2} \tag{5}$$

The lines of a constant value *H* are representing the distribution of the force lines of the magnetic field (a segment of one is plotted in **Figure 1b**), and they are sections of the equipotential surfaces through a plane perpendicular to conductors' axes, the real one and its mirror image. Because of between the conductor and its mirror image the magnetic field intensity is the sum of contributions $H = H' + H''$, the total value *H* will be obviously higher than that one induced by a single conductor with the same flow of electric current on the same radius, and the highest value will be in a connecting line of their axes. Applying the iterative method to calculate the spatial distribution of the electric current and magnetic field intensities according to Sakane and others [4, 13, 14], without an intentional employing the mirroring, the values obtained under the model conditions are very close to those computed using Eqs. (1)–(5).

2.2. Calculation of magnetic field distribution in horizontal EMM

The method of HEMM design presented is based on the calculation of the magnetic field distribution, which is based on the interaction between the inductor and the surface of the shaped molten metal. The condition of equilibrium between gravitational force and electromagnetic force in the HEMM can be defined according to the equation [3, 15]

$$\rho g h = \frac{\mu H^2}{4}, \text{ when } h = \frac{\mu H^2}{4\rho g} \tag{6}$$

where h is the height of liquid metal (m), ρ is the density of molten metal (kg m^{-3}), H is the magnetic field intensity (A m^{-1}), and μ is the permeability of a medium, when $\mu \approx \mu_0 = 4\pi \times 10^{-7}$ H m^{-1}.

The analysis of the specific situation when the molten metal is brought closer to the inductor's face will be conducted with the inductor which has a cross section with an external central peak as shown in **Figure 2**.

As it was mentioned above, the interaction of fields of the inductor and its mirror image moves the centers of the electrical current flow closer together and increases the current density at the inductor's face adjacent to the liquid metal. In such a case, the conductor with radius R through which the current I passes, at a distance l from the electric conductive face of the molten metal, will generate a magnetic field equal to the one generated by a conductor of zero thickness

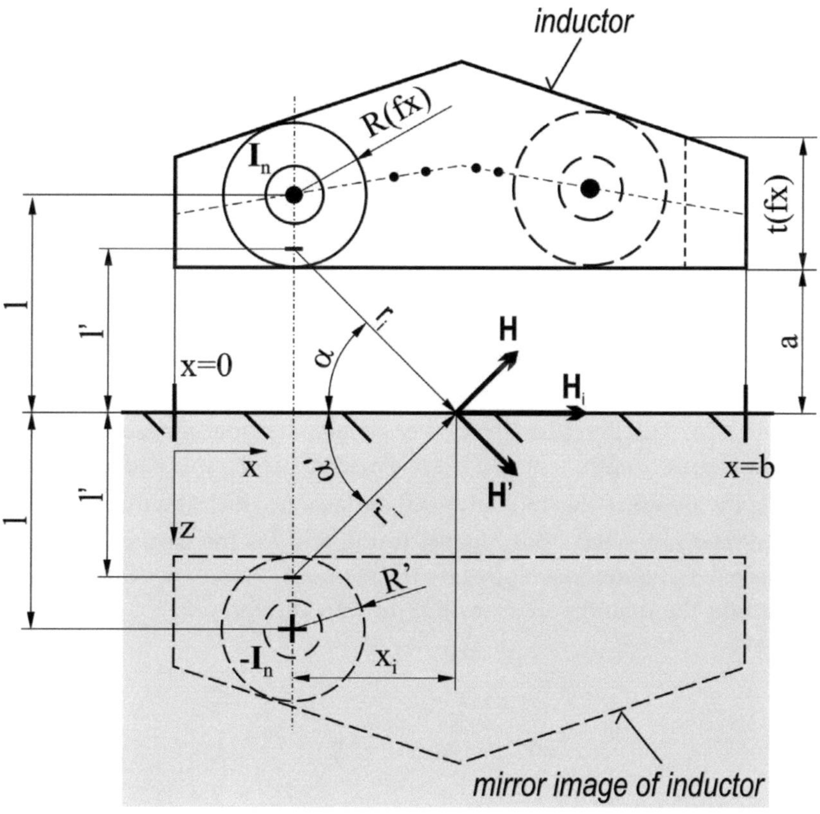

Figure 2. Inductor with a central peak.

having the values of the current I the same, but at the distance of l' from the surface different, that is, $l^2 = l'^2 + R^2$ according to Eq. (4). Then the magnetic field intensity H_i on the molten metal surface will be the vector sum of both components from the conductor itself H and from its mirroring image H'. If the inductor is at a distance a from the liquid metal, when $l = a + R$, the addition of intensity H_i from the selected element at a distance x_i from the perpendicular to the surface will be

$$H_i = \frac{I}{2n\pi r_i} \, . \sin \alpha + \frac{I}{2n\pi r'_i} \, . \sin \alpha' \tag{7}$$

The conductor's cross section of geometric shape, as shown in **Figure 2**, introduces another variable into the calculations, namely that when changing the coordinate x, the thickness t also modifies the value of the hypothetical conductor radius $R_{(fx)}$. After dividing the inductor conductor into n elementary conductors, through each one of which an nth share of the total current I passes, $I_n = I/n$. The radius of each of these conductors $R_{(fx)}$ will be a function of x_i or i. The magnetic field intensity H_{x_i}, which is acting on the surface at a point at a distance x_i from the perpendicular drawn from the hypothetical conductor on the liquid metal, can be computed from the value of current I_n passing through the particular elementary conductor, using the following equation:

$$H_{xi} = \frac{I}{n\pi} \cdot \frac{\sqrt{\left(R_{(fx)i} + a\right)^2 - R_{(fx)i}{}^2}}{\left(R_{(fx)i} + a\right)^2 - R_{(fx)i}{}^2 + x_i^2} = \frac{I}{n\pi} \cdot \frac{\sqrt{a^2 + 2aR_{(fx)i}}}{a^2 + 2aR_{(fx)i} + x_i^2} \tag{8}$$

At a common point on the surface at the x coordinate and with different values of R, the total value H_x will be the sum of the n additions H_{x_i} from all n elementary conductors and their n mirror images, that is,

$$H_x = \sum_{i=0}^{i=n} \frac{I}{n\pi} \cdot \frac{\sqrt{a^2 + 2aR_{(fx)i}}}{a^2 + 2aR_{(fx)i} + x_i^2} \tag{9}$$

The relation $N = H^2/2\delta\sigma$ [15] describes the power generated under the surface of a specific unit area in a volume of great depth, where σ is the electrical conductivity of liquid metal (S m^{-1}), $\delta = \sqrt{1/\mu\sigma\pi f}$ is the depth under molten metal surface at which the magnetic field intensity decreases e times and is termed "skin depth" (mm), and f is the frequency (Hz). Looking at Eq. (6), we can see the pressure developed by the magnetic field varies linearly with the value H^2. If we substitute the quantity H from Eq. (6) into the formula for the power produced $N = H^2/2\delta\sigma$, we get

$$N = \frac{H^2}{2\delta\sigma} = \frac{2g\rho h}{\mu\sigma\delta} = 2g\sqrt{\frac{\pi}{\mu}}\,\rho h\sqrt{\frac{f}{g}} \tag{10}$$

Eq. (10) determines the value of the power supplied by the inductor at the unit area of the molten metal surface, which depends on the height of the molten metal column. It represents conditions of equilibrium between the power of the magnetic field and the metallostatic pressure, taking physical properties (the electrical conductivity σ and the density ρ) of the liquid metal, and also the frequency f into consideration. The alternating magnetic field performs only negligible mechanical work, and by way of induced currents, almost all the energy is transmitted to the slab as heat, where the power N in watts is numerically equal to the quantity of the generated heat Q in J s^{-1}.

Taking the active power N as a criterial quantity to determine the equilibrium conditions between the mechanical and electric quantities for the liquid metal levitation can be considered the very applicable decision. It provides the correct information about the amount of heat required for the thermal analysis and eliminates substantially the need for measuring the electric quantities, as voltage and the electric current in an inductor, its inductance, the finding cos φ, and the challenging calculations. As it follows from Eq. (10), at the given physical properties of liquid metal and a height of the molten metal column h, the power required to generate the levitation will be directly proportional to the square root of the working frequency.

3. Role of operating frequency of EMM

After entering the liquid metal, the alternating magnetic field induces an electrical current in it, which generates its own magnetic field, but the opposite one. As a result, the field decays exponentially when penetrating, and both interacting fields produce the force acting in the liquid metal in the direction from the inductor. At any depth z, in every element of the volume having an area of 1×1 m^2 and the depth dz, the increase of force df arises, and is numerically equal to the increase in the pressure dp (N m^{-2}). At any depth z under the surface, we calculate the total value of pressure p_m using the integration of all the additions in an interval from the surface, where $z = 0$, to the depth z, when we get [4, 16].

$$p_m = \int_0^z \left(\frac{\mu H^2}{2\delta}\right) e^{-2z/\delta} dz = \left(\frac{\mu H^2}{4}\right)\left[e^{-2z/\delta}\right]_0^z \tag{11}$$

If $z = \infty$ (or $z > 3\delta$) then exp $- 2z/\delta$ approaches 1, and the expression of p_m is identical with the expression for electromagnetic force in Eq. (6). As follows from the principle of generation of electromagnetic forces, they do not act on the surface itself but arise in the subsurface layer, on which the surface is supported. The center of their action lies at the depth z_c, which divides the molten metal volume into two parts in which an equal portion of the total pressure is created. According to Eq. (11), 50% of the generated pressure will be acting in the subsurface layer, where exp $- 2z_c/\delta$, thus $z_c \approx 0.3466\delta$. At this depth, the surface is supported and, therefore, the operating frequency of the current flowing through the inductor is the important parameter [7, 17]. The frequency needs to be both high enough so that the surface layers do not flow vertically down from the faces of the shaped liquid metal, and as low as possible because any increase also increases electric energy consumption. Therefore, during shaping of a radial profile, surface tension forces are always involved. These, together with the electromagnetic force, produce the equilibrium with the metallostatic pressure of the liquid metal column on the molten metal surface. The surface tension τ determines that the shape of the fluctuated liquid metal on the surface is semicircular with radius r, in which a pressure is created which is able to overcome the metallostatic pressure produced by the molten metal column h. Under the gravity acceleration g, the equilibrium force can be written thus

$$\rho g h = \frac{2\tau}{r} \tag{12}$$

When appraising the function of the surface tension for shaping the free surface, we will start with the situation presented in **Figure 3**, which can arise when, as a result of an unintended fluctuation in the molten metal, a bulge arises on its surface, and the surface tension force forms it into a hemispherical shape. The analysis is based on model conditions when the uniform magnetic field is acting, thus ignoring the increase in its intensity caused by reducing the air gap.

At the point of the jut creation, the metallostatic pressure is balanced by the pressure developed by the magnetic field and the centers of its action lay at a depth of 0.3466δ under the surface in the direction of the propagating field. The array of these centers of action has the same hemisphere shape with the radius r_v, equal to that of the bulge, but its center is shifted under the plane of the liquid metal at a depth of 0.3466δ. At the axial cross section of the bulge, the surface of the spherical cap above the liquid metal plane is shaped like an arc of a circle, at the center of which the total pressure developed by the field is acting. The center of an arc of a circle according to analytic geometry lies on its axis at a distance of $r_v t/l$ from the center where t is the chord length and l is the length of this arc.

Gravitational force acts on the hemisphere's center which is located at a distance greater than $3/8r_v$ from the plane of the molten metal pool. In order to prevent the hemisphere from sliding

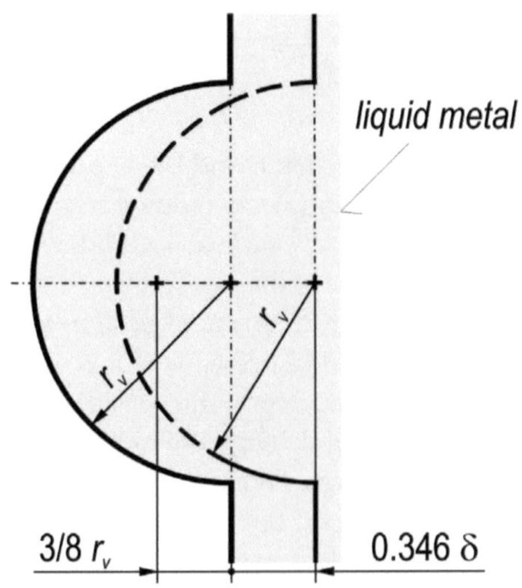

Figure 3. A bulge on a free surface.

http://dx.doi.org/10.5772/intechopen.70681

in the direction of gravity, the center of the electromagnetic force acting on it has to be located at a distance greater than 3/8 r_v from the molten metal plane. In a threshold situation, both centers will be identical $3/8r_v+0.3466\delta=r_vt/l$, which is fulfilled when $r_v=0.816\delta$. Inside the bulge the metallostatic pressure of the liquid metal column of the height $h=2r_v$, which is in equilibrium with the surface tension force, is experienced. If we substitute into Eq. (12) for $h=2r_v$ and the dependence $r_v=0.816\delta$, under these conditions the centers of electromagnetic and gravitational forces have a congruent location, then the expression relating to the minimum working frequency f_{min}, after transforming takes the form:

$$f_{min} = 0.816^2 \frac{g}{\mu\pi} \frac{\rho}{\sigma\tau} = 1.654.10^6 \frac{\rho}{\sigma\tau} \tag{13}$$

Replacing the density ρ, the surface tension τ, and the specific electrical conductivity σ of the particular shaped liquid metal into Eq. (13), we can compute the lowest working frequency f_{min}, at which the EMM is still functional. The working frequency of the current passing through the inductor is an important parameter for EMM functioning. It has to be high enough to ensure the stability of the free surface and at the same time, as low as possible because increasing the frequency increases energy consumption.

Experiments carried out with liquid Al alloys or PbSn in the inductor with a working frequency of 10 kHz have shown that small volumes of the liquid metal placed in a ceramic crucible in the inductor's working space had a tendency to spontaneously move from side to side and their free surface to ripple. This was irrespective of whether the crucible was orientated vertically or horizontally. When increasing the pool dimension, adding more than three times the magnetic field skin depth of metal in the direction of the magnetic field penetration, a marked decline in the oscillations in the liquid metal surface could be observed. After substituting the empirical condition $d = 3\delta$ for the skin depth expression, we can determine the lowest applicable working frequency f_d limited by the pool's thickness d, when $f_d=2.277\,.\,10^6/\sigma d^2$. The working frequency of the EMM has to be higher than or equal to the highest one of the values limited by the surface tension, the minimum frequency f_{min}, and the dimensions of the liquid metal pool f_d. To introduce the dependence of the minimum pool dimension d_{min} on the working frequency into the conditions defined by Eq. (10), we substitute $h=3\delta$ for the column height. Under this condition, the value of the power density N_d will be constant and proportional to the density and the specific electric resistance of liquid metal:

$$N_d = 4.684.10^7 \frac{\rho}{\sigma} \tag{14}$$

At the edge of the shaped molten metal, the surface tension creates the radius r_τ, and hence in the case of the levitating molten metal pool, its height has to be $h \geq r_\tau$, because the edge radius r_τ defines the minimum possible thickness of the shaped strip $d_\tau=2r_\tau$, which depends on the height of molten metal column h and determined from the surface to the strip center [8]. For the particular alloys AlSi8.5Cu (ρ = 2.4 × 10^3 kg m^{-3}; σ = 2.56 × 10^4 S m^{-1}; τ = 0.865 N m^{-1}) and AlSn20 (ρ = 3.4 × 10^3 kg m^{-3}; σ = 2.07 × 10^4 S m^{-1}; τ = 0.8 N m^{-1}) [10], the conditions of the equilibrium between the power generated by the magnetic field and the level of the molten metal pool formulated in Eq. (10) are processed in **Figure 4**. The set of parameters (**Figure 4**) is

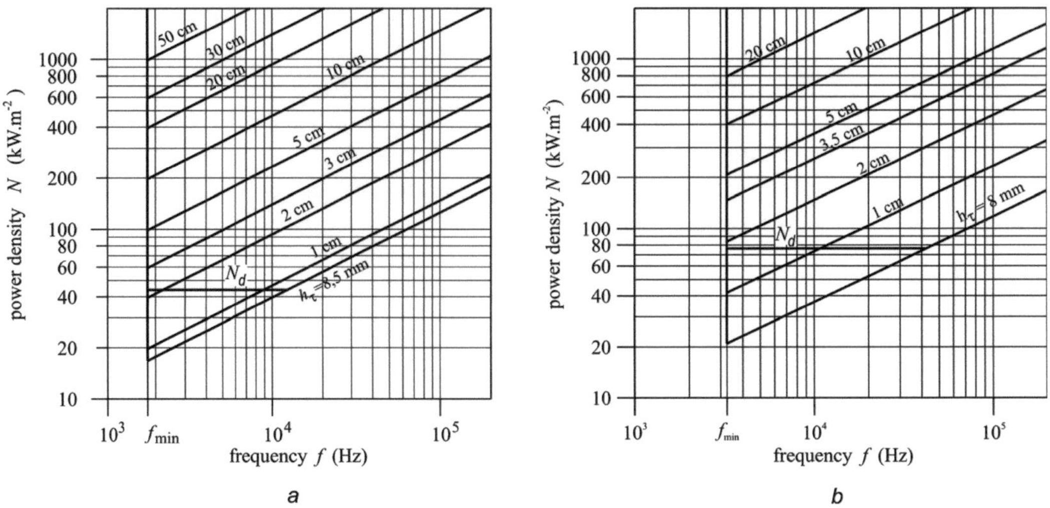

Figure 4. Dependence of power per area unit on a frequency at different heights of molten metal column of alloys AlSi8.5 (a) and AlSn20 (b).

limited by power of 2000 kW m^{-2}, because using parameters above this value entails problems with the removal of generated heat [10, 18].

In a co-ordinate system with a logarithmical scale, where the values of power density N are on the vertical axis and frequency values f are on the horizontal one, the conditions of the equilibrium N-h are shown as a system of lines for the separate values of the height of molten metal column. The boundary conditions at values $r_\tau = h_\tau = h_{min}$ and f_{min} are lines which limit the interval of the parameter values under which the conditions of reliable EMM performance are met. The line of values N_d, where the condition $h = 3\delta$ holds, allows us to read the minimum thickness of the pool $d_{\min}$ for a given frequency. For the given thickness d, the line of value $h = d$ intersects the line N_d at the frequency f_d, which is the lowest value which can still be used to cast the strip of thickness d. Bellow the line N_d, at the height interval $r_\tau \leq h \leq 3\delta$, there are a range of parameters under which the performance of the vertical EMM is maintained.

4. Testing the horizontal inductor of rectangular cross-section

At the development of horizontal electromagnetic mould, much significance has been attached to experiments, the aim of which was to find out a dependence of magnetic field intensity H on the value of a gap between the molten metal surface and the inductor in inductor's working space [10]. The experimental measurements were carried out in the inductor, which had the simple rectangle cross section of the 10 × 40 mm (**Figure 5a**). That inductor has provided a good approach to the working space when taking a measurement. In **Figure 5b**, there are presented the experimental found out courses of magnetic field intensity H depending on gap's size and inductor's shape when alternating current of frequency of 10 kHz and intensity of 3200 A (rms value of 2260 A) were imposed.

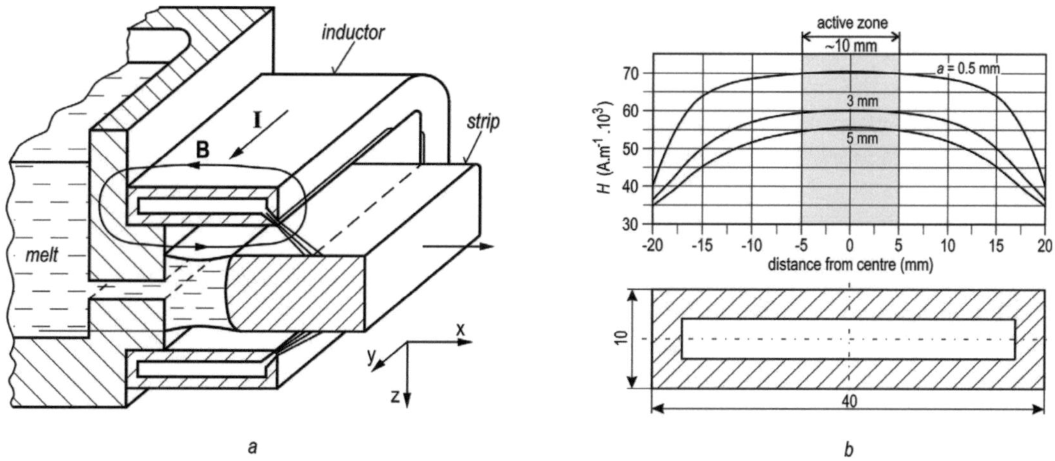

Figure 5. Horizontal EMM with a inductor of rectangular cross section (a) and measured magnetic field intensity *H* under imposed alternating current of 10 kHz frequency and intensity of 3200 A (rms value of 2260 A) (b).

From the shape of curves, it is seen that inside the inductor's operating space the magnetic field is not homogenous, and at the edges, it drops markedly. Therefore, to effectively stabilize the molten metal position against the inductor, it is better to choose the smaller air gaps. Then even the small deviation in the position will cause sufficiently strong changes of magnetic field intensity, which effectively can adjust the incidental deflection of the liquid metal position produced by changes of metallostatic pressure or resulting from the movement fluctuation.

Laboratory HEMM enabled to take measurements at gap up to 10 mm, and the molten metal of the height up to 70 mm, which was kept in a corundum crucible or free levitated. Part of experimental measurements was performed under model conditions in the vertically oriented working space when the molten metal pool was supported from below by a fireclay board. The solid aluminum alloy was inserted into the inductor, and after its meltdown, the desired height of the molten metal was reached by adding an extra metal. Experimentally attained sets of *H-a* values have provided data for next analytic works and development of expression which enable to determine the value of H_x at any point of the molten metal surface having x-coordinate:

$$H_x = \sum H_i = \frac{I}{n\pi} \sum_{x=-\frac{xn}{b}}^{\frac{xn}{b}-n} \frac{\sqrt{a^2 + 2aR}}{a^2 + 2aR + \left(\frac{b}{n} i\right)^2} \tag{15}$$

A plain shape of the inductor enabled to substitute the variable x for multiple of a length of the molten metal surface part being of the length b/n; then, $x_i = ib/n$, when i is integer. The performed experiments have shown that accuracy of calculation H_x for the individual gap values at dividing $n \geq 100$, and the defined geometric shape of the inductor depends namely on the accuracy of determining the value of alternating current intensity and on a method of gap measurement, because the surface of the levitated molten metal rippled moderately having an amplitude approximately ± 0.5 mm.

Using the liquid AlSi8.5Cu1 alloy, for the gap of 3 mm being calculated under the rms current of 2260 A, there was experimentally measured the value of approximately 3–3.2 mm. Differences between the calculated and measured values did not exceed 10%. A validation has shown there is not necessary to use a visual assignment of values from experimentally attained curves *H-a*. It is more advantageous to employ the accurate calculation of *H-a* values which is based on interactions between the inductor and its mirror image known from the theory of induction heating [11, 12] and electromagnetic forming [19].

Based on experiments which were aimed at determining the lower power limit of the EMM and HEMM parameters, the hypothesis was developed that the fundamental limiting factor is the internal motion in the liquid metal produced by the non-potential action of the electromagnetic forces which would inevitably impact on the primary structure formation [20]. To verify this assumption, an experiment was carried out where in the vertical-oriented inductor a cone of Al alloy was melted down on a thermo-insulating insert, the center of which was cooled by immersing a water-cooled copper pipe of outer diameter of 10 mm and bent to form an arch with a radius of 20 mm. After stabilizing the heat balance, most of the liquid metal was solidified and after switching off the device, the block of solidified metal containing the cooling pipe was taken out of the inductor.

As shown in **Figure 6**, at the edge of the sample adjacent to the inductor where the cooling rate was the lowest, and however, where the level of the effect from the alternating magnetic field and the electric current flow was the highest, the gradual transition of an eutectic morphology from finely dispersed at the edge to the typical dendritic structure on the length corresponding to the skin depth δ can be observed.

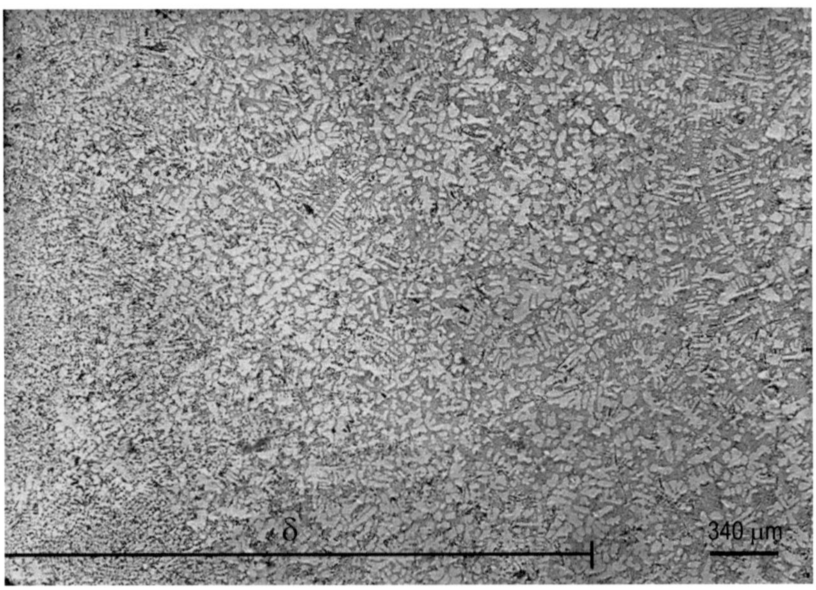

Figure 6. Microstructure of the AlSi8.5Cu alloy from the sample's edge to a distance of 1δ.

The structure consists of fine dendrites of the primary solid solution with no clear orientation. Under the surface at a depth greater than $\delta \approx 3.2$ mm, the microstructure is almost the same as was found over most of the section in central zones of the sample. It consists of the primary solid solution and eutectic (α + Si) of the transition morphology type. The fact that more marked changes in the morphology occur only at a distance about the skin depth δ from the surface, can most likely attributed to the internal motion in the subsurface layer.

5. Design of industrial horizontal EMM

Experiments [9, 10] conducted with the laboratory HEMM comprising one coil inductor having a conductor section of 10 × 40 mm for casting the profile of 150 × 20 mm, as illustrated in **Figure 5a**, proved the feasibility of electromagnetic shaping the strip in the horizontal arrangement. The unit was powered by a generator intended for ultrasonic application with a power output of 10 kW and operating frequency of 10 kHz. The inductor used was characterized by a high drop in the magnetic field intensity in the direction away from central zone, and levitation of the liquid metal could be produced over a length of ±5 mm from the center (**Figure 5b**). At a larger distance, the free surface of the liquid metal rippled uncontrollably [9, 10].

The experiments carried out have shown that this length of active zone, provided by such an inductor, while being adequate for levitation generated by the uniform alternating magnetic field, was at the limit of application. It was evident that to extend the zone of active shaping in the direction of the exiting strip would be most easily achieved by enlarging the inductor, but such a solution would entail an increase in energy consumption. Therefore, the research was focused on the problem of how to extend the levitation zone without enlarging the inductor. Changing the wire cross section from a simple rectangular shape (**Figure 5**) to a more complex shape (**Figure 2**) was proved to be the solution.

The next step in the development of a HEMM prototype highlighted the group of problems which have to be worked out before the prototype can be tried under production conditions. These include determining the boundary parameters of the shaping process, the air gap size, and at both horizontal surfaces the edge radiuses of the shaped strip, and with respect to the different conditions of shaping of the upper and lower surfaces of strip, devising the suitable engineering designs of the inductor [21]. Two designs of inductors capable of shaping the lower and upper surfaces, together with the distribution of the magnetic field, are presented in **Figure 7**.

Research undertaken with the laboratory HEMM with the inductor of a 10 × 40 mm section has shown the solidification zone has to be located in the distance of ±10 mm from the middle, where the value of the air gap is changed about 1 mm. The difference in air gap values from 20 to 21 mm above the upper surface was chosen to ascertain over what length of the inductor in the direction of exiting it is possible to keep this range of air gaps. From the distribution of the magnetic field intensity (**Figure 7b**), it follows that at the nominal value of magnetic field intensity of 32 kA m^{-1}, the arrangement with auxiliary inductors placed 6 mm below the main inductor ensures that the air gap will be ranged from 20 to 21 mm (a_1 = 8 and 9 mm) between

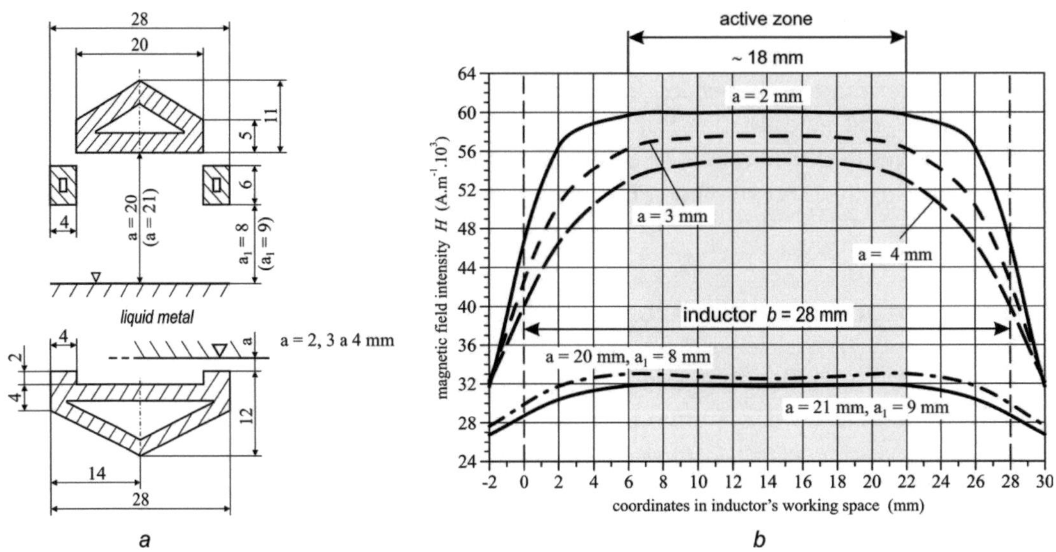

Figure 7. Arrangement of inductors for shaping of both surfaces (a), and the distribution of the magnetic field at an electric current of a peak value of 2520 A (b).

coordinates of 2 and 26 mm, which means over the length up to 24 mm. Over this interval, the internal movements of molten metal will be only negligible. Positioning an auxiliary inductor at the outlet makes it possible to locate a barrier directly at the edge of the working space of the inductor to prevent water entering into the solidification zone. It can be a curtain, baffle or roller, made from silicon rubber. The design can be seen as successfully developed and was used as the universal solution to shaping the upper surface of the strip, while the total width of 28 mm became the standard for inductors intended for the lower surface shaping.

The configuration of the inductor, which is intended for shaping a strip's lower surface, is also based on generating additional intensity of the magnetic field through the use of two auxiliary inductors which are integrated into one unit with the main inductor. **Figure 7a** shows the cross section of the inductor which has 2 mm raised edges, together with the distribution of its magnetic field intensity at air gap values of $a = 2$, 3, and 4 mm at the raised edges (**Figure 7b**). For every value of $a = 2$, 3, and 4 mm at the edge, the values of the air gap at the central plane of the inductor are about 2 mm higher, hence the distance of inductor's central part from the surface of the molten metal will be 4, 5, and 6 mm. On the lower surface with a magnetic field intensity of 56 kA m^{-1}, a satisfactory stability of molten metal can be expected between coordinates of 2 and 26 mm, that is, over the length of 24 mm, where the air gap a will range from 2 to 4 mm. Under working conditions, it is desirable to locate the solidification zone l/s in the interval between coordinates 6 and 22 mm of the inductor's width in between the auxiliary inductors. It is possible to extend the interval to the x coordinate of 4 mm, which will result in increasing the strip's thickness by about 1 mm, or up to the x coordinate of 2 mm, when the strip's thickness will rise to approximately 1.6 mm, whilst retaining the functionality of the HEMM.

Both design arrangements, which meet requirements for the industrial use of the HEMM, are presented in **Figure 8** with a tank of molten metal. In **Table 1**, the HEMM's parameters for continuous casting of a 20 mm thick strip of AlSi8.5Cu and AlSn20 alloys are listed.

The raised edges of the HEMM's working space, which ensure a proper distribution of the magnetic field in the lower surface of the strip, also form a channel from which the cooling water splashed from the cooling system or condensed on the inductor has to be drained. Similarly at the end of the casting process, when the liquid metal no longer exerts a lifting force, the channel serves to drain the surplus molten metal. A steady run-off of water can be ensured by the gentle 1–2° slope of the HEMM or can be created with a sloped surface from

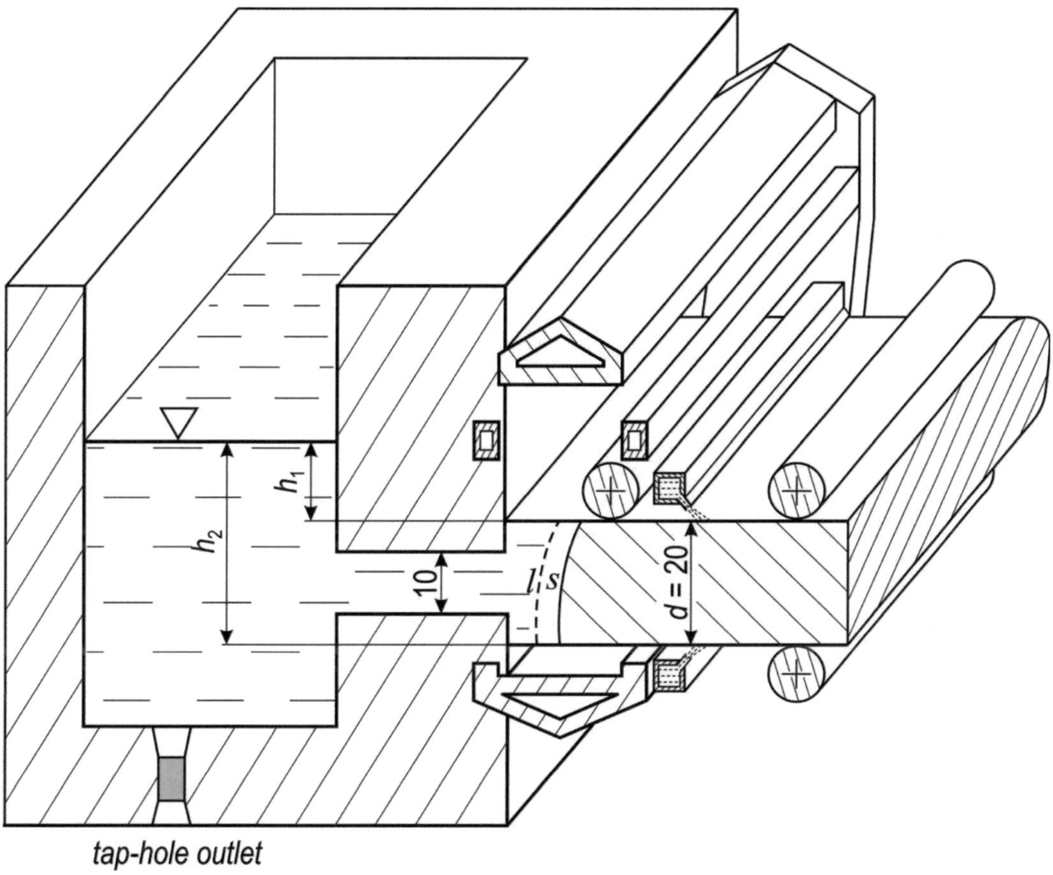

Figure 8. Arrangement of the HEMM unit for continuous casting of a 20 mm thick strip from Al alloys.

Material	H_2 (kA m^{-1})	h_2 (mm)	H_1 (kA m^{-1})	h_1 (mm)	ΣQ (J s^{-1})	I_{ef} (A)	δ (mm)	ρ (kg m^{-3})
AlSi8.5Cu	47.32	29.83	27.04	9.76	879	1507	3.17	2.4×10^3
AlSn20	56.30	29.84	32.16	9.74	1282	1791	3.26	3.4×10^3

Table 1. Parameters of horizontal EMM for continuous casting of Al alloys.

organosilicate or moister-proof ceramics over the whole width of the inductor to the lateral gap. The slope area could also function as an emergency run-off for the liquid metal.

One of the technological features of the HEMM, as seen in **Figure 8**, is the distinct positioning of a cooling zone for both the upper and lower surfaces. Values for the intensity of the magnetic field of 47.32 and 27.04 kA m^{-1}, when shaping a 20 mm thick strip of AlSi8.5Cu alloy in the inductor (**Figure 7a**) with current I flowing at 2132 A ($I_{ef} = I\sqrt{2} = 1507$ A), correspond to 139 kW m^{-2} of power density generated at the upper surface, and to 46 kW m^{-2} at the lower surface, according to **Figure 4** and Eq. (10). When the AlSn20 alloy is shaped with current I = 2533 A (I_{ef} = 1791 A), the magnetic field intensity of 56.28 and 32.16 kA m^{-1} correspond to values of the generated power density of 204 and 66 kW m^{-2}, respectively. Better conditions for the heat removal and less intensive induction heating of the upper surface result in the solidification zone being about 4–5 mm closer to the outlet in the upper part than in the lower one. Analysis of temperature conditions [18] has shown that axial heat removal will prevail, isosolidus of which has a shape indicated in **Figure 8**. This creates the adequate conditions for directional solidification and additional feeding when compensating for the loss of volume during solidification. In order to finish the process of continuous casting, it is necessary to reduce the level of or completely to evacuate the liquid metal from the tank as quickly as possible. This reduces to a minimum the end segment of the continuous cast strip having a reduced thickness.

At this point, it is necessary to understand that if the height of the levitating molten metal falls below 10 mm, then the pool will disintegrate as a consequence of its internal movement. From a technological point of view, the most favorable method for evacuating the tank is by "tapping," that is, by creating an opening in its bottom which is filled with for example clay bonded moulding sand, through which the molten metal can be channeled into a collecting crucible. Breaking an opening has to involve simultaneous cutting off of the supply of liquid metal to the tank. It follows that at the expected withdrawal speed of 1.5 m min^{-1}, the strip covers the distance of 10 mm between the solidification zone and the area of the strongest cooling in 0.4 second, and the whole length of the working space between the tank lining and the outlet plane of the inductor in approximately 1 s.

The inductor in the form as seen in **Figure 8** enables the solidification zone to be located on the upper surface only approximately 3 mm from the inductor's edge, and the cooling zone to be close to the edge. This is possible because under the auxiliary inductor, there is ample space to put a roller, curtain, or baffle to prevent the water leaking into the operating space. On the lower surface, the conditions are more complicated because the small gap enables the zone of strong cooling to be located 2–3 mm in front of the face of the inductor, and the solidification zone then has to be at least 5.5 mm behind the inductor's face. In this case, the distance between the solidification and cooling zones will be 7.5–8.5 mm or, with some allowance, 10 mm. Determining a speed of withdrawal was the subject of research [18] which has pointed out the crucial effect of three factors: the efficiency of cooling of the shaped strip by water spraying; the amount of Joule heat produced in the process of shaping; and the positioning of the solidification zone being limited by the magnetic field distribution. Work [18] states that when pouring AlSi alloy it is possible to achieve an exit speed up to 2.0 m min^{-1} when the area

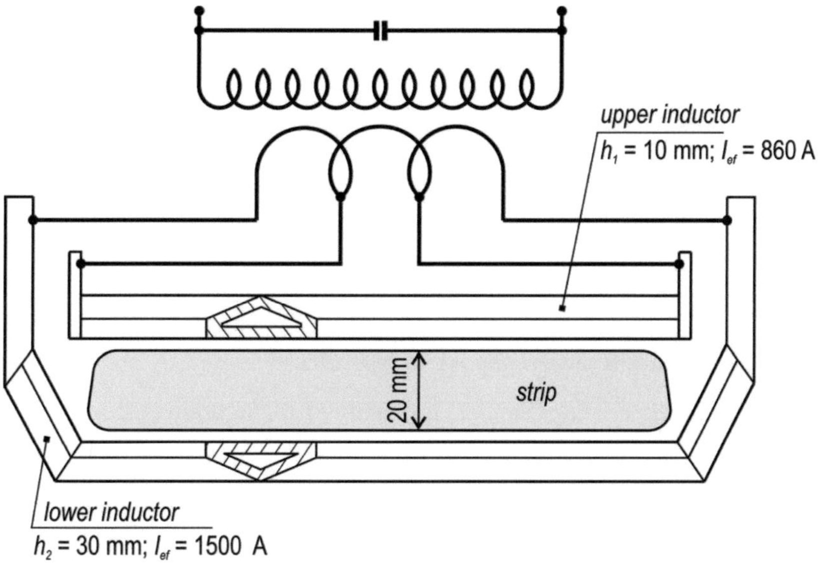

Figure 9. Horizontal EMM consisting of two separate inductors connected parallel for shaping a strip of 20 mm thickness.

of optimal bilateral cooling of the strip is 10 mm away from the solidification zone, and with certain allowance, a speed of about 1.5 m min^{-1} can be achieved for common cooling.

Another possible concept of the horizontal electromagnetic inductor is presented in **Figure 9**. The devised electromagnetic mould consists of two separate inductors with the same cross section of a base, the upper and bottom ones, which are connected parallel and passed with the electric current of different intensities [22]. This requirement can be assured by the separated feeding of the both inductors, or by connecting a shunt to the upper inductor. Applying an air gap of 2–6 mm for both peripheries of the strip, the inductor confining the bottom surface has to be passed by the substantially higher current of 1500 A than the upper inductor that is passed by 860 A.

6. Conclusions

The horizontal EMM is a promising technical instrument, the development of which has already reached a stage when it is possible to expect its adoption by industry. When undertaking the research, the author based it on a critical analysis of a prototype unit and primarily on the design of a laboratory inductor, which enabled the experiments with the HEMM to be carried out, and provided crucial suggestions for the design of an industrial unit prototype. The theoretical analyses supported by the extended experimental work has shown that conditions for confining and shaping the upper and bottom surfaces of liquid metal required the distinct spatial arrangement of inductors. The presented way of designing the electromagnetic moulds is based on the analysis of effects of changes in inductor's shape on the magnetic field distribution and subsequent targeted use of inductor's shape adjustment.

Acknowledgements

The author appreciates the financial support provided by Slovak Research and Development Agency for the projects APVV-0857-12 and APVV-16-0485.

Author details

Marcela Pokusová

Address all correspondence to: marcela.pokusova@stuba.sk

Faculty of Mechanical Engineering, Slovak University of Technology in Bratislava, Bratislava, Slovak Republic

References

[1] Getselev ZN. Casting in an electromagnetic field. Journal of Metals. 1971;**23**(10):38-39. DOI: 10.1007/BF03355734

[2] Yoshida M. Electromagnetic casting up to date and future. Tetsu-to-Hagane. 1987;**73**(3): 403-410. DOI: tetsutohagane1955.73.3_403

[3] Zhu XR, Harding RA, Campbell J. Calculation of the free surface shape in the electromagnetic processing of liquid metals. Applied Mathematical Modelling. 1997;**21**(4):207-214. DOI: https://doi.org/10.1016/S0307-904X(97)00008-5

[4] Sakane J, Li BQ, Evans JW. Mathematical modeling of meniscus profile and melt flow in electromagnetic casters. Metallurgical Transactions B. 1988;**19**(3):397-408. DOI: 10.1007/BF02657737

[5] Lavers JD. Computational methods for the analysis of molten metal electromagnetic confinement systems. ISIJ International. 1989;**29**(12):993-1005. DOI: http://doi.org/10.2355/isijinternational.29.993

[6] Roche JR, Canelas A, Herskovits J. Shape optimization for inverse electromagnetic casting problems. Inverse Problems in Science and Engineering. 2012;**20**(7):951–972. DOI: http://dx.doi.org/10.1080/17415977.2011.637206

[7] Ren Z, Deng K, Zhou Y, Zhu S, Jiang G. Basic study of horizontal electromagnetic levitation casting of aluminum sheet. ISIJ International. 1996;**36**:S194-S196. DOI: http://doi.org/10.2355/isijinternational.36.Suppl_S194

[8] Murgas M, Pokusova M. Electroamgnetic crystallizer for continuous casting the AlSi8.5 alloy strip. Solidification of Metals and Alloys. 1997;**33**:20

[9] Pokusova M, Murgas M. Horizontal electromagnetic caster for continuous casting of the Al-alloys. In: Electromagnetic Processing of Materials; 26–29 May 1997; Paris-la-Defense. Paris: Centre Francais De L'Electricite; 1997. pp. 73–78

[10] Pokusova M. Electromagnetic Processing of Aluminium Alloys. 1st ed. Trnava: AlumniPress; 2007. p. 100

[11] Brown GH, Hoyler CN, Bierwirth RA. Induction Heating Handbook. New York: D. Van Nostrand Company, Inc.; 1947. p. 370

[12] Langer E. Theory of Induction and Dielectric Heat. Praha: Academia; 1964. p. 291

[13] Vives C, Ricou R. Experimental study of continuous electromagnetic casting of aluminium alloys. Metallurgical Transaction B. 1985;**16B**(2):377-384. DOI: 10.1007/BF02679730

[14] Yamane H, Bhamidipati JR, El-Kaddah N. Analysis of electromagnetic confinement and solidification of a vertical molten metal sheet in electromagnetic strip casting. In: International Symposium on Electromagnetic Processing of Materials—EPM94. 25 – 28 October 1994; Nagoya. Tokyo: The Iron and Steel Institute of Japan; 1994. p. 35-40

[15] Asai S. Birth and recent activities of electromagnetic processing of materials. ISIJ International. 1989;**29**(12):981-992. DOI: 10.2355/isijinternational.29.981

[16] Zhang Z, Xun J, Shi L. Study on multiple electromagnetic continuous casting of aluminum alloy. Journal of Materials Science & Technology. 2006;**22**(4):437-440

[17] Nakata H, Etay J. Meniscus shape of molten steel under alternating magnetic field. ISIJ International. 1992;**32**(4):521-528. DOI: http://doi.org/10.2355/isijinternational.32.521

[18] Taraba B, Behulova M. Modelling of the continuous casting of pure aluminium in electromagnetic crystallizer. In: CO-MAT-TECH 93; 26–27 October 1993; Trnava. Trnava: STU in Bratislava; 1993. pp. 148–150

[19] Xu D, Xuesong L, Fang K, Fang H. Calculation of electromagnetic force in electromagnetic forming process of metal sheet. Journal of Applied Physics 2010;**107**:124907. DOI: http://dx.doi.org/10.1063/1.3437201

[20] Shen J, Li J, Fu H. Electromagnetic confinement and shaping for plate-form part. Journal of Materials Science & Technology. 2000;**16**(1):27-32

[21] Pokusova M, Murgas M. Using electromagnetic levitation for continuous casting of thin aluminium alloys strip. ISIJ International. 2015;**55**(8):1669–1676. DOI: http://dx.doi.org/10.2355/isijinternational.ISIJINT-2014-843

[22] Pokusova M, Murgas M, Tehlar P. Method of horizontal continuous casting of metals and assembly inductors of horizontal electromagnetic crystallizer. Slovakia; Patent application No. SK50272015 (A3), 2017

Chapter 4

Effects of Modification Techniques on Mechanical Properties of Al-Si Cast Alloys

Williams S. Ebhota and Tien-Chien Jen

Additional information is available at the end of the chapter

http://dx.doi.org/10.5772/intechopen.70391

Abstract

Mechanical, physical and chemical properties of a part depend on the size, morphology and dispersion of the constituents of the microstructure of the part's material. Therefore, this chapter discusses the different processes of altering the microstructure of Al-Si–based alloys to desired functional properties. These processes, commonly called modification methods, were broadly categorised into three: chemical, thermal and mechanical methods. Chemical method, which involves the addition of some elements, in trace levels, to alloys to be modified, is the best modification option. The elements for modifying are called modifiers. The three commonly used modifiers (sodium, Na, strontium, Sr and antimony, Sb) are discussed. The chapter, however, notes that for optimal alloy's mechanical attributes, thermal treatment is usually combined with both chemical and mechanical modification processes. The thermal method involves rapid cooling of alloy for modification, while the mechanical method depends on force to break up large α-Al dendrites and plate-like Si phases.

Keywords: aluminium, Al-Si alloys, refinement, chemical modification, mechanical modification, thermal modification, mechanical properties, grain, eutectic silicon, microstructure, heat treatment

1. Introduction

Quite often, the functional requirements of a component are conflicting. For the engineer, this occurrence makes most materials in their natural form deficient. For instance, a part may require both hardness and ductility to function properly in a given working environment. It is more or less impossible to find such material existing naturally. This type of situation has led engineers to explore different production techniques that produce the desired conflicting properties [1]. At times, two different materials may be combined, with each having one of the required properties. The combined materials could be metal and metal, metal and non-metal or non-metal and non-metal, and during merging, they may be in the same state or otherwise. The new material reflects the properties of the different materials that are combined. Several techniques have been established and used to improve the properties of aluminium and its alloys. The concept behind these methods is modification of the material's microstructure, which ultimately alters the properties of the material [2].

The light weight and high corrosion resistance of aluminium make it significant in its applications; however, there are restrictions on aluminium applications, due to its soft and brittle nature [3]. Thus, to optimally exploit its natural attributes in automotive, aerospace and defence sectors requires a high strength-to-weight ratio. Processes have been developed to strengthen and harden this metal. The primary method of strengthening aluminium is alloying, which is the addition of a calculated amount of selected elements to aluminium. Aluminium alloy is a metallic substance and consists of approximately 90–96% aluminium and another or more elements, most commonly silicon (Si), copper (Cu), magnesium (Mg), zinc (Zn) and manganese (Mn) [4]. However, commercially available aluminium alloys have about 0.1–0.4% iron (Fe) by mass, which gives the alloys special qualities. While the iron content could be seen as an unwanted impurity, it depends on the fed raw materials and the electrolytic reduction process. There are other alloy elements that provide special properties, and are usually applied in smaller amount (less than 0.1% by mass), and include elements such as bismuth (Bi), chromium (Cr), boron (B), lead (Pb), zirconium (Zr), nickel (Ni) and titanium (Ti) [5]. The most important and commonly used alloying element of aluminium is Si. Silicon addition to aluminium improves the fluidity of the Al-Si alloy, feeding, hot tear resistance, tensile strength and hardness.

However, alloying has not completely satisfied material engineers' quest to meet the trends of Al-Si alloys functional requirements, due to as-cast mechanical properties limitations. Cast alloys of Al-Si produced by conventional processes of melting, pouring and solidification, without post-process is called as-cast alloys. Studies have shown that the microstructure of as-cast alloys under the conventional solidification conditions consists of coarse flakes of Si that promote *brittleness* within these alloys [6, 7]. The primary Si in the form of a plate in Al-17%Si is shown is **Figure 1**. Consequently, material engineers and scientists have developed several

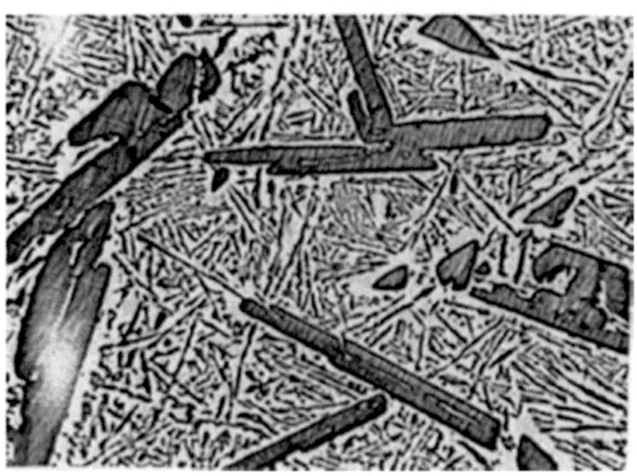

Figure 1. Primary Si in the form of a plate in Al-20%Si [8].

http://dx.doi.org/10.5772/intechopen.70391

processes to enhance Al-Si alloys mechanical and physical properties. The properties of a material are defined by the characteristics of its microstructure. In the case of Al-Si alloys, microstructures can be modified either chemically or mechanically. In a chemical modification, certain elements are added in trace levels to the matrix depending on the needed property. This chapter, therefore, discusses Al-Si alloys modification concept, Al-Si modification methods and their applications.

2. Modification of Al-Si alloy techniques' classification

Generally, modification and refinement processes are used for improving mechanical properties of alloys by altering the alloy's Si morphology and distribution. There are several modification and refinement techniques that can be used and these techniques can be categorised into three:

i. Chemical modification and refinement processes; addition of a calculated amount of nucleation agents

ii. Mechanical modification and refinement processes; ultrasonic, squeeze, stirring, centrifugal and vibration methods

iii. Thermal modification process; superheating, quench and cooling

2.1. Chemical modification and refinement processes

The addition of trace levels of certain additive (modifier) to a molten Al-Si to alter its structure is called modification. Modification reduces the size of eutectic Si particles to enhance the cast's mechanical properties such as ductility and strength. The addition of a modifier such as Sr transforms the Al-Si cast to fine and globular/fibrous morphology. Chemically stimulated modification produces a fine flake-like or fibrous structure. Many elements have been discovered to produce a fibrous eutectic Si structure such as Na, Sr, K, Ce and Ca. These following elements, Sb, As, and Se have been found also to produce a refined flake-like structure. These three elements Sr, Na, and Sb are the most effective modifiers in trace levels of additions and widely used in the foundry industry. The strongest modifiers known are Na and Sr. Other modifying elements are K, Rb, Ba, La, Yb, As and Cd, as presented in **Table 1**.

2.1.1. Sodium: Na-modification

The first commercially modifier applied to Al-Si alloys was Na. It is required in trace levels, usually <0.007%Na, to make the full modification. Advantages of Na-modification include: its effective use for many years; small amount required for modification; short residence time (the time it takes to remove inclusions); minimal surface agitation; and reduced offensive fumes. However, there are several drawbacks: it has about 10–50% volatility recoveries; limitation due to the danger in handling caused by its rapid reaction with moisture; formation of thick oxide skin that hinders fluidity, which may cause entraining in casting; it makes surface appearance of casting scaling; Na attacks mould coatings; and over modification challenges.

Elements	Morphology of eutectic Si
No addition	—
Sodium, Na	Fibrous
Calcium, Ca	Fibrous
Strontium, Sr	Fibrous
Potassium, K	Fibrous
Barium, Ba	Fibrous
Cerium, Ce	Fibrous
Rubidium, Rb	Fibrous
Europium, Eu	Fibrous
Antimony, Sb	Lamellar (or a fine version of acicular)
Ytterbium, Yb	Lamellar (or a fine version of acicular)
Arsenic, As	Lamellar (or a fine version of acicular)
Selenium, Se	Lamellar (or a fine version of acicular)
Cadmium, Cd	Lamellar (or a fine version of acicular)
All rare earth metals and misch metals except Eu including Laa, Cea, Pra, Nd, Sm, Gd, Tb, Dy, Ho, Er, Tm, Yba and Lu	Lamellar (or a fine version of acicular)

Table 1. Modifier elements and their effects on eutectic Si [9–11].

Modification of Al-Si microstructure is performed by the addition of a minute quantity of ternary element such as Na to Al-Si. The use of addition of a trace level alkaline earth metals or alkali metals to AI-Si alloys to alter their structures began several years ago. This alteration enhances the mechanical properties; raise the ultimate tensile strength (UTS) and increases the ductility. The influence of Na on tensile strength and elongation is shown in **Figure 2**. Describing this morphological alteration, conventionally, there are two versions of explaining the principle of Na-modification [12, 13]: (i) based on Na influence on Si growth and; (ii) based on Na influence nucleation of the Si phase. However, Day and Hellawell identified three various modes of eutectic nucleation and growth in Al-Si alloys that are composition and solidification conditions depended. These modes are [13]:

i. Nucleation at or near to the wall and front growth facing the thermal gradient

ii. Nucleation of eutectic on primary dendrites

iii. Heterogeneous nucleation of eutectic on nucleant particles in the interdendritic liquid

The growth model seems to have the widest acceptance, due to the appearance of Si interconnectivity in both unmodified and modified structures. Consequently, the continuous

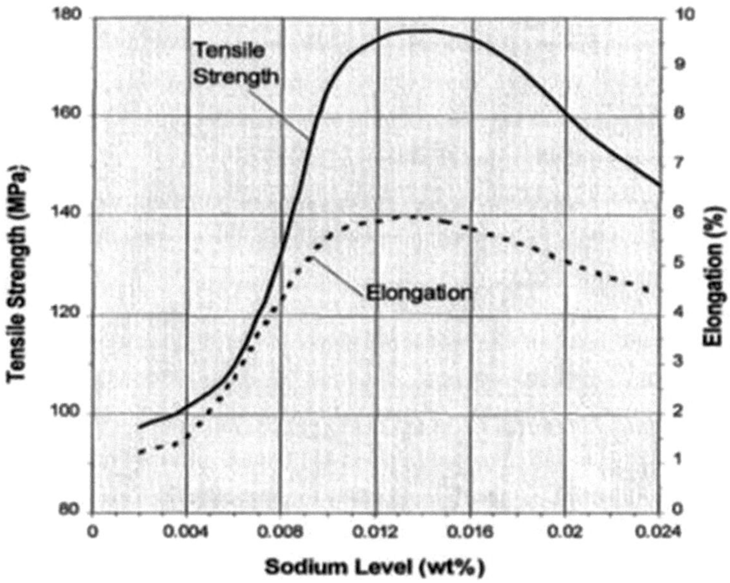

Figure 2. Sodium effect on tensile strength [14].

nucleation of Si has been contested. However, two limitations were observed in the growth model. In a Na-modified casting, a significant change of the microstructure was noted, while little spacing changed and fibrous Si form are produced in the directionally grown specimens. Again, it was noted that Na addition has little influence on the equilibrium liquidus temperature, but there is a huge change in the plateau temperature produced during the thermal investigation of a cast alloy [13]. Enhancement of capacity of Si to branch by Na, in the growth models, has been advocated and this reduces spacing and total undercooling, at a specified growth velocity.

Conversely, Flood and Hunt concluded in their work that nucleation and Si growth are both affected by the addition of Na to Al-Si cast alloys. In the Na-modified casting, two effects were reported that [15]:

i. Sodium presence transforms the Si growth morphology from the plate-like form to the fibrous form.

ii. If the temperature rise in the liquid is small, Na stops nucleation from happening ahead of the eutectic growth front. The lack of nucleation primarily accounts for the finer structure and larger undercooling of modified castings or ingots.

Trace of phosphorous is often found in Al-Si alloys, which causes the formation of a granular structure and aluminium phosphite ($AlPO_3$). Aluminium phosphite accelerates primary Si crystallisation that appears in the microstructure in the form of polyhedral platelets. In Na-modification, Na reacts with phosphorus to form sodium phosphite (Na_3O_3P). The solid solution phase in the form of dendrites crystallises out first before the Si phase.

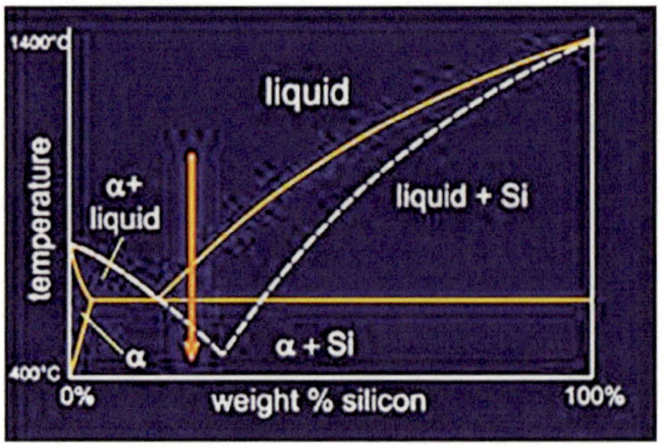

Figure 3. Modified Al-Si alloy phase diagram [16].

Significant change is observed in Si morphology and particle spacing when 0.01 wt.%Na was added. This process transforms solidifying alloy into highly refined Al-Si eutectic and also, reduces temperatures and moves the eutectic point to a higher Si content. The broken line in **Figure 3** shows the modified Al-Si alloy phase diagram.

Addition of Na modifies the eutectic Si growth to an irregular fibrous form instead of the usual coarse flakes. The composition of Al-Si alloy at the new eutectic point due to the Na-modification is *hypoeutectic* instead of hypereutectic, which results in the formation of primary α-Al instead of Si. **Figure 4** shows micrograph of Al-13%wtSi and Al-13%wtSi-0.01%Na respectively. Addition of small of the quantity of 0.01%Na as an impurity to Al-Si alloy modifies its microstructure and improves its properties.

Wessén, Andersson and Granath investigated Na-modification effect on the mechanical properties of a secondary alloy, Al-6%Si-2.5%Cu, produced from rheocasting, applied to thick wall components production [17]. The study revealed noteworthy alterations in the microstructure of the Na-modified; individual Si lamellar could not be identified, while the average size of Si lamellar of the unmodified alloy was 100 μm. The reduction of the quantity of Na from 4.3×10^{-5} wt.% to 3×10^{-5} wt.% did not show a significant change in the structure and shows that a trace level of Na can substantially transform eutectic. The energy disperse X-ray spectroscopy (EDX) analysis shows intermetallic phases Al_2Cu and $Al_{15}(Fe, Mn)_3Si_2$ and these are shown in the SEM image in **Figure 5**.

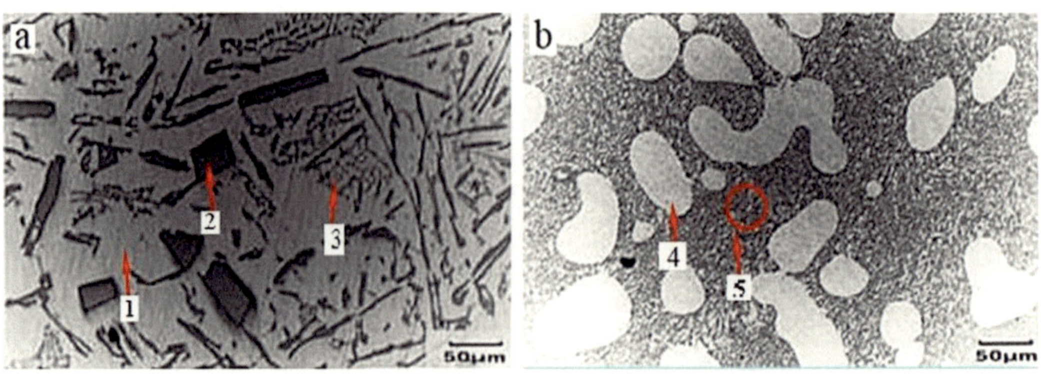

Figure 4. Micrograph of [16] (a) Al-13%Si; (b) Al-13%Si-0.01%Na, where 1 = α-Al dendrite; 2 = primary Si; 3 = *Eutectic* Si; 4 = α-Al; and 5 = fibrous eutectic Si.

http://dx.doi.org/10.5772/intechopen.70391

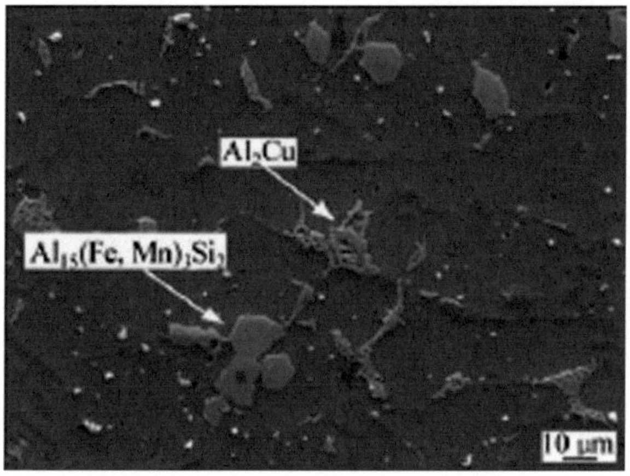

Figure 5. SEM image showing main intermetallic phases in microstructure [17].

2.1.2. Strontium: Sr-modification

Modification by Sr and Na has the similar result of fibrous eutectic Si structure. However, Na is much more volatile than Sr, as is often considered as a semi-permanent modifier. This property and its easy application, coupled with other metallurgical advantages have increased the use of Sr as a modifier in recent years. Other advantages are [18–29]: about 80–90% recovery rate; addition melt easily; wide effective concentration range; last long in the melt during holding times; less delicate to over-modification; produces smooth appearance castings; found in suitable master alloy form; and it does not react with refractories; it has no environmental challenges. In a well Sr-modified hypoeutectic casting microstructure, the estimated average area of the fibrous eutectic Si particle is 3.8 ± 0.6 µm^2 and the aspect ratio is about 1.58 ± 0.29. The impact of Sr concentration on strengths is shown in **Figure 6**.

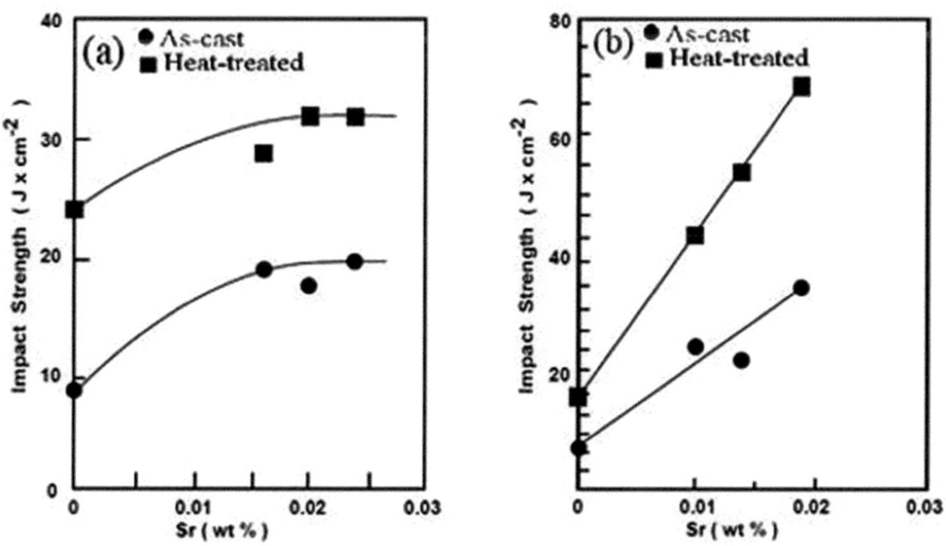

Figure 6. Impact of Sr concentration on strengths of: (a) AA601 cast alloy; and (b) AA401 alloys [20].

There are challenges that are associated with Sr-modification such as the promotion of gas levels in the melt and the cost is comparatively high. Nevertheless, these challenges can be overcome by easier melt treatment practice and other casting quality enhancement processes. Over modification causes the mechanical properties of the alloy to revert to that of a typical unmodified alloy, **Figure 7** shows the microstructures of, unmodified, modified and over modified alloys.

The transformation of eutectic Si morphology in Al–Si casting alloys from coarse plate-like to fine fibrous networks can be achieved by trace addition of Sr the alloy. To further explain the process of Sr-modification, Timpel et al investigated the distribution of Sr in two ways [24]: in nanometre resolution by transmission electron microscopy (TEM) and in atomic resolution by atom probe tomography (APT). The two methods showed that within the eutectic Si phase, there is Sr co-segregation with Al and Si. Two kinds of segregations, type I and type II, were identified:

i. Type I segregation is a nanometre-thin rod-like, accountable for the formation of numerous twins in a Si crystal and facilitate its development in various crystallographic directions.

ii. Type II segregations are more stretched structures that impede the growth of a Si crystal and regulate its branching.

This study agrees with earlier studies of modification mechanisms, which hinged the modification on growth restriction of eutectic Si phase and impurity induced twinning [11, 25]. **Figure 8** shows the optical micrographs of Al–10%Si–0.1Fe alloy for unmodified alloy and 200 ppm Sr-modified alloy.

Atom probe tomography analysis and TEM images of Al and Si interface are presented in **Figures 9** and **10** respectively. The ATP data set does not contain crystallographic information. Therefore, TEM is used to obtain such information with a spotlight on the structural and compositional characteristics of the eutectic Si phase.

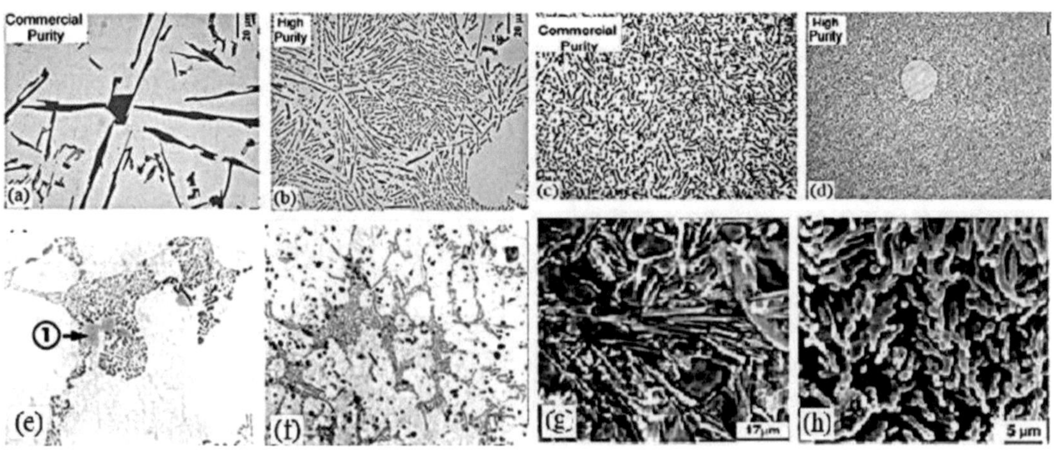

Figure 7. Eutectic microstructures of fully solidified Al-Si alloys: (a) unmodified commercial purity; (b) unmodified high purity; (c) Sr modified commercial purity;(d) Sr modified high purity [21]; over Sr-modification of AA601 alloy: (e) >0.03%Sr formation of Al_4SrSi_2 phase; (f) >0.09%Sr formation of coarse of the eutectic Si [22]; SEM images of etched microstructures of the Al-Si eutectic [23], (g) unmodified Al-Si alloy; (h) Sr-modified Al-Si alloy.

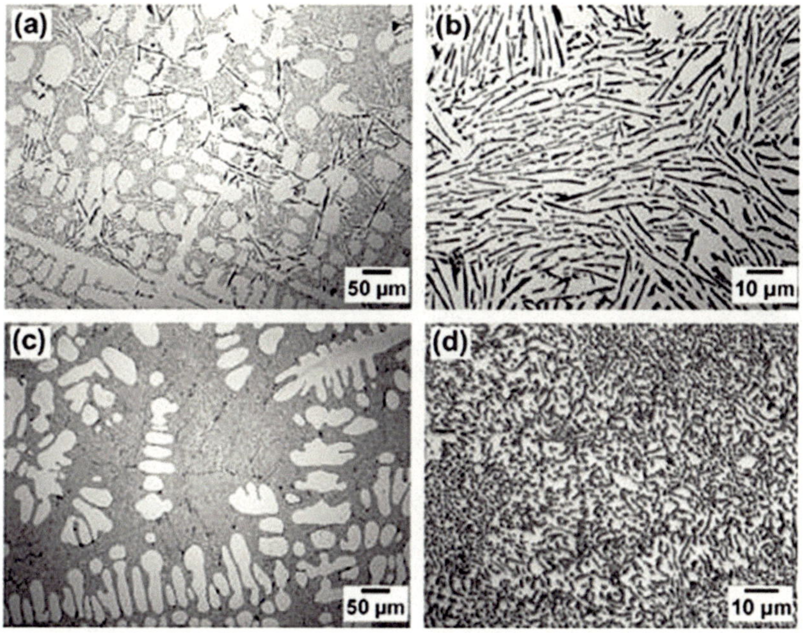

Figure 8. Optical micrographs of Al–10%Si–0.1Fe alloy [24]: (a) and (b) unmodified alloy; (c) and (d) alloy modified by 200 ppm Sr.

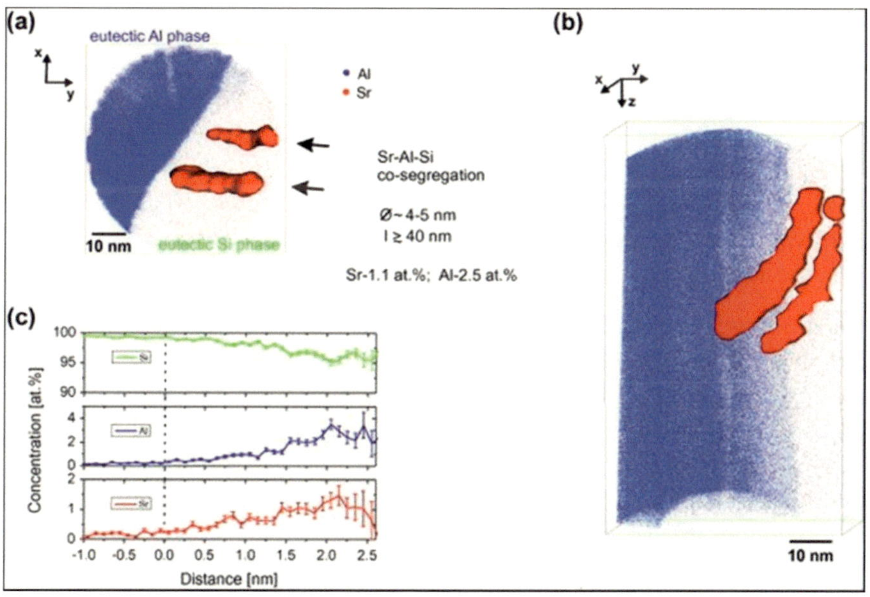

Figure 9. APT analysis of eutectic Al–Si interface of Al–10%Si–0.1%Fe alloy of 200 ppm Sr-modified; (a) iso-density surface; (b) representation of 0.17 Sr atoms/nm^3 in both co-segregations; (c) concentrations of Al, Si and Sr in proxigram, which depend on the distance to the Si/Sr–Al–Si co-segregation interface in (a) and (b) [24].

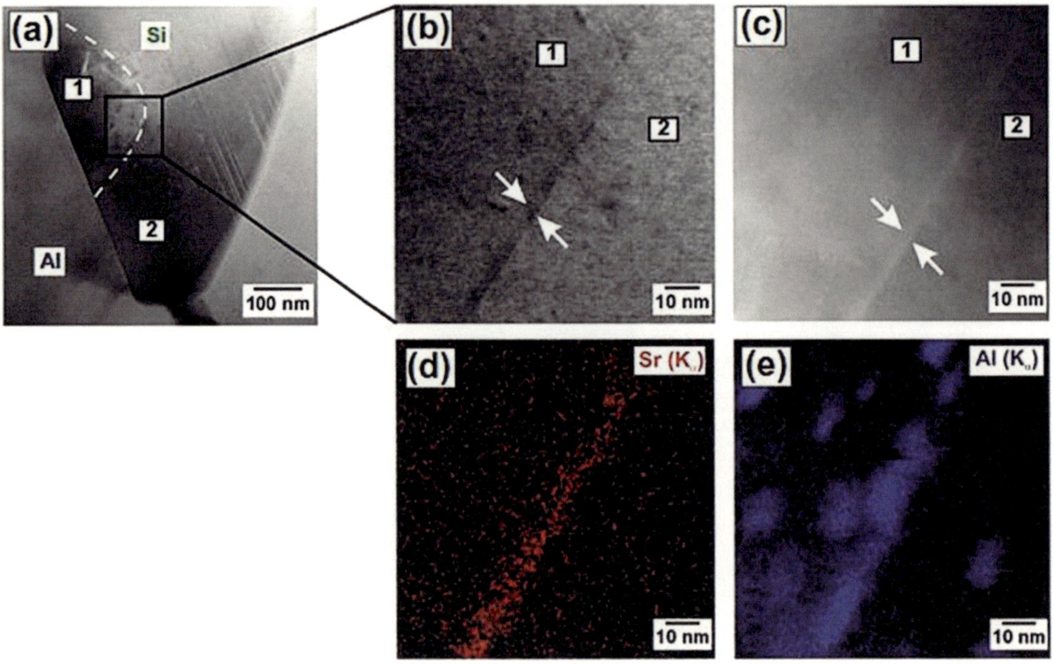

Figure 10. TEM images of eutectic Al–Sr interface of Al–10%Si–0.1%Fe alloy of 200 ppm Sr-modified. (a) Visible internal boundary of eutectic Si phase, along a {1 1 1} plane; (b) bright field scanning transmission electron microscopy (BF-STEM) image of the enlarged rectangular mark; (c) high angle annular dark field (HAADF) image of the enlarged rectangular mark; (d) and (e) energy disperse X-ray spectroscopy (EDX) mapping of Sr and Al [24].

In another study, hypoeutectic Al-Si alloys containing two levels of Fe (0.5 and 1.1 wt.%) and modifier (Sr) in the range of 30–500 ppm, were investigated. Significant reduction in the number of eutectic grains and the formation of polygonal-shaped Al_2Si_2Sr intermetallic were observed in excess addition (100 ppm) of Sr. TEM examination showed that the Al_2Si_2Sr phase is bounded by the P-rich particles, and this infers poisoning or deactivation of nuclei for the Al-Si eutectic. The poisoning is due to the formation of Al_2Si_2Sr phase about the particles. Further reduction in the number of eutectic Al-Si nucleation actions was recorded at 1.1 wt.% Fe due to the formation of pre-eutectic, β-Al_5FeSi platelets.

There is the difference in the nucleation temperatures between unmodified and Sr-modified Al-10%Si alloys. In the Sr-modified alloy, the eutectic-nucleation temperature is depressed with minimum occurrence before recalescence and growth temperature. **Figure 11** shows unmodified and Sr-modified Al-10%Si alloys of low Fe (0.5 wt.%) and high Fe (1.1 wt.%) cooling curves recorded during solidification, respectively. Considering the curves, T_a is α-Al nucleation temperature; T_b is β-Al_5FeSi nucleation temperature; and T_N is Al-Si eutectic nucleation temperature.

2.1.3. Antimony: Sb-modification

Addition of Sb produces a refined flake-like eutectic structure, unlike Sr or Na-modified alloys, which result in a fibrous structure. Sb remains a permanent constituent of the alloy, unlike

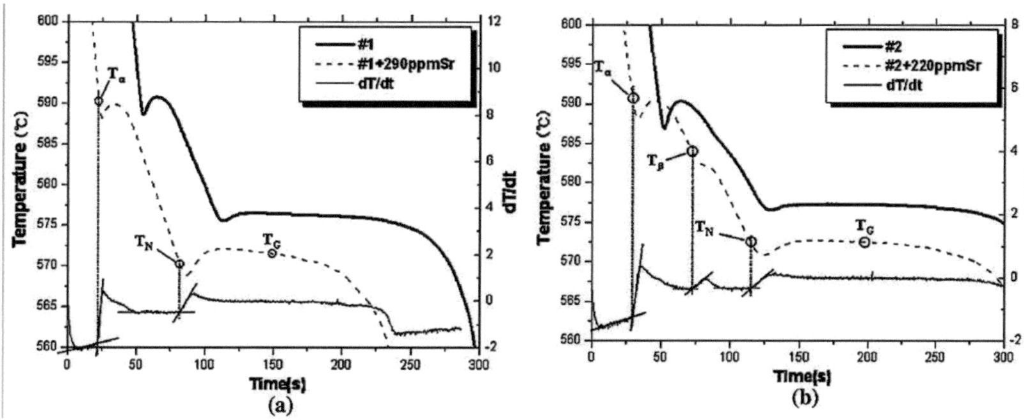

Figure 11. Cooling curves during solidification of Al-10Si alloys in the unmodified and Sr-modified conditions containing (a) low Fe (0.5 wt.%); (b) high Fe (1.1 wt.%) [18].

Sr and Na, which fade away with time. Because of this, the supplier adds it to the foundry ingot. The ultimate tensile strength of unheated treated alloy, $AlSi_6Cu_4$, was improved from 5 to 10% by increasing the addition of Sb from 1000 to 2500 ppm to the alloy respectively [26]. The Sb-modification decreases the size of grain but increases the number. The maximum size of unmodified Al-Si alloy grain is about 220 μm. The size of the eutectic grain is reduced from 156 μm (0 ppm Sb) to 84 μm (1000 ppm Sb) by Sb-modification. Antimony, Sb, is widely used in Japan and Europe and commonly called permanent modifier. Its addition to molten Al-Si alloy is straightforward, as it does not require any special set up, and once it is added, it becomes a permanent part of the alloy. The use of Sb as a modifier has merits such as insensitive to re-gassing; it does not fade; and is appropriate for components that are susceptible to porosity formation. However, Sb-modification has the following drawbacks: Sb reacts and reduces the effectiveness of Sr and Na; Sb may react with hydrogen dissolved in the metal and forms a stabile gas, a toxic material; may cause the slower solidifying regions of casting poor mechanical properties; and Sb is least effective, as lowest level of transformation is achieved compared to Sr and Na [20, 27].

2.2. Effect of modification on casting quality

Despite many years of application, Na and Sr modification and its influence on the gas content of Al-Si alloy melts are still being contested by researchers. Several studies declare that Sr-modification has no effect on the alloy's hydrogen content [28, 29]. The measured hydrogen content in a melt after 0.03% Sr addition, using Al-90% Sr master alloy, to a non-degassed A356 melt at 710°C, is shown in **Figure 12**. Others claim that Sr-modification levels addition above 0.10–0.12% causes gas porosity [30–32]. Porosity formation in alloys during solidification is a major challenge for casting industry due to its negative effect on total elongation and fatigue performance. Jahromi et al. reported that 0.013% Sr and 0.1% Sb were found to be the optimum additions to modified A356 alloy to fibrous structure in a sand casting. More porosity developed in Sr-modified than Sb-modified [29]. Denton and Spittle reported that hydrogen content of Al-Si alloys increased significantly in the addition of Sr to Al-Si melts, at elevated temperatures [33].

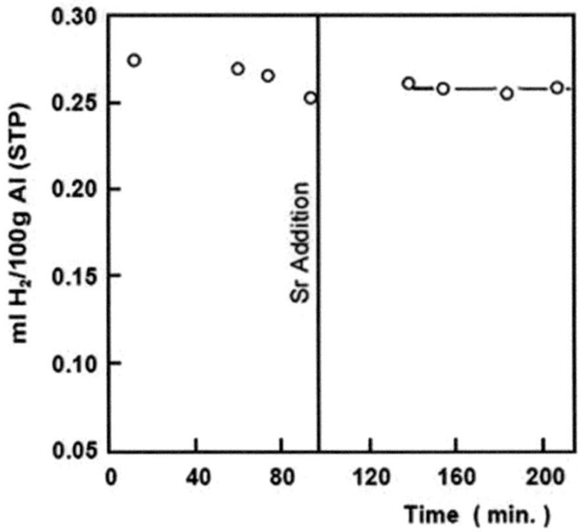

Figure 12. Effect of addition of 0.03% Sr to a non-degassed A356 melt at 710°C [22].

3. Rapid cooling (quenching) and mechanical modification

Coarse columnar grain structures are developed by Al-Si cast alloys under normal casting conditions. But these structures can be transformed into fine grain structures and uniform distribution in the alloy by mechanical modification and through rapid cooling. Rapid cooling results in a fine dendritic structure in the alloy [34, 35]. Large dendritic structure is due to undercooling during solidification. Rate of cooling affects the size of critical nuclei, and subsequently, the effective number of nuclei that will ultimately produce fine-grained structures.

Cooling rate can be expressed by this relation:

$$d = C v^{-n} \tag{1}$$

where d—secondary dendrite arms spacing, SDAS, (μm); C and n—are constants; and v—cooling rate (°K/s).

The local solidification times (t_f) can be calculated in terms of SDAS measurements through the following expression [36]:

$$SDAS = 5.5 \left[\frac{\Gamma_{sl} D_l \ln\left(\frac{C_{eut}}{C_o}\right)}{m_l(1-k_o)(C_{eut}-C_o)} t_f \right]^{\frac{1}{3}} \tag{2}$$

where Γ_{sl}—Gibbs-Thomson coefficient; D_l—diffusion coefficient in liquid; m_l—liquidus curve slope; k_0—coefficient of partition; C_{eut} and C_0—are the eutectic composition and the initial alloy concentration respectively.

The removal of superheat and latent heat from a liquid at a cooling rate of 10^2–10^6 K/s to form solid is called rapid solidification. To accomplish this, there are certain conditions that must be satisfied [37]. There should be impressive:

i. High undercooling before solidification occurs

ii. High solidification front speed during continuous solidification

iii. Rapid cooling during solidification

Duwez and his team at the California Institute of Technology developed an innovative method in 1960 to increase solid solubility and to yield metastable crystalline in some simple binary eutectic alloy structures [38]. A similar process was performed earlier by a Russian researcher, Salli, in 1958 [7]. In their modification technique, a gun was applied to deposit a small droplet of molten metal, at high velocity, on a freezing surface. This resulted in the formation of an irregular solidified splat of metal. The estimated cooling rate range of the process reported was 10^5 to 10^6 K/s as against the conventional cooling rate of 10^2 K/s or less. Other metal rapid quenching systems, with varied solidification effects, have been introduced since the development of fast cooling by Duwez. However, the common principal aims of these systems are to: increase solid solubility limits; decrease grain size; create metastable crystalline phases; form metallic glasses; and increase chemical homogeneity. Recent studies have shown that rapid cooling systems such as atomisation, melt spinning and splat quenching are effective in the modification of Si phase in Al-Si alloys. In an investigative study, the microstructure and mechanical properties of A356 alloy prepared from a copper mould cooled by a phase-transition medium [34]. The study reported that a cooling rate of 10^2 K/s was obtained using this method and this method was described as a fast-cooling technology. The study indicated that:

i. Variation in the quantity of cooling medium controlled the cooling rate to a certain extent.

ii. The primary and SDAS were better refined by this technology compared to the use of conventional casting technique

iii. Increase in cooling rate decreases SDAS while strength and microhardness increase correspondingly.

The studies essentially focused on: characterisation of rapidly solidified Al-Si alloys microstructure; and the determination of retained-Si amount in α-Al by X-ray diffraction methods, which depends on lattice parameter.

3.1. Al-Si alloy quenching media

There are several quenching fluids (quenchants) used in the quenching of high strength Al-Si alloys. There is no an ideal all-purpose quenchant, their applications depend on some factors such as composition, cast thickness, etc. Quenchants that are commonly used for different aluminium alloys are water spray, cold water immersion, hot water immersion, air blast, still air, glycols/polymers, fast quenching oil, liquid nitrogen and brine solutions [39]. Water is the most common quenchant, with the advantages of being cheap, readily available and providing the fast cooling rate needed to produce the required properties. Furthermore, the temperature of water can be altered to produce a wide range of quenching characteristics. Boiling water is used in many aluminium alloys quenching operations due to sufficient cooling rate. Colder water or polymers may be used in the case of premium property requirements, such as in A357 and A201 castings alloys [40].

In 2014, Abdulazeez et al. observed that microhardness of Al-Si-Mg alloys is affected by quenchants differently [41]. Water quenching was said to give higher microhardness compared to polymers. This was attributed to faster cooling rate, restraining elements from solid solution (α-Al) diffusion and grain boundary precipitation by water. The generally acceptable water quenching temperatures ranges for various aluminium alloys are presented in **Table 2**.

Type	Alloy/temper	Water temperature (°C)			
		21–32	54–65.6	60–71	65.6–100
Casting	C355		✓		
	A356		✓		
	AA356 premium	✓	✓		
	AA357 premium	✓			
	AA201	✓			

Table 2. Normal water quenching temperatures for some aluminium casting alloys.

In the heat treatment sequence, quenching is the next vital process, and its purposes are as follows: to suppress precipitation; to preserve the maximum amount of hardening elements precipitates in solution to develop a supersaturated solid solution at low temperatures; and to confine several vacancies [42]. Quench rate limit is 4°C/s, above this, the yield strength increases slowly. To maximise vacancy confinement concentration and minimise part deformation after quenching, optimal quenching rate is required. A slow quenching rate reduces residual distortion and stresses in parts, but it causes harmful effects such as precipitation during quenching; reduction in grain boundaries; increase tendencies for corrosion; localised over-ageing and leads to a response to ageing treatment reduction. Optimal cooling rate should be established, and the optimum combination of ductility and strength depends on rapid cooling.

4. Mechanical modification techniques

Modification of Al-Si alloy can as well be achieved through the use of mechanical techniques, such as centrifuge, sonic and ultrasonic vibration, squeezing, etc. These mechanical means have been identified to cause grain refinement, density increase, shrinkage, degassing, change size, shape and distribution of the second phase [43, 44]. Refinement is produced through some mechanical means of breaking up newly developed dendrites, as in case of semisolid-metal (SSM) casting technique. The size, distribution and morphology of α-Al particles govern die filling, and subsequently, control the ability to produce thin wall castings.

4.1. Vibration energy

Refinement of primary austenite structure was achieved by Chernov in 1868 by applying vibration during solidification [45]. Since then, several other researchers have investigated and applied the beneficial effects of vibration energy to treat many alloys of aluminium, zinc, brass, etc. during solidification [46, 47]. Vibration influences the structure of a solidifying alloy by suppressing the growth of columnar and the formation of equiaxed grains. The effect of vibration on unmodified and Na-modified alloy has been reported. A schematic of a mould mounted on the vibrator or shaker is shown in **Figure 13**.

These beneficial effects include nucleation promotion, which reduces as-cast grain size; decreases shrinkage porosities; and stimulates the formation of a more homogeneous metal structure with cracking susceptibility decrease. Therefore, due to vibration, alterations occur to the morphologies and dispersion of eutectic and dendritic phases in the microstructure. Pillar's study revealed that eutectic Si of unmodified Al-Si alloys was modified by vibration at a frequency of 12 Hz and amplitude of 10 mm [48]. Contrarily, it was found that vibration coarsened the eutectic Si in Na-modified Al-Si alloys and this was attributed to fine eutectic Si agglomeration. Abu-Dheir et al. observed that at constant frequency of 100 Hz and varying amplitude resulted in different degree of breakup of the dendrites and eutectic Si phase [49]. Micrographs of Al-12.5% Si castings of without vibration and with vibration at 100 Hz are shown in **Figure 14**.

It was seen in the optical micrographs that degree of fragmentation is a function of the amplitude, which shows proportionality. However, the study indicated that there is a limit of

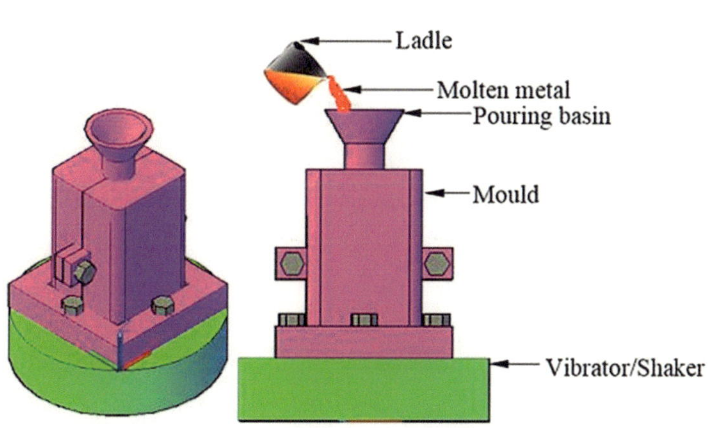

Figure 13. Schematic of mould assembly mounted on the vibrator.

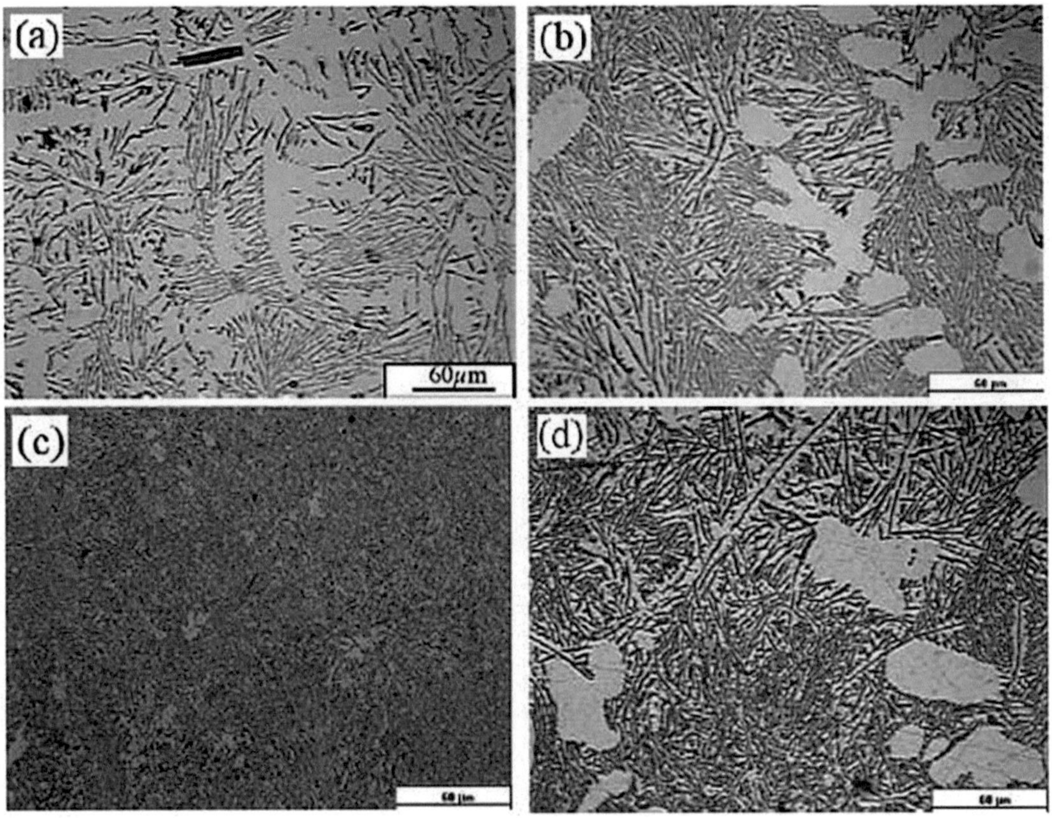

Figure 14. Micrographs of Al-12.5% Si castings (a) casting without vibration; (b) with vibration at 100 Hz and 18 μm; (c) with vibration at 100 Hz and 149 μm; (d) with vibration at 100 Hz and 199 μm [49].

amplitude above which the size of fragmented dendrites and eutectic Si start to form coarse flakes due to agglomeration. The microstructural characteristics of Al–12.5% Si casting without and with vibration at constant frequency (100 Hz) and varying amplitude are presented in **Table 3**.

Casting condition	Lamellar spacing (μm)	Si flake length (μm)	Notes
No vibration	2.5	27	Si cuboids, large dendrites
18 μm	2	15	Broken dendrites
49 μm	2.77	31	Refined broken dendrites
149 μm	N/A	N/A	Fibrous Si observed
199 μm	1.5	10.5	Coarse Si flakes

Table 3. Microstructural characteristics of Al–12.5% Si casting without and with vibration at constant frequency (100 Hz) and varying amplitude [49].

There are different methods of applying vibration. Electromagnetic vibration is one of the non-contact methods used to generate vibration in the solidifying alloy. The vibration is produced by using an orthogonal static magnet and alternating electric fields [50]. It was observed that the collapse of the cavities created by this method was accountable for the refinement of the microstructure for Al–7% Si and Al–17% Si [51].

4.2. Centrifugal casting technique

Centrifugal casting technique (CCT) is a casting production process that involves rotation of a mould during pouring and solidification of the casting. The schematic in **Figure 15** shows the forms and the major components of a centrifugal casting machine [52].

The attributes of material depend not only on the composition chemistry but also on the morphologies and distribution of the microstructural features present in the microstructure. These microstructural features include the primary and eutectic Si phases, α-Al dendritic and intermetallics. Low solidification rate results in large flakes of Si, large dendritic cells and large inter-dendrite arm spacing of α-aluminium dendrites. Centrifugal casting process increases cooling rate and consequently, produces small dendritic cells, small inter-dendrite arm spacing and small flakes of Si. Alloys of Al-Si by CCT are morphologically transformed from acicular to fibrous [52]. Micrographs of A390-5%Mg alloy as-cast fabricated by gravity casting and CCT are shown in **Figure 16**.

Speed of rotation is a parameter that controls the rate at which centrifugal casting process affects cooling. Some studies have reported that mould rotational speed range of 1200–1500 RPM as the optimum [53] and other relevant processing parameters are pre-heating and pouring temperatures [52]. The effect of CCT on casting can be classified into three: centrifugal pressure, inherent vibration of the process and fluid dynamics.

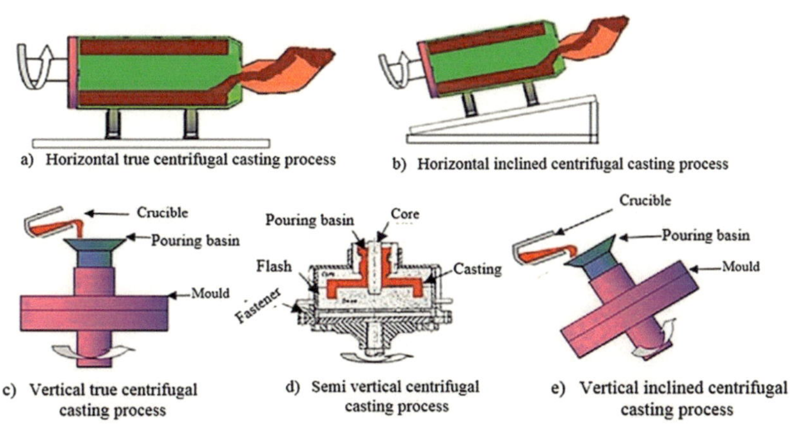

Figure 15. Classification of centrifugal casting methods [52]. (a) horizontal true centrifugal casting process; (b) horizontal inclined centrifugal casting process; (c) vertical true centrifugal casting process; (d) semi vertical centrifugal casting process; (e) vertical inclined centrifugal casting process.

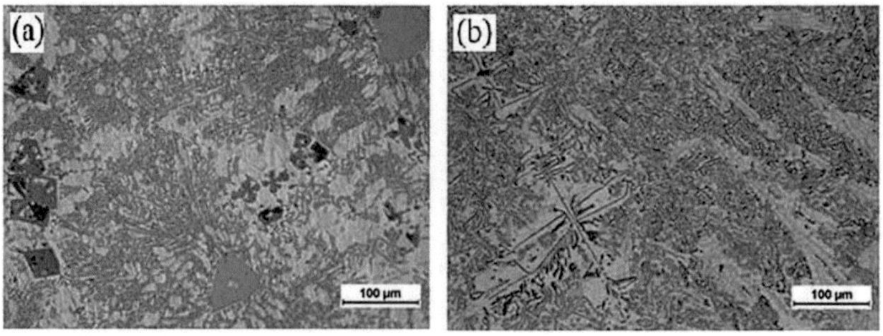

Figure 16. Micrographs of A390-5%Mg alloy as-cast (a) by gravity casting and (b) by CCT.

5. Conclusion

Components made of materials in their natural forms often do not satisfactorily meet functional requirements, due to harsh and extreme working environments. Scientists and engineers have continuously modified these natural occurring materials, using different production processes, to suit their harsh working environments. Material's mechanical, physical and chemical properties depend on the size, morphology and dispersion of the constituents of the microstructure of the material. Aluminium is one of such natural materials that have evolved into several alloys and composites. The mechanical properties of aluminium-based alloys and composites have further been improved by microstructural alteration processes, termed as modification processes. These modification techniques can be classified into three:

i. Chemical modification processes; addition of a calculated amount of nucleation agents

ii. Mechanical modification processes; ultrasonic, squeeze, stirring, centrifugal, and vibration methods

iii. Thermal modification process; superheating, quench and cooling

Chemical modification, which is the addition of trace levels of certain elements, such as Na, Sr and Sb, into aluminium alloys, is most effective. However, optimal modification occurs when thermal modification process is combined with chemical or mechanical process.

Author details

Williams S. Ebhota* and Tien-Chien Jen

*Address all correspondence to: willymoon2001@yahoo.com

Mechanical Engineering Department, University of Johannesburg, Johannesburg, South Africa

References

[1] Peter I, Rosso M. Light alloys—From traditional to innovative technologies. In: Ahmad Z, editor. New Trends in Alloy Development, Characterization and Application. Rijeka: InTech; 2015. p. Ch. 01

[2] Ebhota WS, Inambao FL. Principles and baseline knowledge of functionally graded aluminium matrix materials (FGAMMs): Fabrication techniques and applications. International Journal of Engineering Research in Africa. 2016;**26**:47-67

[3] Sukiman NL, Zhou X, Birbilis N, Hughes AE, Mol JMC, Garcia SJ, et al. Durability and corrosion of aluminium and its alloys: Overview, property space, techniques and developments. In: Ahmad Z, editor. Aluminium Alloys—New Trends in Fabrication and Applications. ed Rijeka: InTech; 2012. p. Ch. 02

[4] Kaufman JG. Introduction to Aluminum Alloys and Tempers. Materials Park, OH 44073-0002: ASM International; 2000

[5] Støren S. Understanding Aluminium as a Material. Oslo: The Norwegian Institute of Technology, Trondheim and by Skanaluminium; 1994

[6] Das S, Yegneswaran AH, Rohatgi PK. Characterization of rapidly solidified aluminium-silicon alloy. Journal of Materials Science. 1987;**22**:3173-3177

[7] Uzun O, Karaaslan T, Keskin M. Production and structure of rapidly solidied Al-Si alloys. Turkish Journal Physics. 2001;**25**:455-466

[8] Atasoy OA, Yilmaz F, Elliott R. Growth structures in aluminium-silicon alloys I. The coupled zone. Journal of Crystal Growth. 1984;**66**:137-146

[9] Gruzleski JE, Closset BM. The Treatment of Liquid Aluminum-Silicon Alloys. USA: American Foundrymen's Society; 1999

[10] Chang JY, Ko HS. Twin probability of eutectic si in rare earth modified Al-7wt%Si alloy. Journal of Materials Science Letters. 2000;**19**:197-199

[11] Lu SZ, Hellawell A. The mechanism of silicon modification in aluminum-silicon alloys: Impurity induced twinning. Metallurgical and Materials Transactions A. 1987;**18A**:1721-1733

[12] Thall BM, Chalmers B. Modification in aluminium silicon alloys. Journal of the Institute of Metals. 1950;**77**:79-97

[13] Day MG, Hellawell A. The microstructure and crystallography of aluminium-silicon eutectic alloys. Proceedings of the Royal Society of London. Series A, Mathematical and Physical Sciences. 1968;**305**:473-491

[14] Shilvock WD. The effect of alloy and impurity variation on the treatment, casting and physical properties of aluminium-silicon eutectic alloys [PhD]. New Zealand: University of Canterbury Christchurch; 1995

[15] Flood SC, Hunt JD. Modification of Al-Si eutectic alloys with Na. Metal Science. 1981;**15**:287-294

[16] Warren A. Phase Diagrams: Solidification of Al-Si Alloys. 1997. Available: https://www.southampton.ac.uk/~pasr1/al-si.htm#page1 [17/04/2017]

[17] Wessén M, Andersson NE, Granath O. Effect of sodium modification on microstructure and mechanical properties of thick-walled AlSi6Cu2.5 rheocast component. Transaction of Nonferrous Metals Society, China. 2010;**20**:643-1648

[18] Cho YH, Lee HC, Oh KH, Dahle AK. Effect of strontium and phosphorus on eutectic Al-Si nucleationand formation of β-Al5 FeSi in hypoeutectic Al-Si foundry alloys. Metallurgical and Materials Transactions A. 2008;**39A**:2435-2448

[19] Krishna SG. Effect of strontium modification and heat treatment on microstructure of Al-319 alloy. International Journal of Mechanical and Production Engineering. 2013;**1**:30-33

[20] Comalco. Modification of Foundry Al-Si Alloys. Brisbane, Queensland, Australia: Comalco Aluminium Limited; 1987

[21] McDonald SD, Nogita K, Dahle AK. Eutectic nucleation in Al–Si alloys. Acta Materialia. 2004;**52**:4273-4280

[22] Gruzleski JE, Closset BM. The treatment of liquid aluminium-silicon alloys. Des Plaines, Illinois, USA: American Foundryman's Society Inc.; 1990

[23] Ludwig TH. Trace elements in Al-Si foundry alloys [PhD]. Trondheim: Materials Science and Engineering, Norwegian University of Science and Technology (NTNU); 2013

[24] Timpel M, Wanderka N, Schlesiger R, Yamamoto T, Lazarev N, Isheim D, et al. The role of strontium in modifying aluminium–silicon alloys. Acta Materialia. 2012;**60**:3920-3928

[25] Clapham L, Smith RW. Segregation behaviour of strontium in modified and unmodified Al-Si alloys. Journal of Crystal Growth. 1988;**92**:263-270

[26] Farkašová M, Tillová E, Chalupová M. Modification of Al-Si-Cu cast alloy. FME Transactions. 2013;**41**:210-215

[27] Prasada Rao AK, Das K, Murty BS, Chakraborty M. On the modification and segregation behavior of Sb In Al-7Si alloy during solidification. Materials Letters. 2008;**62**:2013-2016

[28] Fommei F. Modification treatments of Al-Si alloys. Alumino. 1977;**46**:121-135

[29] Jahromi SAJ, Dehghan A, Malekjani S. Effects of optimum amount of Sr and Sb modifiers on tensile, impact and fatigue properties af A356 aluminum alloy. Iranian Journal of Science & Technology, Transaction B. 2004;**28**:225-232

[30] The fluidity of sodium and strontium modified sand-cast aluminium-silicon foundry alloys. Cast Metals. 1989;**2**:109-112

[31] Hess PD. Sr as a Modifying Agent for Aluminum-Silicon Alloys—II. Progress Report 12E501111978

[32] Fuoco R, Correa ER, Correa AVO. Effect of modification treatment on microporosity formation in 356 Al alloy. AFS. 1995;**103**:379-387

[33] Denton JR, Spittle JA. Solidification and Susceptibility to hydrogen absorption of Al–Si alloys containing strontium. Materials Science and Technology. 1985;**1**:305-311

[34] Zhang LY, Jiang YH, Ma Z, Shan SF, Jia YZ, Fan CZ, et al. Effect of cooling rate on solidified microstructure and mechanical properties of aluminium-A356 alloy. Journal of Materials Processing Technology. 2008;**207**:107-111

[35] Maube SE, Wang'ombe DN, Maranga SM, Kihiu JM. Effect of cooling rate and heat treatment on the microstructure and impact resistance of recycled aluminium sand cast alloy. Journal of Sustainable Research in Engineering. 2014;**1**:24-30

[36] Manente A, Timelli G. Optimizing the Heat Treatment Process of Cast Aluminium Alloys, Recent Trends in Processing and Degradation of Aluminium Alloys. Rijeka, Croatia: InTech; 2011. p. Ch. 09

[37] Lavernia EJ, Ayers JD, Srivatsan TS. Rapid solidification processing with specific application to aluminium alloys. International Materials Reviews. 1992;**37**:1-44

[38] Duwez P, Willens RH, Klement W Jr. Continuous series of metastable solid solutions in silver-copper alloys. Journal of Applied Physics. 1960;**31**:1136-1137

[39] Sarmiento G, Bronzini C, Canale A, Canale L, Totten G. Water and Polymer Quenching of Aluminum Alloys: A Review of the Effect of Surface Condition, Water Temperature, and Polymer Quenchant Concentration on the Yield Strength of 7075-T6 Aluminum Plate. Journal of ASTM International. 2009;**6**:1-18

[40] Croucher T. Water Quenching Aluminum. 2010. Available: http://www.croucher.us/water-quenching [11/07/2017]

[41] Abdalla Hassan MJ, Abdulazeez MS, Jassim AM, Abdalla HMJ. Effect of quenching media, heat treatment and alloying elements on properties of Al-Si-Mg alloy. Journal of Babylon University. 2014;**22**:803-815

[42] Apelian D, Shivkumar S, Sigworth G. Fundamental aspects of heat treatment of cast Al-Si-Mg alloys. AFS Transactions. 1989;**97**:727-742

[43] Selivorstov V, Dotsenko Y, Borodianskiy K. Influence of low-frequency vibration and modification on solidification and mechanical properties of Al-Si casting alloy. Materials. 2017;**10**:560

[44] Limmaneevichitr C, Pongananpanya S, Kajornchaiyakul J. Metallurgical structure of A356 aluminum alloy solidified under mechanical vibration: An investigation of alternative semi-solid casting routes. Materials & Design. 2009;**30**:3925-3930

[45] Fisher TP. Effects of vibrational energy on the solidification of aluminium alloys. British Foundryman. 1973;**66**:71-84

[46] Campbell J. Effects of vibration during solidification. International Metals Reviews. 1981;**26**:71-108

[47] Richards R, Rostoker W. The influence of vibration on the solidification of an aluminum alloy. Transactions of the ASM. 1956;**48**:885-901

[48] Pillai NR. Effect of low frequency mechanical vibration on structure of modified aluminum-silicon eutectic. Metallurgical Transaction. 1972;**3**:1313

[49] Abu-Dheir N, Khraisheh M, Saito K, Male A. Silicon morphology modification in the eutectic Al–Si alloy using mechanical mold vibration. Materials Science and Engineering: A. 2005;**393**:109-117

[50] Tamura T, Amiya K, Rachmat RS, Mizutani Y, Miwa K. Electromagnetic vibration process for producing bulk metallic glasses. Nature Materials. 2005;**4**:289-292

[51] Radjai A, Miwa K. Effects of the intensity and frequency of electromagnetic vibrations on the microstructural refinement of hypoeutectic Al-Si alloys. Metallurgical and Materials Transactions A. 2000;**31**:755-762

[52] Ebhota WS, Karun AS, Inambao FL. Centrifugal casting technique baseline knowledge, applications, and processing parameters: Overview. International Journal of Materials Research. 2016;**107**:960-969

[53] Ebhota WS, Karun AS, Inambao FL. Investigation of functionally graded aluminium A356 alloy and A356-10%SiCp composite for hydro turbine bucket application. International Journal of Engineering Research in Africa. 2016;**26**:30-46

Chapter 5

Friction Stir Welding of Aluminium Alloys

Rajkumar Vijayakumar, Venkatesh Kannan and
Arivazhagan Natarajan

Additional information is available at the end of the chapter

http://dx.doi.org/10.5772/intechopen.70233

Abstract

This chapter investigates on the characterization of friction stir welded dissimilar aluminium alloys AA2024 with AA5052, AA2024 with AA6061 and AA 5052 with AA6061. Five tool designs were employed with first two dissimilar combinations to analyze the influence of rotation and traverse speed over microstructural and mechanical properties. H13 tool steel was used as tool material with various pin profiles which includes cylindrical, cylindrical-threaded, squared, tapered and stepped types. In the dissimilar welding of AA 2024 with AA 5052, sound welds were produced with stepped pin tool. In the dissimilar welding of AA 2024 with AA 6061, ratio between tool shoulder to diameter of tool pin was the most influential factor. Welded joints failed in the Heat affected zone (HAZ) of 6061 where the hardness values were comparatively less. In dissimilar welding of AA 5052 with AA6061, cylindrical pin tool was used at a constant speed of 710 rpm and at different feed rates of 28 and 40 mm/min. Micro structural examination showed variation of grain size in every zone and their influence on mechanical properties. Correlating mechanical and metallurgical properties, the optimized process parameters of speed and feed were identified to be 710 rpm and 28 mm/min respectively for all attempted dissimilar combinations.

Keywords: aluminium alloy, friction stir welding, dissimilar welding, AA2024, AA 5052, AA6061

1. Introduction

Aluminium alloys are widely used in variety of applications ranging from basic to complex such as in the making of aircraft bodies. Due to varied service conditions, there are scenarios where different series of aluminium alloys are to be joined. The most extensively used non precipitation hardenable series of Al alloys in aeronautical applications are 3xxx, 5xxx. Some of the precipitation hardenable series of Al alloys include 2xxx, 6xxx and 7xxx series. Merits of 2xxx series are that they possess excellent mechanical properties at high temperatures whereas 7xxx series show good mechanical properties at low temperatures and also exhibit high corrosion resistance properties [1]. Identification of appropriate joining process and process parameters employed becomes important in service performance of these materials and in its actual service conditions. aluminium alloys are lighter yet possess good strength and ductility. This makes the material a preferable candidate to work in varied working environments.

AA2024, AA5052 aluminium alloys are widely used in, automotive, aerospace and shipbuilding industries [2]. Fusion welding of dissimilar aluminium alloys is very challenging mainly due to the formation of low melting eutectics by the constituent elements resulting in weld solidification cracking. Hot cracking in aluminium alloys are extremely sensitive to weld metal compositions [3]. Hence solid state joining process becomes more suitable for welding aluminium alloys, since this process does not involve melting. Hence the defects like weld solidification cracking, porosity, segregation, liquid cracking on heat effected zone and brittle inter metallic formation could be avoided using this technique [2, 4]. In FS welding techniques, tool geometry plays a major role in obtaining desirable metallurgical and mechanical properties [5–7].

In this study AA2024 and AA5052 were fabricated using friction stir welding process using five different tool pin profiles. Four traditional pin profiles namely cylindrical, threaded, squared, tapered pin and a newly designed stepped pin profile were employed in this research. Al-Cu dissimilar metals were welded as in reference [4] and found that better tensile strengths were achieved in the plates which was welded at 710 rpm, where the tested speed range was 600–1000 rpm. They also concluded that the defect-free joints could be obtained when the plates that had superior mechanical properties were fixed on the advancing side. The author reported that the pin transfers the material layer by layer, while the shoulder transfers the material in bulk and forges it. FS welding of AA 6061 has experimented as in reference [8] by changing ratio of shoulder to pin diameter and reported that the defect free welding can be obtained when keeping the D/d ratio is 3:1. Studies on the FS welding of AA 2219-AA5083 alloy as in reference [3] by changing various pin profiles and found that, cylindrical threaded pin has produced defect free welding with good tensile strength. From extensive literature survey it can be deduced that the newly developed stepped pin profile has not been incorporated in the friction stir welding of aluminium alloys.

Aluminium alloy AA6061, is extensively used in marine industries and in the construction of storage tanks and pipelines. Joining process for dissimilar materials are considered quite challenging as compared to joining similar metals, due to change in chemical composition of base metals and their mechanical properties [9–11]. Fusion welding of nonferrous metals is tedious due to high heat inputs. Formation of secondary phase in friction stir welding (FSW) process is absent since the temperature reached in this process is well below the melting point of parent metals.

Thermal dissipation emanates local isothermal stresses. This thermal gradient developed has important and adverse effect on the metallurgical properties and in turn on the mechanical properties of the joint, precisely in the formation of soft zones. This microstructural change affects the performance in service conditions of the weld joints, since mechanical properties decreases with reference to the base material. Valuable details can be obtained by understanding weld thermal cycles precisely in conjunction with transformation curves and their microstructural effect in mechanical properties [12].

Literatures indicate that optimal parameters for joining of dissimilar aluminium alloys are rotation speeds being 600–1000 rpm, D/d ratio of 3:1 traverse speeds around 15–40 mm/min. Out of many materials processed using FSW, the most commendable results obtained were for aluminium and all of its alloys including cast and wrought conditions. Most influential parameters in the FSW process were researched to be tool geometry, travel speed, rotational speed, rotational direction, rotational axis, eccentricity [13]. Studies show that out of two base metals employed, when harder material is placed on the advancing side and softer material in retreating side, better mechanical properties are obtained. Different tool profiles are investigated by researchers, as geometry of tool pin plays a major role in FSW. Threaded, squared and triangular tool pin profiles are efficient to transfer the material from top to bottom of the joint and vice versa by stirring action. The analysis of the results during certain experiments by researchers revealed that during the time of mechanical tests performed, the crack initiation may be significant, at least for welded joints with relative lower stress concentrations and low to moderate loads [14]. Certain researchers have attempted various welding techniques on aluminium alloys. Gas metal arc welding (GMAW), shielded metal arc welding (SMAW) and gas tungsten arc (GTA) welds have been attempted for welding aluminium alloys. Important morphological characteristics have been observed on MIEW process when compared to that of GMA welds. When MIEW welds were fabricated, solidification process tended to promote heterogeneous nucleation. Therefore, auto refinement of grain size is promoted. On the other hand, when multi pass GMA welds were employed, columnar epitaxial solidification happens resulting in increase in grain size [15, 16].

Aluminium alloys AA 5052 and AA 6061 are FS welded with specific tool contours and using two welding parameters of variable feed rate and constant speed. This chapter also discusses on characterization of metallurgical and mechanical properties of the above combination to evaluate the performance characteristics of FS welded joints.

2. Experiments and procedures

2.1. Materials and methods

The materials used in this study are commercially available AA 2024-T4 (Al-Cu alloy), AA 5052 and AA 6061-T4 (Al-Mg-Si alloy). Plates of 5 mm thick AA2024, AA5052 and AA6061 were friction stir welded in certain combinations using conventional milling machine employing a specially designed fixture. The chemical compositions of the samples are tabulated in **Table 1**.

D/d ratio (shoulder/pin) was kept as 3, where shoulder diameter is 16 mm and pin diameter being 6 mm. The aluminium plates were made into coupons of 100 mm × 50 mm where the welding was carried out using milling machine with necessary fixtures. AISI H13 tool steel which has high thermal fatigue resistance was used in this study.

Material	Mg	Mn	Cu	Fe	Si	Cr	Zn	Ti	Al	% of Elongation
AA2024	1.5	0.6	4.35	0.5	0.5	0.10	0.25	0.15	Rem	137
AA5052	1.4	0.14	0.14	0.4	0.26	0.15	0.08	–	Rem	330
AA6061	1	0.15	0.27	0.7	0.6	0.19	0.6	0.15	Rem	75

Table 1. Chemical composition of base metals.

2.2. Mechanical characterization

Tensile test was carried out using INSTRON 8801 UTM according to ASTM E8/E8M standards of sub size specimen and the tests were carried out at a strain rate of 0.5 mm/min. Micro hardness measurements were carried out at a load of 100 gf with dwell time of 10 s and distance of 0.25 mm interval across the weldment.

2.3. Microstructural characterization

To study the microstructure of the weldments of these dissimilar aluminium alloys Keller's reagent (150 ml water + 3 ml of Nitric Acid, 6 ml of hydrochloric and hydrofluoric acid) was used. In order to analyze the constituents in thermo-mechanically affected zone (TMAZ) and weld, scanning electron microscopy (SEM) with energy-dispersive spectroscopy (EDS) was used.

2.4. Process parameters

In dissimilar welding of AA2024 with AA 5052 and AA 2024-T4 (Al-Cu alloy) with AA 6061-T4 (Al-Mg-Si alloy) welding was carried out by placing AA2024 in the advancing side. It was due to higher mechanical strength of AA 2024. AA 5052 and AA6061 were placed on the retreating side. Different pin profiles viz. cylindrical, threaded, squared, tapered and stepped pin were used in this study. The length of tool pin is kept constant at 4.8 mm. Experiments were conducted at a feed rate of 40 mm/min and 28 mm/min against two different speeds of 710, 1000 rpm. The tensile test results welded with all tool profiles and their Ultimate tensile strength (UTS) values along with place of fracture is tabulated in **Table 2**.

The process parameters used in the welding of AA 2024-T4 (Al-Cu alloy) and AA 6061-T4 (Al-Mg-Si alloy) is given in **Table 3**.

S No	Tool pin profile	UTS (MPa)	Percentage strength	Fracture spot
1	Threaded	259	78	TMAZ of 5052
2	Squared	200	60	Weld
3	Stepped	297	90	TMAZ of 5052
4	Cylindrical	195	59	Weld
5	Tapered	202	61	Weld

Table 2. Tensile test results of AA 2024 and AA 5052 welded specimens.

S No	Tool design	Rotational speed	Traverse speed (mm/min)	Tilting angle (deg)
1	Threaded pin	710 and 1000 rpm	28	2
2			40	
3	Squared pin	710 and 1000 rpm	28	
4			40	
5	Tapered pin	710 and 1000 rpm	28	
6			40	
7	Cylindrical pin	710 and 1000 rpm	28	
8			40	
9	Stepped pin	710 and 1000 rpm	28	
10			40	

Table 3. Welding parameters of AA 2024 and AA 5052 alloys.

The dissimilar alloys AA 5052 and AA6061 were butt welded using cylindrical pin tool with 2 threads. AA 5052 was kept at the advancing side and AA6061 in the retreating side. The two specimens were welded using two parameters: 710 rpm at 28 mm/min and 710 rpm at 20 mm/min.

3. Results and discussion

3.1. Dissimilar friction stir welding of AA 2024 - AA5052 alloys

3.1.1. Metallurgical characterization

The different types of pin profiles used in this work along with the corresponding tool dimensions can be seen from **Figure 1(a–e)**.

It is observed that macroscopic defects appeared when using the cylindrical pin tool. The cross sectional macrograph is represented in **Figure 2(a–e)**.

It clearly shows void defect while using cylindrical and tapered pin tool. This could be due to inferior metal flow during the welding process while using the corresponding profile tools. On the other hand, the joints fabricated using threaded, squared and stepped pin appeared to be free from defects which could possibly be due to effective mix-up and proper inter diffusion of elements from both base metals. It could also be due to optimized heat input within the weld nugget while using these profiled tools. This may be due to lack of heat input during the welding process and the nature of pin profile. Comparing all tool pin profiles, threaded pin and stepped pin resulted in uniform mix-up of elements from both base metals. The micrographs of FS welded samples obtained using these tool profiles are represented in **Figure 3(a–f)**.

It is observable from **Figure 3(a** and **d)** that the material flow from advancing side of AA2024 to the weld nugget is vivid in both threaded and stepped pin profiles. Further refined grain structures appeared on the weld nugget as well as on the AA2024 side.

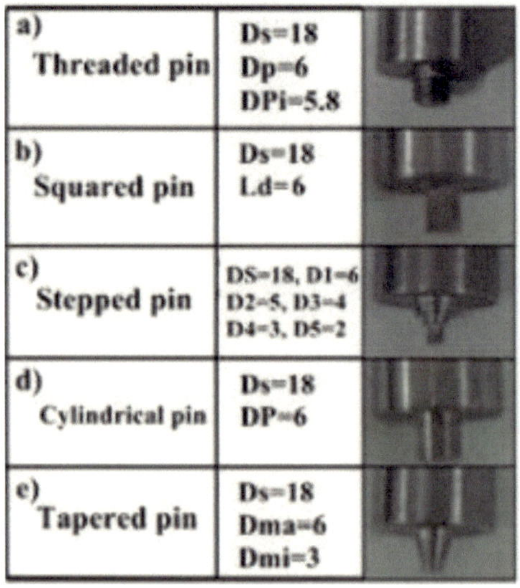

Figure 1. Various tool pin profiles used and their dimensions.

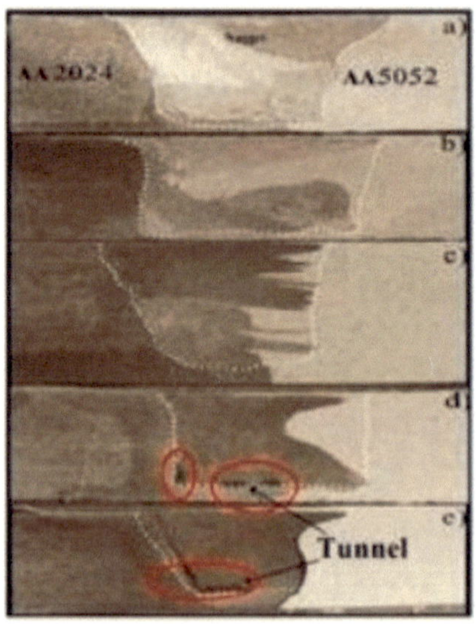

Figure 2. Cross sectional macrostructures of welded samples.

3.1.2. Mechanical characterization

The tensile test has been carried out for the FS welded samples and the averaged results are represented in **Table 2**. Two samples were tested on weldments fabricated with each tool pin profiles to ensure the repeatability of weld consistency and testing. Samples welded with threaded pin gave tensile strength of 259 MPa which is 78% of the strength of AA5052. Interestingly the stepped pin profile yielded tensile strength of 297 MPa. The fracture also occurred at the TMAZ of AA5052 and not in the weld. Squared pin, cylindrical pin, and tapered pin fractured in nugget with 60, 59, and 61 percentage of base strength of AA 5052. Stress-strain graphs plotted for both samples are depicted in **Figure 4**.

http://dx.doi.org/10.5772/intechopen.70233

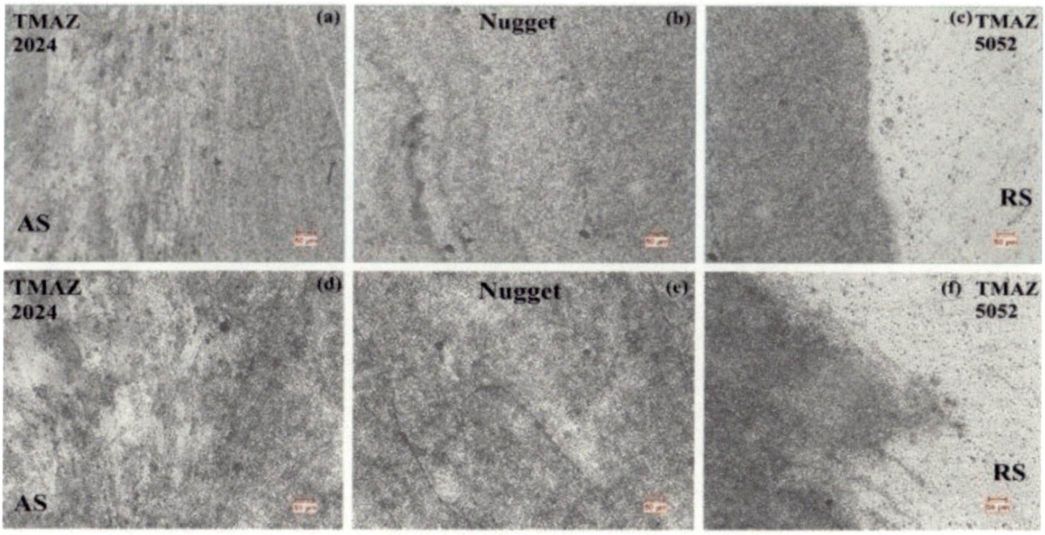

Figure 3. Microstructures captured at various zones of AA 2024 and AA5052 welds.

The hardness profiles of the samples welded with traditional tool pins were found to have lower range of hardness value in the weld nugget and heat affected zone. In both the cases, it is apparent the hardness values in the weld nugget seems to be higher than that of the base metal of AA 5052 and less that of AA2024. Hardness plot for threaded and stepped pin profile weldments can be seen from **Figure 5**.

Sudden drop in the hardness value from weld zone to TMAZ in retreating side led to the fracture at that point as micro hardness values are directly proportional to the strength of weld which is also evident from the tensile test. The hardness profile noticed in the weld region of the sample welded with stepped pin is comparatively higher than that of the hardness achieved in

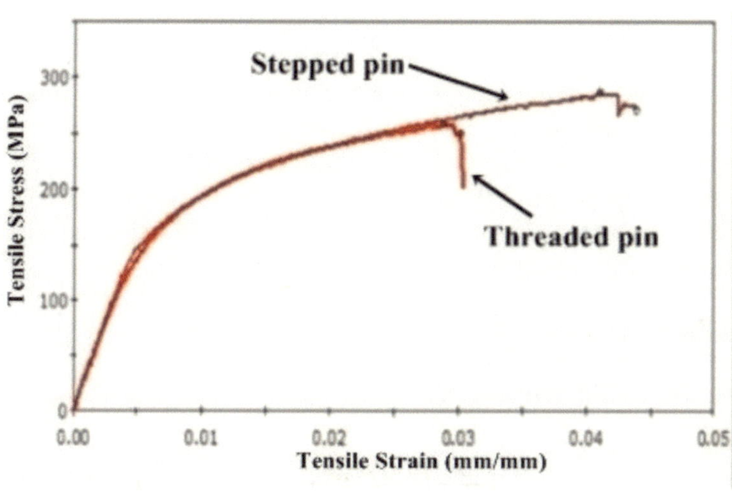

Figure 4. Stress strain plots of AA 2024 and AA 5052 welded samples.

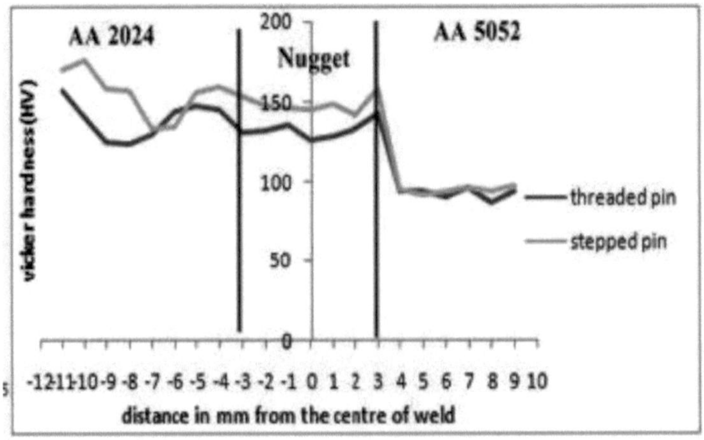

Figure 5. Hardness plots of threaded and stepped pin welded samples.

joints obtained using other tool profiles eventually proving high strength. Average micro-hardness value of 150 HV was obtained in the welded joints produced through stepped pin tool.

3.2. Dissimilar friction stir welding of AA2024-AA6061 alloys

The surface morphologies of the weld fabricated using cylindrical, threaded and squared pin tool profile are shown in **Figure 6**(**a** and **b**).

The traverse speed and rotational speed for threaded and squared pin were fixed at 28 mm/min at 710 rpm and 40 mm/min at 1000 rpm for respective samples. By varying the process parameters and tool geometry, no defects were found in the welds except for tapered pin. Defective surface morphologies and improper heat diffusion were observed on using tapered pin (**Figure 6(b)**).

3.2.1. Macrostructures

From the macro-graphic studies, different regions of weldments are identified and it represents the effective stir of both the base material in the nugget zone **Figure 6**(**c**, **d**).

The presence of AA2024 constituent elements in the nugget is more when compared to AA6061. This is because AA2024 was placed on advancing side of the weld. Welds without voids were fabricated using cylindrical threaded pin with parameters of 710 rpm, 28 mm/min and using squared pin with parameters of 1000 rpm, 40 mm/min. For the same process parameters, few macroscopic defects occurred on using cylindrical, tapered and stepped pin. This could be attributed to absence of vertical traverse of metal in nugget region.

3.2.2. Microstructural analysis

The microstructures at different regions of the welded dissimilar materials are shown in **Figure 7**(**a–g**).

There are no substantial changes in the base metal microstructure, though the weld nugget undergoes considerable amount of thermal changes. It is evident from the microstructure

http://dx.doi.org/10.5772/intechopen.70233

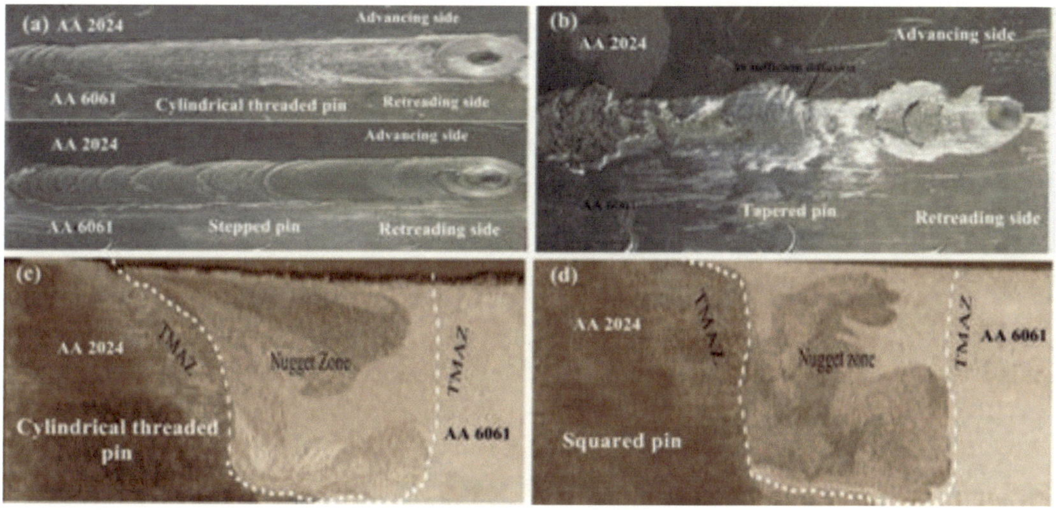

Figure 6. Surface morphology of AA2024 and 6061 welds and macrographs of threaded and squared pin profiles.

(**Figure 7**(**b** and **f**)) that the thermal cycle has significantly influenced the heat affected zone. Yet there is no plastic deformation occurred in this region. There is considerable grain growth in the thermo-mechanically affected zone (TMAZ). This could be due to the plastic deformation and low heat levels encountered during welding. A distinct boundary separates the recrystallized zone (weld nugget) from the TMAZ, which is apparent from the micrographs. The dynamically recrystallized zone is the stirred zone. Severe plastic deformation has taken place in this zone resulting in fine equiaxed grains. Stirred zone is a commonly used phrase in friction stir processing which denotes that substantial volume of material is being processed. Weld nugget micrographs (**Figure 7(d)**), depicts highly refined and increased grain boundaries, which has enhanced the weldment strength.

3.2.3. Tensile test

Maximum weld strength of 194 MPa and 209 MPa were obtained for cylindrical threaded and squared pin, respectively. As shown in **Figure 8**, for both the tool pin geometries, fracture occurred at the HAZ of 6061 alloy.

For the other tool pin geometry fracture has occurred at the stirred zone. The weld region shows lower strength compared to both base metals. Joint efficiencies obtained for cylindrical threaded pin and squared pin are 80 and 87%, respectively.

3.2.4. Hardness measurements

Vickers hardness tests were conducted across the various regions of the weld spacing of (0.25 mm) shown in **Figure 9**.

Average hardness value across the weldments obtained for cylindrical threaded pin and for squared pin were 105.15 HV and 135.6 HV respectively. The hardness of weld nugget was considerably less than that of AA2024, whereas the hardness was comparatively higher than base metal of 6061 and TMAZ.

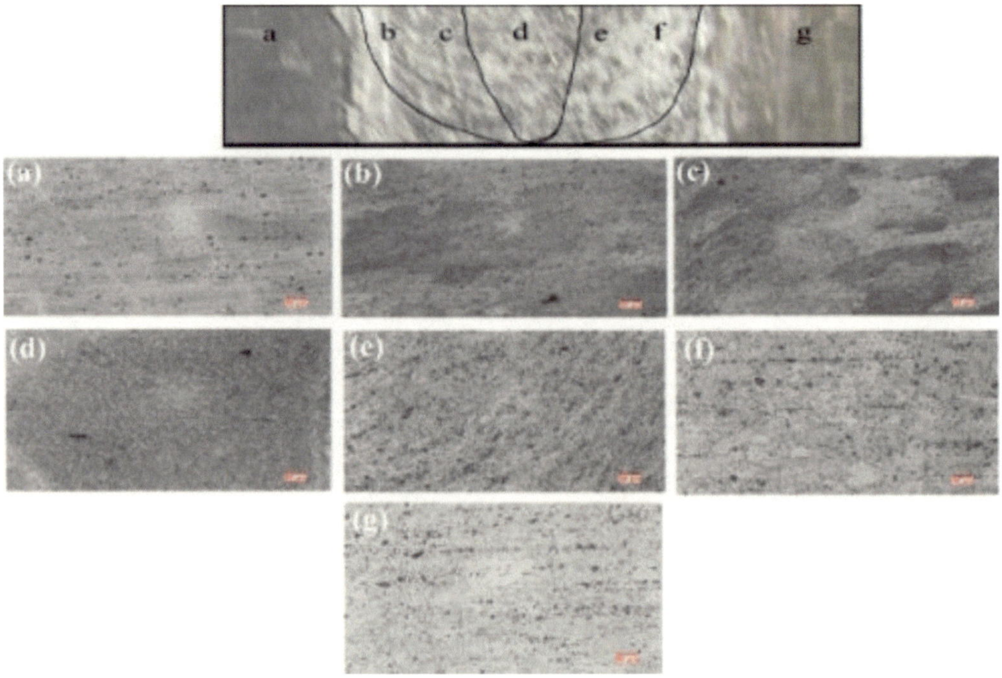

Figure 7. Microstructures captured at various zones of AA 2024 and AA 6061 weldment.

3.3. Dissimilar friction stir welding of AA 5052 and AA 6061 alloys

Evolution of grain structures, their texture, temperature gradient, recrystallization techniques and precipitation of inter – metallic constituents are some of the factors included in micro structural studies. These factors indeed influence the quality and strength of any welds when

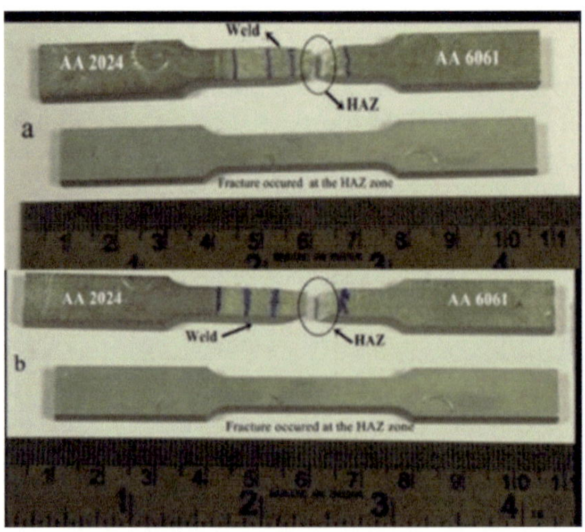

Figure 8. Tensile tested AA 2024 and AA 6061 samples welded with (a)Cylindrical threaded pin (b) Squared pin.

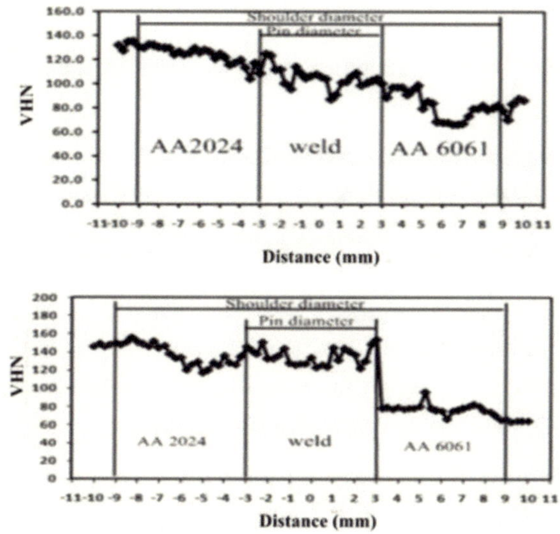

Figure 9. Hardness plots of samples welded using threaded and stepped pin profiles.

assessed through tests namely tensile, creep, fatigue etc. [17]. The tool dimension and the tool used for welding are shown in **Figure 10(a** and **b**).

3.3.1. *Cross sectional macrographs*

The cross sectional macrograph of weldments showed proper material flow from advancing side to retreating side. This is evidence of the fact that the weld process parameters considered are optimal. A micro void was visible in the sample welded at 40 mm/min (**Figure 10(d)**). This may be due to improper fusion involved in the weld nugget.

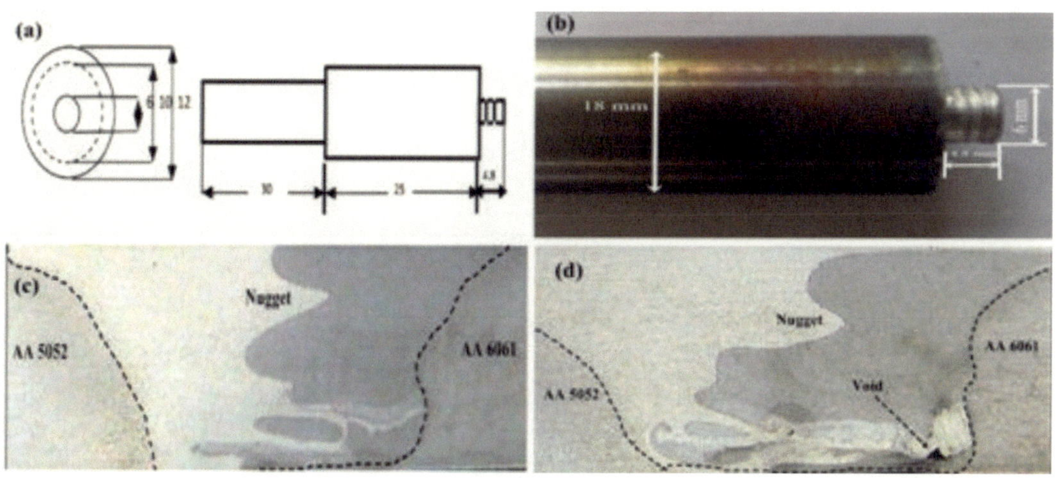

Figure 10. Dimensions of cylindrical threaded tool pin and cross sectional macrostructures of welded samples.

3.3.2. *Microstructural analysis*

The microstructures taken at various regions of the weldments of both the samples are shown in **Figure 11**.

The microstructures of the base metals of AA 5052 and AA 6061 are shown in **Figure 11**(**a** and **b**) respectively, whereas **Figure 11**(**c** and **d**) shows the micrographs at TMAZ. **Figure 12**(**a** and **b**) shows the microstructures of weld region and **Figure 12**(**c** and **d**) depicts the microstructures at the interfaces.

It can be observed from (**Figure 11**(**e** and **f**)) that grain formation is refined in the weld zone as compared to TMAZ of both alloys. This should be due to optimality of process parameter that was selected and effective cooling rate. In addition, on comparing the both TMAZ's, AA 6061 side has considerable grain size increase (**Figure 11(f)**).

This may have led to deterioration of tensile properties in this zone. It has been analysed in reference [6] that the change of grain size is due to the effect of elevated temperatures and that the strain rates were insignificant. It was also concluded that the decrease in the flow stress at high temperatures was primarily due to the thermally activated dislocation lines.

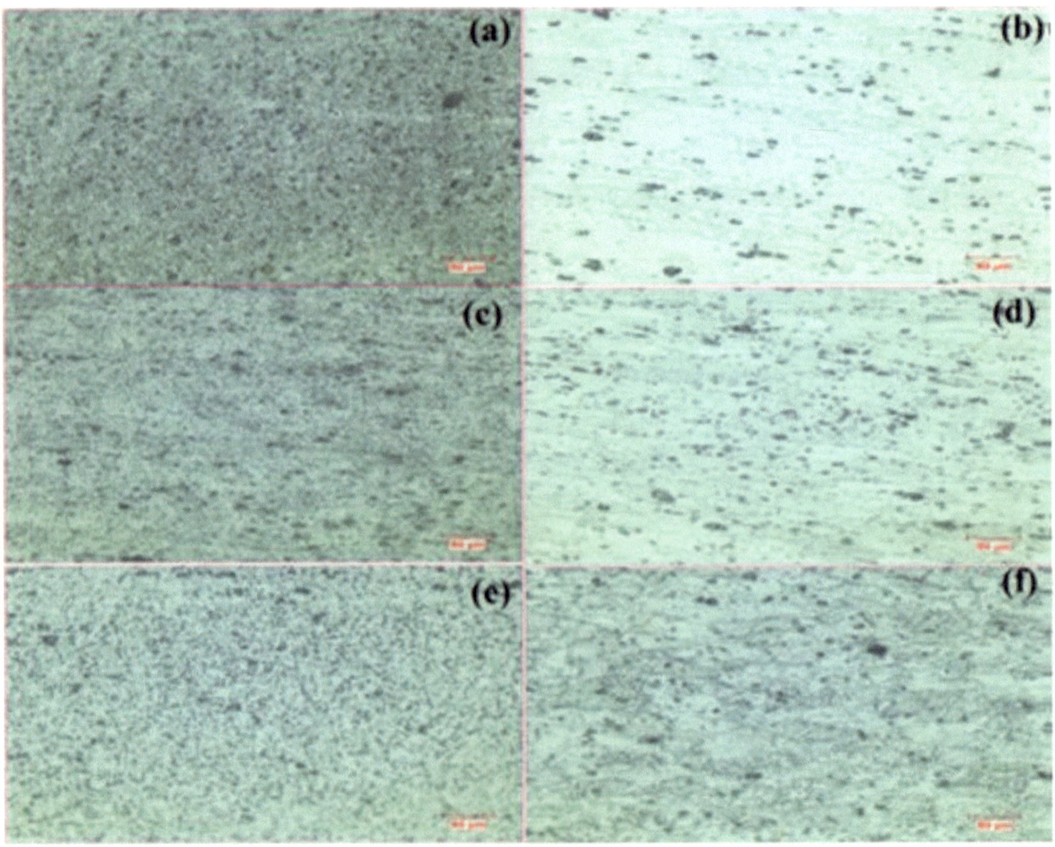

Figure 11. Microstructures of AA 5052 and 6061 sample welded at feed rate of 28 mm/min.

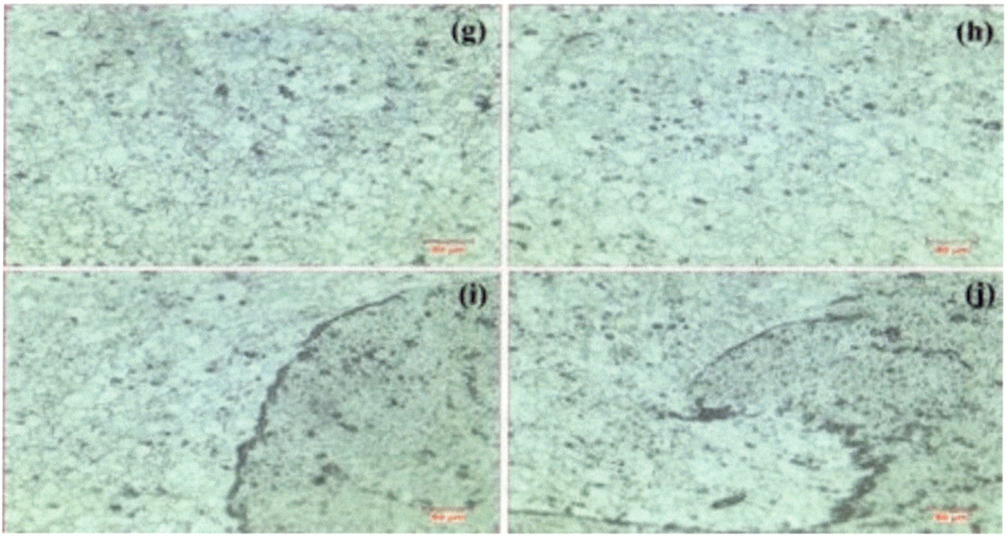

Figure 12. Micrographs of AA 5052 and AA 6061 samples captured at (g & h) weld regions and (I & j) at interface regions.

3.3.3. Tensile test

Tensile samples were wire cut using electrical discharge machining (EDM) and prepared according to the ASTM standards of sub-size dimensions. It is vivid from **Figure 13(a)** that in both cases, the fracture has occurred in the TMAZ of the 6061 alloy. This affirms that the weld nugget possesses better strength than base metals.

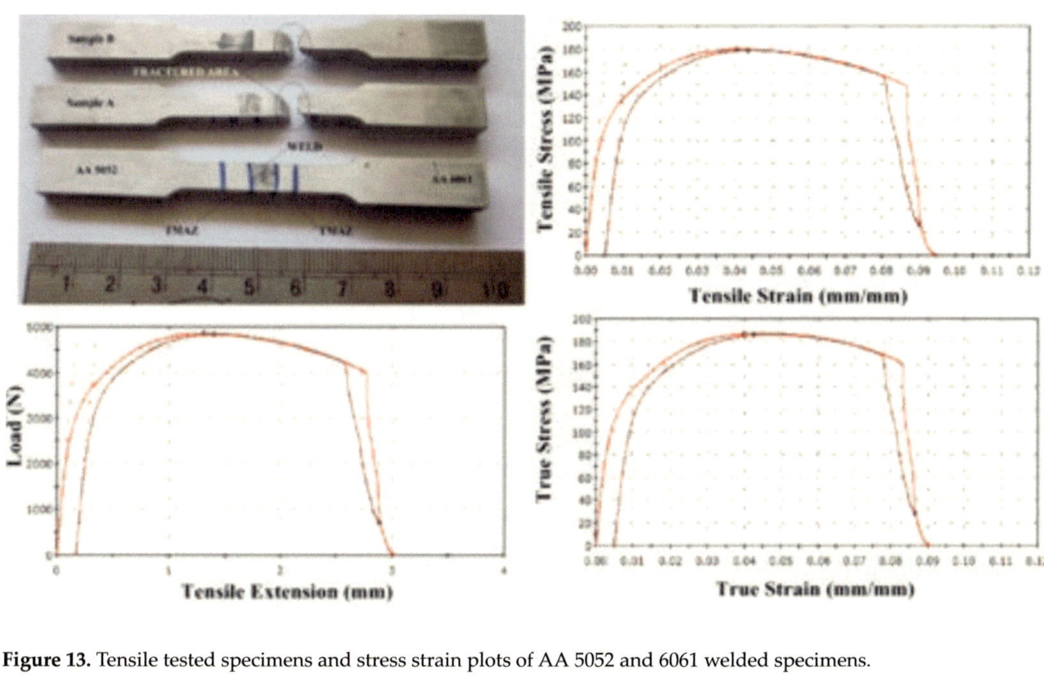

Figure 13. Tensile tested specimens and stress strain plots of AA 5052 and 6061 welded specimens.

There is no much difference between UTS of both the samples. However, the tensile stress at break point of Sample B is proportionately higher than Sample A. This means Sample B has yielded suddenly after the elastic limit, whereas Sample A has yielded after demonstrating high ductility. The load at break point adds to the same. Results of tensile test are being tabulated in **Table 4**.

Welding parameters have been optimized as in reference [2] by design of experiments and the optimum level of settings concluded from their experiments. Corresponding parameters chosen for the studies are rotational speed of 700 rpm, transverse speed of 28 mm/min and D/d ratio being 3 respectively. The performance of cylindrical threaded pin tool profile was better among others considered. D/d ratio plays a vital role and contributes to an overall efficiency of about 60%. It is found in reference [17] that the hardness variations in the FS weld zone of Al 5083 alloy could not be supported by study of grain size in the weld region alone. Hence, in any study, correlation of mechanical performances with that of metallurgical factors becomes important in analysing the nature and reason for the fracture.

3.3.4. Hardness measurements

The hardness profile obtained for the samples are shown in **Figure 14**.

From the plots, it can be understood that the feed rate has only little influence on the hardness of the weldments. From the advancing side (AA 5052) to the retreating side (AA 6061) of the weldment, the hardness values show a decreasing trend and again a gradual increasing trend. It is worthy to note that the fracture has also occurred in the TMAZ of AA 6061 (minimum

Sample	Maximum load	UTS (MPa)	Tensile strain at break (%)	Tensile stress at break (MPa)	Load at break (kN)	Elongation (%)
Sample A	4853	180	10.5	2.75	0.07	10.5
Sample B	4837	179	8.4	27.06	0.73	8.33

Table 4. Tensile test report of AA 5052 and AA 6061 welded samples.

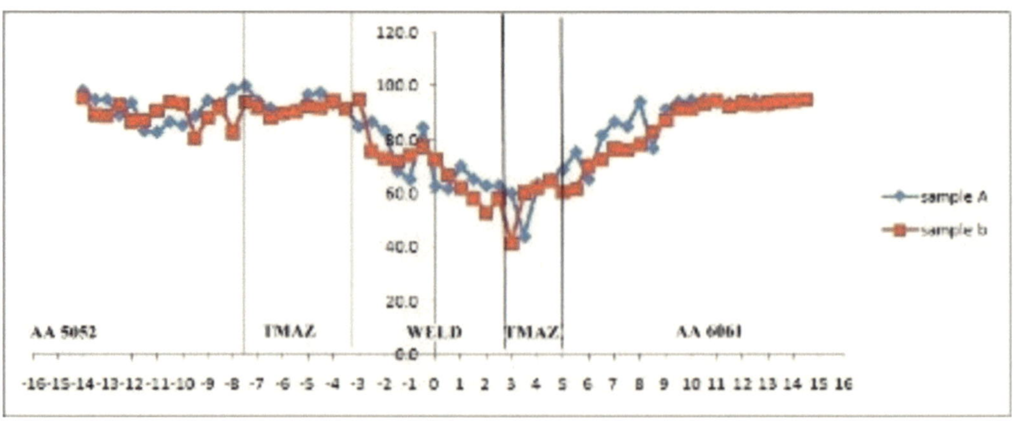

Figure 14. Hardness plots of AA 5052 and AA 6061 welded samples.

hardness zone). It can be noted from the graph that hardness values of TMAZ in the retreating side are comparatively less. This could be due to constituent gradient in mixing/diffusion of AA 6061 with material from advancing side and weld nugget. As evident from nature of the micrograph, this could be a reason for the failure of tensile specimen in the TMAZ of AA 6061 side. Moreover, it may be due to the grain coarsening effect that the hardness values in the weld nugget shows lower trend when compared with parent metals.

4. Conclusions

- In the FSW of AA 2024 with AA5051, sound welds were produced using a new stepped pin tool, at 710 rpm and 28 mm/min. From morphological studies, there is quantifiable reduction in grain size of weld nugget compared to parent metals. Maximum tensile strength of 297 MPa was achieved on samples welded using stepped pin tool. This strength is comparatively higher than that of the ones that were welded with other profiled pins.
- From the FSW of AA 2024 with AA6061, it is deduced that for cylindrical pin, D/d ratio of 3, rotational speed of 710 rpm, traverse speed of 28 mm/min, were the best optimal parameters. Further, better mechanical properties were observed for the same. In addition, it is concluded that squared pin and cylindrical threaded tool profiles perform better than rest of tool profiles considered.
- FSW between AA 5052 and AA 6061 alloys sounds promising. Cylindrical threaded pin has imparted excellent bondage between both alloys (AA 5052 and AA 6061) by effective FS joining. Both the samples have exhibited nearly equal ultimate strength. In terms of ductility, Sample B (with 710 rpm and 28 mm/min feed rate) outperformed Sample A (with 710 rpm at 40 mm/min).

Acknowledgements

The authors wish to extend their sincere thanks to SNR Sons Charitable Trust, Coimbatore and the management of VIT University, Vellore for providing necessary support in executing this research.

Author details

Rajkumar Vijayakumar[1]*, Venkatesh Kannan[2] and Arivazhagan Natarajan[2]

*Address all correspondence to: vrajkumarphd@gmail.com

1 Department of Mechanical Engineering, Sri Ramakrishna Institute of Technology, Coimbatore, India

2 School of Mechanical Engineering, VIT University, Vellore, India

References

[1] Schneider R, Heine B, Grant RJ. Mechanical behaviour of commercial aluminium wrought alloys at low temperatures. In: Monteiro WA, editor. Light Metal Alloys Applications. Croatia: InTech; 2014. pp. 61-76. DOI: 10.5772/57069

[2] Sheikhi S, Bolfarini C. Preliminary study on the microstructure and mechanical properties of dissimilar friction stir welds in aircraft aluminium alloys 2024-T351 and 6061-T4. Journal of Materials Processing Technology. 2007;**6**:132-142

[3] Koilraj M, Sundareswaran V, Vijayan S, Koteswara Rao SR. Friction stir welding of dissimilar aluminium alloys AA 2219 to AA 5083 – Optimization of process parameters using Taguchi technique. Materials and Design. 2012;**42**:1-7

[4] Xue P, Ni DR, Wang D, Xiao BL, Ma XY. Effect of friction stir welding parameters on the microstructure and mechanical properties of the dissimilar Al-Cu joints. Materials Science and Engineering A. 2011;**528**(13):4683-4689

[5] Li Y, Murr LE, Mcclure JC. Flow visualization and residual microstructures associated with the frictions stir welding of 2024 aluminium to 6061 aluminium. Materials Science and Engineering A. 1999;**271**:213-223

[6] Buffa G, Huaa J, Shivpuri R, Fratini L. Design of the friction stir weld using the continuum based FEM model. Materials Science and Engineering A. 2006;**419**(2):381-388

[7] Mishra RS, Ma ZY. Friction stir welding and processing. Materials Science and Engineering: R: Reports. 2005;**50**(1):1-78

[8] Arora A, Debroy T. Toward optimum friction stir welding tool shoulder diameter. Scripta Materialia. 2011;**64**:9-12

[9] Heidarzadeh A, Khoaverdizadeh H, Mahmoudi A, Nazari E. Tensile behaviour of friction stir welded AA 6061-T4 aluminium alloy joints. Materials and Design. 2012;**37**:166-173

[10] Rao P. Microstructure and mechanical properties of friction stir lap welded aluminium alloy AA 2014. Journal of Materials Science and Technology. 2011;**28**(5):414-426

[11] Kumbhar NT, Sahoo SK, Samajdar I, Dey GK, Bhanumurthy K. Microstructure and micro textural studies of friction stir welded aluminium alloy 5052. Materials and Design. 2011;**32**:1657-1666

[12] Ambriz RR, Jaramillo D. Mechanical behavior of precipitation hardened aluminium alloy welds. In: Monteiro WA, editor. Light Metal Alloys Applications. Croatia: InTech; 2014. pp. 36-59. DOI: 10.5772/58418

[13] Vilica P, Telmo G Santos. Non destructive techniques for detecting imperfections in friction stir welds of aluminium alloys. In: Kvackaj T, editor. Aluminium Alloys Theory and Applications. Croatia: InTech; 2011. pp. 94-114. DOI: 10.5772/14743

[14] Ribeiro AS, de Jesus AMP. Fatigue behaviour of welded joints made of 6061-T651 aluminium alloys. In: Kvackaj T, editor. Aluminium Alloys Theory and Applications. InTech; 2011. pp. 136-156. DOI: 10.5772/14489

[15] Ambriz RR, Mayagoitia V. Welding of aluminium alloys. In: Recent Trends in Processing and Degradation of Aluminium Alloys. Croatia: InTech; 2011. pp. 63-86. DOI: 10.5772/18757

[16] Mahabunphachai S, Koc M. Investigations on forming of aluminium 5052 and 6061 sheet alloys at warm temperatures. Materials and Design. 2010;**31**:2422-2434

[17] Sato Y, Urata M. Hall-Petch relationship in friction stir welds of equal channel angular presses aluminium alloys. Materials Science and Engineering A. 2003;**35**(4):298-305

Chapter 6

Formability and Performance of Al-Zn-Mg-Cu Alloys with Different Initial Tempers in Creep Aging Process

Heng Li, Chao Lei and He Yang

Additional information is available at the end of the chapter

http://dx.doi.org/10.5772/intechopen.70849

Abstract

The initial temper may directly affect the deformation behavior and material performance in creep age forming (CAF) process. Five heat treatment states are selected as the initial tempers for CAF, namely, solution, peak-aging (T6), over-aging (T73), retrogression and re-solution. The formability and performance of an Al-Zn-Mg-Cu alloy with the above initial tempers in creep aging process are investigated via using creep and stress relaxation aging tests, mechanical property tests, corrosion resistance tests and microstructure analysis. The differences of formability are attributed to the inhibitions of different distributed matrix precipitates (MPts) on the dislocation movement, namely, the more coarsening the MPts is, the easier the dislocation movement. During creep aging process, the mechanical properties are improved for the solution, retrogression and re-solution tempers with fine MPts, but reduced for the T6 and T73 tempers due to coarsening of MPts. Since the distribution of grain boundary precipitates (GBPs) becomes discontinuous, the corrosion resistances of the creep aged specimens are enhanced for all initial tempers. Taking both mechanical properties and corrosion resistances into account, the re-solution temper may be a preferable choice to achieve high performance of the components beyond the precise shape in CAF.

Keywords: Al-Zn-Mg-Cu alloy, creep age forming, initial temper, precipitate, formability, mechanical property, corrosion resistance

1. Introduction

With the increasing demands for high-performance and light-weight in the components of aircraft and automobile, the Al-Zn-Mg-Cu (7xxx) series alloys have been widely used due to their high strength and relatively low density. The precipitation reaction is the leading strengthening mechanism in this class of alloys, and the precipitation sequence has been generally identified as supersaturated solid solution → Guinier-Preston (GP) zones → η' → η ($MgZn_2$) [1]. As a novel and promising forming technology, creep age forming (CAF) has been developed and applied to manufacture large integral panel components, such as the upper wing panels of Airbus A380, and both the upper and lower wing skins of B-1B′ Long Range Combat Aircraft [2]. Compared to traditional manufacturing techniques, CAF has one of the most prominent advantage, namely, due to the simultaneity of creep deformation and age hardening, the shape forming process and the material strengthening process can be integrated into one single process, thus reducing the cost of manufacturing.

The complexity of CAF lies in the strong interaction coupling between creep and aging. Specifically, the precipitates generated in aging process will significantly hinder dislocation movement. Inversely, the dislocation density variation induced by creep deformation will affect the nucleation and growth of precipitates. Therefore, it is very difficult to precisely control the product shapes and material properties. Furthermore, it is well known that the Al-Zn-Mg-Cu alloys have many heat treatment states, and any one of the states can be the initial temper of the material used in CAF. Various initial tempers inevitably cause different couplings between creep and aging, and then make it difficult to precisely control the CAF process [3]. In order to acquire the high performance of formed components beyond the precise shape, it is very necessary to study the dependences of formability and performance on the initial tempers of Al-Zn-Mg-Cu alloys during CAF.

A lot of experimental studies mainly aimed at finding out the effects of processing parameters on CAF. Arabi Jeshvaghani et al. [4, 5] studied the effects of time and temperature on microstructure evolution of 7075 aluminum alloy sheet during CAF, and observed the transmission electron microscope (TEM) bright field images of matrix precipitates (MPts) and grain boundary precipitates (GBPs) in the formed samples after different forming periods, 6, 12 and 24 h, respectively. Guo et al. [6] found that the external elastic tensile stress promotes the formation of precipitates and shortens the aging period of an Al-Zn-Mg-Cu alloy. Lin et al. [7] found that the main precipitates of Al-Zn-Mg-Cu alloys during creep aging process are metastable η' and stable η phases, and both of them are sensitive to the external factors, such as applied stress and aging temperature. Chen et al. [8] found that the applied stress induces the coarsening of precipitates in 7050 aluminum alloy during creep aging process, and both yield strength (YS) and tensile strength (TS) of the creep aged samples are lower than those of the stress-free aged samples. Lin et al. [9] studied the relationship between exfoliation corrosion (EXCO) sensitivity and creep aging process conditions, and found that the EXCO resistance of an Al-Zn-Mg-Cu alloy first increases and then decreases with the increase in stress and temperature. Li et al. [10] investigated the creep aging behaviors of an Al-Cu-Li alloy with different initial tempers, namely, solution, under-aging and peak-aging, and revealed there is a close relationship between creep mechanism and initial temper.

In view of the large springback occurred in the unloading process, many numerical studies mainly aimed at predicting springback after CAF to improve the forming precision. Based on the damage theory by Kowalewski et al. [11] and the conventional creep damage model [12], Ho et al. [13] proposed a unified creep constitutive model, and simulated the whole CAF process including loading, forming and unloading by using this constitutive model combined with the commercial finite element solver ABAQUS. Based on this work, Jeunechamps et al. [14] suggested a unified creep aging constitutive model considering the age hardening during

CAF, in which the nucleation and growth of precipitates were related to the creep deformation. Lin et al. [15] introduced an integrated process to model stress relaxation, creep deformation, precipitation hardening and springback in CAF. Zhan et al. [16] developed a microstructure internal variables based creep aging constitutive model, which can describe the YS evolution through integrating the contribution of each hardening mechanism, namely dislocation hardening, age hardening and solution hardening.

It is noted that the initial tempers of aluminum alloys in above experiments and modeling are mostly solution temper, and the effects of various initial tempers on deformation and properties of Al-Zn-Mg-Cu series alloys in CAF are rarely reported. Therefore, in order to clearly understand the interaction between creep and aging, and achieve both high-precision forming and mechanical property improvement, this study focuses on investigating the effects of initial tempers on formability and performance of Al-Zn-Mg-Cu series alloys. In terms of creep deformation and stress relaxation, the formability of an Al-Zn-Mg-Cu alloy with various initial tempers under CAF conditions is evaluated firstly. Then, the evolution of microstructures and performance are revealed. Finally, based on the relations among creep deformation, material properties and microstructure evolution, the effects of initial tempers on formability and performance are discussed.

2. Experimental procedures

2.1. Specimen preparation and heat treatment for various initial tempers

A typical commercial 7050 aluminum alloy (AA7050) hot rolled plate with 30 mm thickness was used in this work, which belongs to the heat-treatable high strength Al-Zn-Mg-Cu series alloys. This alloy was provided by Northeast Light Alloy Co., Ltd., Harbin, China, and its chemical composition (wt.%) was verified by a SPECTRO MAXx direct-reading spectrometer, as listed in **Table 1**. The as-rolled Al-Zn-Mg-Cu alloys have been certified as containing coarse constituent particles (Al_7Cu_2Fe and Mg_2Si) and fine intermetallics ($MgZn_2$ and Al_2CuMg). The fine intermetallics can be dissolved by subsequent solution treatment, but the coarse constituent particles are quite stable and insoluble [17, 18]. The creep aging test specimens with the gauge length of 50 mm and thickness of 3 mm were machined out from the hot rolled plate by wire electrical discharge machining (WEDM) along the rolling direction, and their geometry and dimensions are shown in **Figure 1**.

In this work, five heat treatment states were selected as the material initial tempers in creep aging tests. The solution temper is the most widely used material initial temper in industrial CAF. The peak-aged (T6) and over-aged (T73) tempers are the most common delivery conditions of AA7050. Since the retrogression and re-aging (RRA) treatment can help Al-Zn-Mg-Cu

Zn	Mg	Cu	Zr	Ti	Fe	Si	Al
6.02	1.97	2.23	0.12	0.03	0.10	0.07	Bal.

Table 1. Chemical composition (wt.%) of as-received AA7050.

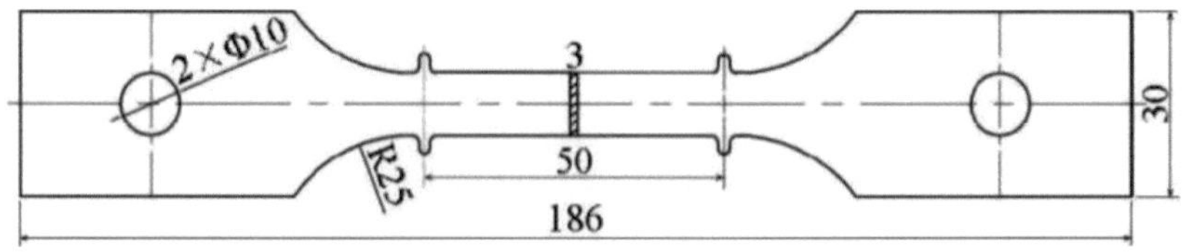

Figure 1. Geometry and dimensions of creep aging test specimens (unit: mm).

alloys to simultaneously achieve high mechanical properties and excellent corrosion resistance [19, 20], the retrogression temper is selected to be an initial temper. That is to say, the creep aging process will play the role of re-aging in RRA. Nevertheless, Lin et al. [21] pointed out that the narrow temperature-time window in RRA does not apply to the thick plate. Thus, the re-solution and re-aging (RSRA) treatment with wider temperature-time window has been developed in response to the similar effect as RRA, and the re-solution temper is chosen as one of the initial tempers in this study [22]. **Figure 2** shows various heat treatment procedures conducted to obtain the designed initial tempers of AA7050.

- For the solution temper, the as-rolled material was subjected to a solution treatment at 470°C for 50 min and subsequent water quenching.
- The T6 temper was obtained by a peak-aging treatment at 120°C for 24 h.
- For the retrogression temper, the specimen with T6 temper had endured a retrogression treatment at 200°C for 5 min in a salt bath.
- The T73 temper was obtained by a two-step aging at 115°C for 8 h and then 165°C for 16 h.
- For the re-solution temper, the specimen with T73 temper went through a re-solution treatment including hot insulation at 470°C for 50 min and rapid water quenching.

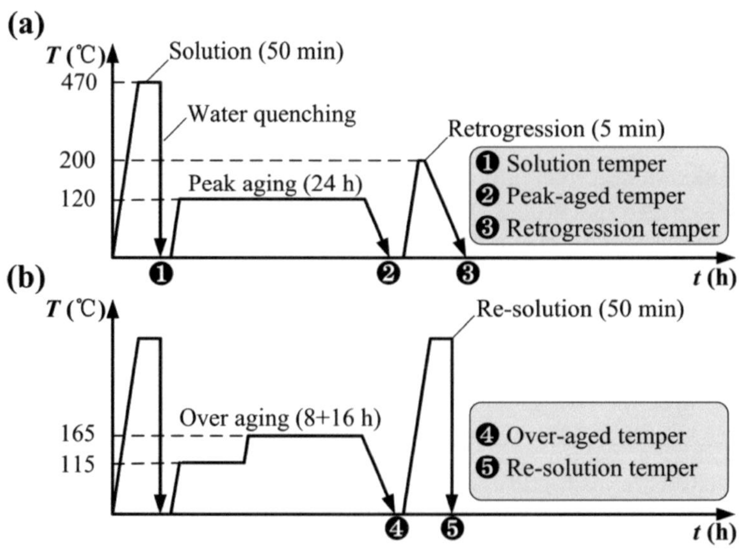

Figure 2. Heat treatment procedures of AA7050 for obtaining various initial tempers: (a) solution, T6 and retrogression; (b) T7 and re-solution.

2.2. Creep aging and stress relaxation aging tests

The creep aging and stress relaxation aging tests were carried out on a 100 kN electronic creep testing machine with special thermal environmental chamber. In the chamber, the temperature fluctuation can be controlled to less than 1°C. The tests were conducted under the aging temperature of 165°C to obtain more significantly creep deformation than those under the conventional T6 treatment temperature of 120°C. The applied stress of 250 MPa was within the typical stress range in the actual CAF process [23], and loaded on the specimen according to a loading speed of 0.5 mm/min. Meanwhile, in order to avoid the serious over-aging, the testing time is reduced to 18 h from the T6 treatment time of 24 h [24].

The installation position of the specimen was located in the middle of the chamber. The specimen was heated to 165°C, and then holds for 10 min before loading. The temperature was monitored at three positions around the specimen, namely, the two ends and the middle of the specimen. The change of gauge length on the specimen was measured by a high-precision (5×10^{-4} mm) grating line displacement transducer.

- In creep aging tests, the applied stress of 250 MPa remained the same throughout the creep deformation process. When the tests went on for 6 h, some tests were stopped as interrupted tests to investigate the evolution of microstructure and performance during CAF process.
- In stress relaxation aging tests, the applied stress of 250 MPa was a starting stress to obtain a given strain which remained unchanged during the stress relaxation process.

2.3. Microstructure characterizations and performance tests

The nano-sized precipitate microstructures were analyzed using the TECNAL G2 F30 transmission electron microscope (TEM). A slice was cut from the creep aged or stress relaxation aged specimen by WEDM, and then sanded to 60 μm thickness using sandpaper. Some 3 mm diameter round pieces as the TEM specimens were blanked from the slice and twin-jet electro-polished in a solution of 20% perchloric acid and 80% ethanol (in volume) at −20°C and 20 V. The quantitative analysis of the MPts in TEM bright field image was conducted by the Image-Pro Plus 6 software with no less than five images counted for each specimen.

The performance of the creep aged and stress relaxation aged specimens examined in this study is YS, TS, elongation, hardness, electrical conductivity, EXCO resistance and stress corrosion crack (SCC) resistance. The YS, TS and elongation were tested by the tensile tests conducted on a MTS CMT5205 electronic universal testing machine under the tensile speed of 2 mm/min. The hardness tests were carried out using a digital micro-hardness tester. Also, 10 tests were performed on each specimen to calculate an average. The electrical conductivity was measured by the eddy current method. The SCC resistances were evaluated using the slow strain rate tests (SSRT) in a corrosive environment of 3.5% NaCl. The accelerated EXCO tests were carried out in a classic etchant consisted of 4.0 M NaCl, 0.5 M KNO_3 and 0.1 M HNO_3. The specimen surfaces of EXCO tests were studied using a JEOL JCM-6000 scanning electron microscope (SEM).

3. Results

3.1. Creep deformation, stress relaxation and formability

Figure 3 shows the creep deformation and stress relaxation behaviors of AA7050 with different initial tempers during creep aging and stress relaxation aging tests. In the constant-stress creep aging tests under 165°C and 250 MPa, the specimens with four initial tempers of solution, T6, retrogression and re-solution all exhibit typical creep behaviors including primary and steady stages, but the specimen with initial temper of T73 represents primary and tertiary stages, as shown in **Figure 3(a)**. The primary creep stage lasts about 0.5 h and has a decrease of creep rate. After that, the strain hardening caused by creep and the softening caused by structural recovery gradually achieve a balance, and then the creep process enters into the second stage, namely steady creep stage. Four creep strain curves almost overlap with each other before the first half of steady stage. The bifurcation on those curves appears about 6 h later. After creep aging tests of 18 h, the largest creep strain is obtained in the T73 temper, and the least creep strain appears in the re-solution temper. However, the specimens with initial tempers of T6 and T73 apparently enter the tertiary creep stage. It means that the interior of material has generated a lot of microdefects, which is no longer applicable to the manufacture of aviation components. Using the amount of creep strain to evaluate the formability, the consequence of the creep aging formability of the specimens with three initial tempers except T6 and T73 is retrogression > solution > re-solution. The final creep strain for the retrogression temper is about 1.21 and 1.34 times than that for the solution and the re-solution tempers, respectively.

A practical CAF process usually involves loading, forming, and unloading stages. The workpiece is fitted closely with the tool surface after loading stage, and its shape will remain unchanged in subsequent forming stage until unloading. In the forming process, part of the elastic deformation is transformed into an irreversible plastic (creep) deformation, and the stress gradually decreases. Thus, the CAF process is more like a stress relaxation process [25]. After the forming stage, the stress cannot be released completely. The residual stress causes the workpiece to produce a larger springback in the unloading stage. In general, the larger stress relaxation signifies the better formability of CAF. **Figure 3(b)** shows the stress relaxation curves of AA7050 with different initial tempers during the stress relaxation aging tests under

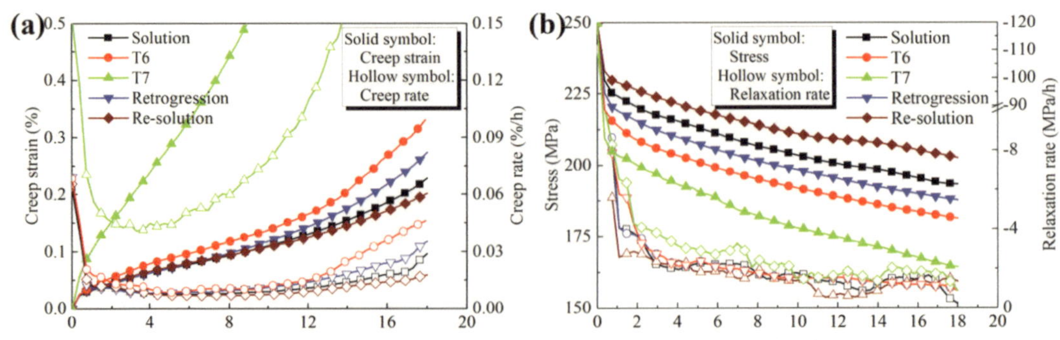

Figure 3. Creep deformation and stress relaxation behaviors of AA7050 with different initial tempers: (a) curves of creep strain and creep rate; (b) curves of stress relaxation and relaxation rate.

165°C and initiating stress of 250 MPa. The stress decreases for the five initial tempers are very sharp in the beginning, and turn to slow in the later period. The T73 temper has the largest relaxed stress, and the re-solution temper has the smallest relaxed stress. Comparing the magnitude of relaxed stress, it can be seen that the order of formability of the specimens with five initial tempers is T73 > T6 > retrogression > solution > re-solution. This result is consistent with the above conclusion of creep aging tests. An inference can be drawn that the root cause of stress relaxation is the occurrence of creep deformation.

3.2. Evolution of microstructures

3.2.1. Matrix precipitates

In the precipitation reaction of the Al-Zn-Mg-Cu alloys under the CAF condition, the metastable η' phase and stable η phase are main precipitates and formed in the supersaturated aluminum matrix. **Figure 4** shows the MPt microstructures of AA7050 with different initial tempers before the creep aging tests. Both η' and η phases can affect the creep behavior, so there is no special distinction between the types of precipitates. **Figure 4(a)** shows that there is no obvious precipitate in the alloy with solution temper. **Figure 4(b)** displays many fine precipitates with average radius of 2.8 nm and number density of 1.7×10^{18} cm^{-3} evenly distributed in the matrix of alloy with T6 temper. **Figure 4(c)** represents some coarse precipitates with average radius of 4.3 nm and number density of 6.3×10^{17} cm^{-3} uniformly

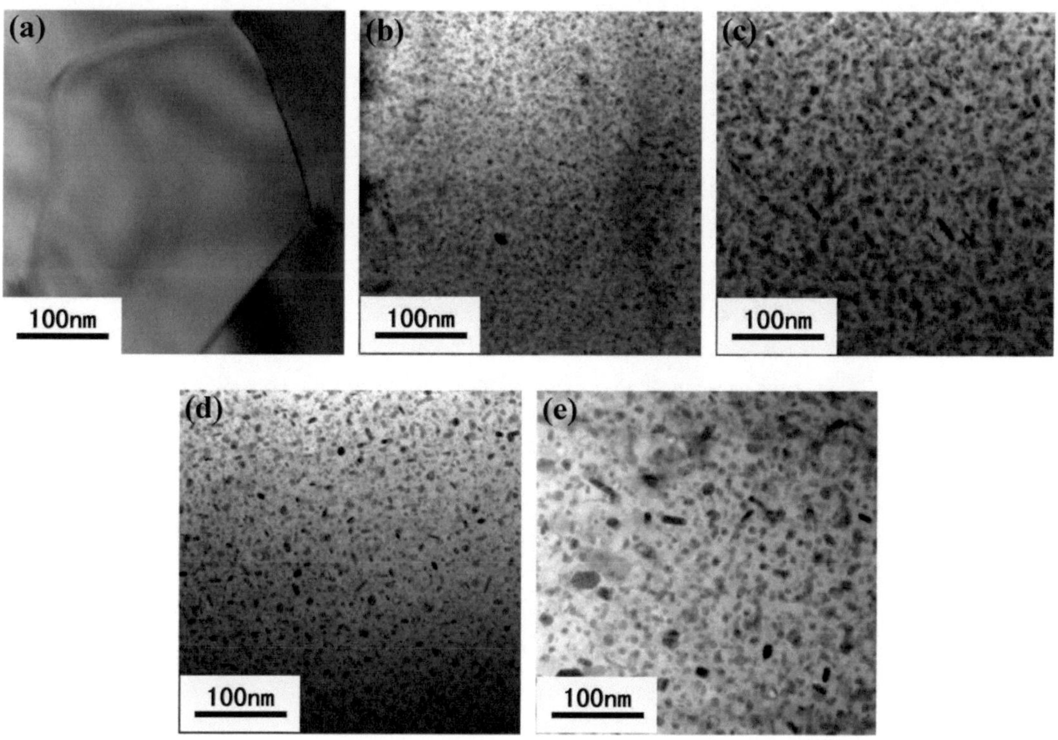

Figure 4. TEM images of MPts of AA7050 with different initial tempers before creep aging tests: (a) solution; (b) T6; (c) T73; (d) retrogression; (e) re-solution.

distributed in the matrix of over-aged alloy. Through retrogression treatment for the peak-aged alloy, parts of MPts has been dissolved, and the residual MPts have the average radius of 3.2 nm and number density of 5.5×10^{17} cm^{-3}, as shown in **Figure 4(d)**. Therefore, it can be seen that the short-time retrogression treatment cannot dissolve the whole precipitates. **Figure 4(e)** shows the sporadic MPts distributed in the matrix of re-solution treated alloy, and the average radius and number density of these precipitates are 4.8 nm and 7.9×10^{14} cm^{-3}, respectively. This suggests that due to the higher temperature and longer insulation time of the re-solution treatment, plenty of precipitates in over-aged alloy are dissolved, and only few precipitates are remained and coarsened.

In **Figure 3**, some creep curves are almost overlapping in the early stage of the creep aging process. It can be inferred that the initial microstructures have little effect on the creep deformation. The different creep rates appear in about 6 h, and the reason can be attributed to the different precipitate microstructures of AA7050 with different initial tempers after creep aging tests of 6 h. It also can be seen that, in the creep aging tests, the minimum creep rates appear at around 6 h, which indicates that the precipitate microstructures have the strongest effect on creep deformation at this time.

In the creep aging process, due to the precipitation reaction, the size of the MPts is greatly changed. **Figure 5** shows the MPt microstructures of AA7050 with different initial tempers after

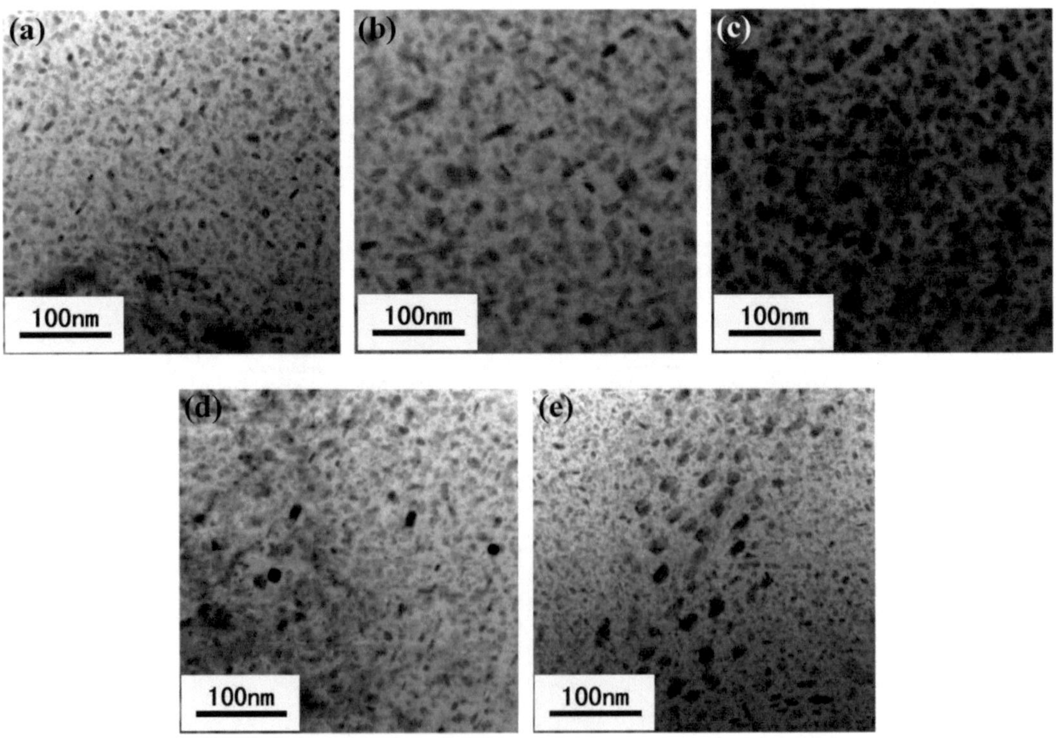

Figure 5. TEM images of MPts of AA7050 with different initial tempers after creep aging tests of 6 h: (a) solution; (b) T6; (c) T73; (d) retrogression; (e) re-solution.

the creep aging tests of 6 h. As shown in **Figure 5(a)**, there are a lot of fine precipitates with an average radius of 3.5 nm. **Figure 5(b)** and **(c)** presents that the MPts in the alloy with T6 and T73 tempers grow, respectively, to 6.3 and 7.4 nm. **Figure 5(d)** represents the microstructure evolution of the retrogression treated specimen, and all previous MPts are coarsened to an average radius of 4.6 nm. This image clearly shows that, after the creep aging tests of 6 h, the size of MPts in the alloy with retrogression temper is bigger than that with solution temper. In **Figure 5(e)**, since most of MPts in T73 temper are dissolved by the previous re-solution treatment, after the creep aging tests of 6 h, a lot of new precipitates with small size are re-precipitated in the matrix of the alloy with re-solution temper accompanied by the coarsening of previous residual precipitates. In calculating the size of the new precipitates, the initial average radius of previous residual precipitates is set to be a threshold value. Excluding the precipitates which are bigger than the threshold radius of 4.8 nm, the average radius of new precipitates is 2.9 nm. It can be seen that the size of new MPts of re-solution tempered specimen is the smallest in the specimens with five initial tempers after the creep aging tests of 6 h.

3.2.2. Grain boundary precipitates

Figure 6 shows the GBP microstructures of AA7050 with different initial tempers before the creep aging tests. As shown in **Figure 6(a)**, there is no obvious precipitate on the grain

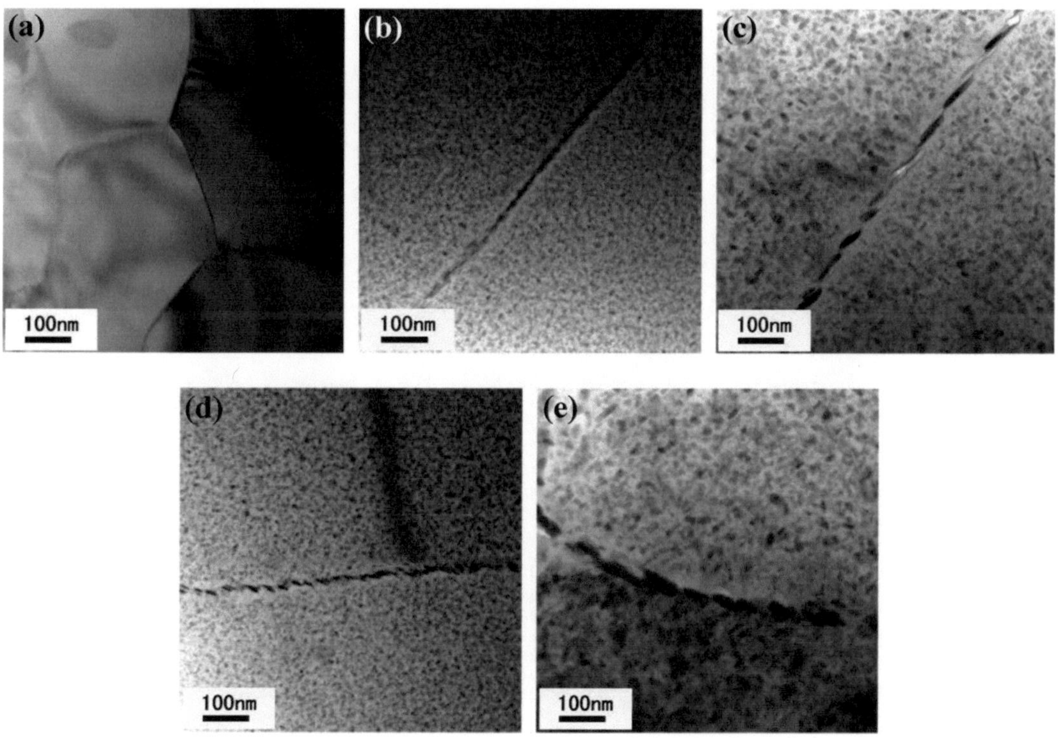

Figure 6. TEM images of GBPs of AA7050 with different initial tempers before creep aging tests: (a) solution; (b) T6; (c) T73; (d) retrogression; (e) re-solution.

boundary of the specimen with solution temper. This indicates that the intermetallics of as-rolled AA7050 have been dissolved by the solution treatment. **Figure 6(b)** shows that the specimen with T6 temper has continuously distributed GBPs. By comparing **Figure 6(b)** and **(d)**, it can be found that the alloy with retrogression temper has slightly discontinuous GBPs than that with T6 temper. This indicates that the role of short-time retrogression treatment on the dissolution of precipitates is limited. **Figure 6(c)** shows that the specimen with T73 temper has discontinuous GBPs with large spacing. The comparison between **Figure 6(c)** and **(e)** shows that, through re-solution treatment, the GBPs become smaller in size. It demonstrates that the re-solution treatment not only dissolve the MPts but also the GBPs.

Figure 7 represents the GBP microstructures of AA7050 with different initial tempers after the creep aging tests. The GBP microstructures of the specimen with initial temper of solution is similar to that of T6 temper, namely continuously distributed GBPs, as shown in **Figure 7(a)**. **Figure 7(b)** shows that the GBPs in the specimen with T6 temper translate from continuous to discontinuous in the creep aging tests. **Figure 7(c)–(e)** reveal the growth of previous GPBs in the specimens with initial temper of T73, retrogression and re-solution, resulting in more discontinuous grain boundary structures. Among them, the specimen with T73 temper has the largest sized GBPs with the widest spacing.

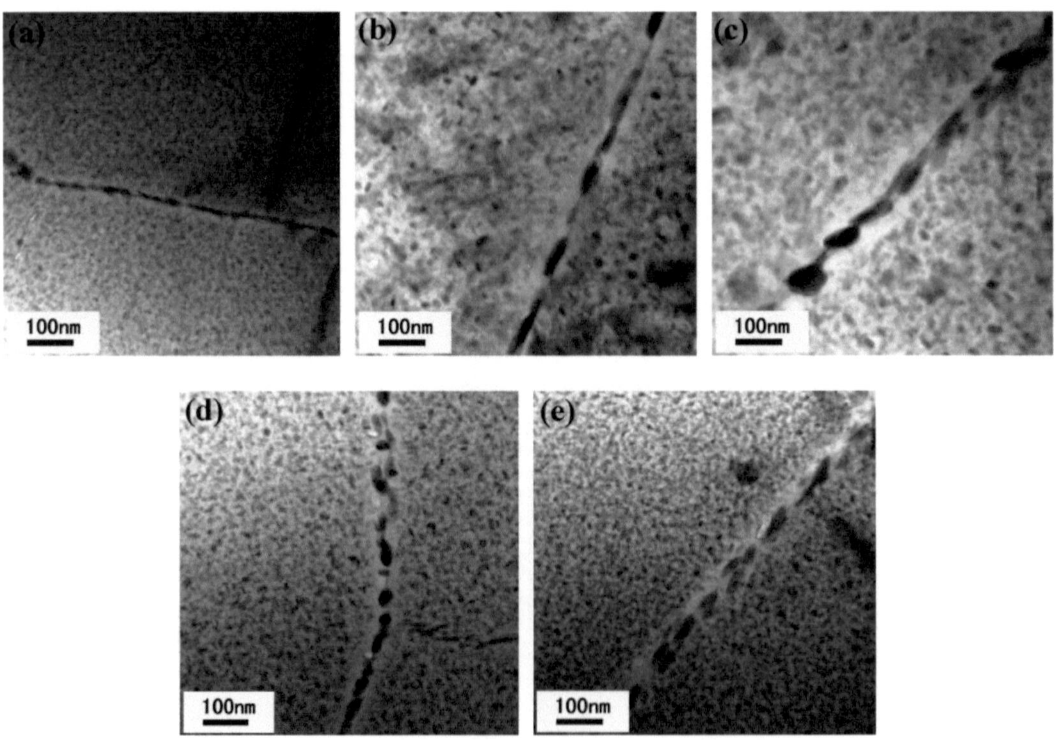

Figure 7. TEM images of GBPs of AA7050 with different initial tempers after creep aging tests: (a) solution; (b) T6; (c) T73; (d) retrogression; (e) re-solution.

3.3. Evolution of material performance

3.3.1. Mechanical properties and electrical conductivity

Figure 8(a) illustrates the YS, TS and elongation of AA7050 with different initial tempers before and after creep aging tests. It can be seen that the mechanical strength of the specimen with initial temper of solution obtains the maximized promotion in the creep aging tests. On the contrary, the mechanical strengths of the specimens with initial tempers of T6 and T73 reduce. Compared with solution and re-solution tempers, the specimen with initial temper of retrogression has the highest strength before creep aging. However, the mechanical strength of this specimen is significantly lower after creep aging. Its YS and TS are approximately 18% and 9% lower than those of the creep aged specimen with initial temper of solution. The YS and TS of the creep aged specimen with initial temper of re-solution are only 4% and 2% lower than those of solution temper. In addition, the sorting of elongations is the opposite to that of mechanical strengths, namely T73 > T6 > retrogression > re-solution > solution.

Figure 8(b) shows the hardness and electrical conductivity of AA7050 with various initial tempers before and after the creep aging tests. The contrast situation of hardness is identical to that of strength. Tsai and Chuang [26] determined that the electrical conductivity of aluminum alloy may serve as an indicator of SCC resistance, so the electrical conductivity tests were carried out as the circumstantial evidence for subsequent SCC resistance tests. It can be see that, for each initial temper, the electrical conductivity of specimen increases with time. At the end of creep aging process, the sequence of electrical conductivity is T73 > T6 > re-solution > retrogression > solution.

3.3.2. SCC and EXCO resistances

In the SSRT, the index r_{tf} is commonly used to evaluate the SCC resistance and calculated by [22]:

$$r_{tf} = \frac{t_{fe}}{t_{fc}} \tag{1}$$

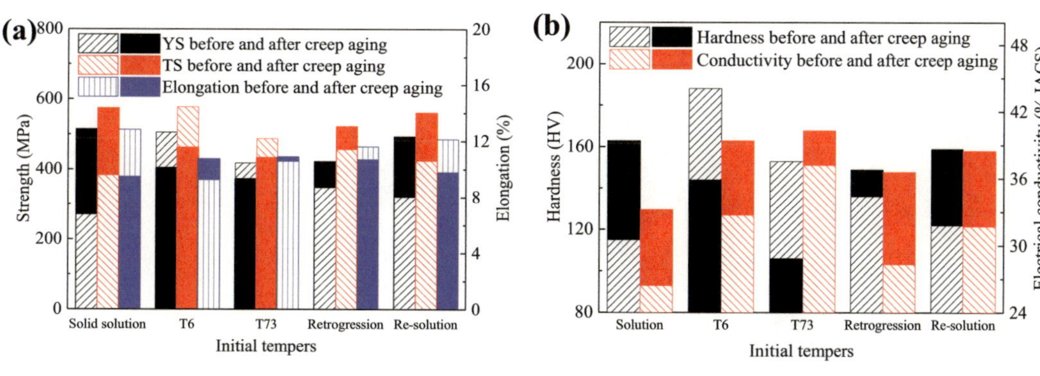

Figure 8. Performance change of AA7050 with different initial tempers before and after creep aging tests: (a) strength and elongation; (b) hardness and electrical conductivity.

where t_{fe} is the time of specimen fracture in the corrosive environment and t_{fc} is the corresponding time in the atmospheric environment. The closer the r_{tf} value is to 1, the better the SCC resistance.

Table 2 lists the r_{tf} values of creep aged AA7050 with different initial tempers. After creep aging process, the specimens with initial tempers of T6, T73 and re-solution have better SCC resistance than those with initial tempers of solution and retrogression.

In the accelerated EXCO tests, the corrosion products gradually break away from the specimen surface. **Figure 9** displays the surface morphology of the creep aged specimens continuously immersed in the EXCO solution for 48 h. The creep aged specimen with initial temper of solution has been seriously corroded, and a mass of corrosion products is detached from the metal surface. There are negligible corrosion phenomena which occur in the creep aged specimens with initial tempers of T6 and T73. For the retrogression temper, there are some cracks appeared on the metal surface. Conversely, some tiny cracks can be found on the surface of the creep aged specimen with initial temper of re-solution.

Using EXCO grade to evaluate the degree of corrosion, the total EXCO behavior can be classified into six grades, namely free of obvious corrosion (expressed as N), slight pitting (P), growing pitting (EA), slight surface crack (EB), corrosion peeling and blister (EC) and corrosion product

Initial temper	**Solution**	**T6**	**T73**	**Retrogression**	**Re-solution**
r_{tf} value	0.724	0.972	0.987	0.903	0.956

Table 2. r_{tf} values of AA7050 with different initial tempers after creep aging tests.

Figure 9. EXCO morphologies of creep aged AA7050 with different initial tempers: (a) solution; (b) T6; (c) T73; (d) retrogression; (e) re-solution; (f) corresponding EXCO grade evolution.

shedding (ED). **Figure 9(f)** shows the EXCO grade evolution of the creep aged AA7050 with various initial tempers during the EXCO tests, and it can be found that the degree of EXCO aggravates with immersion time.

4. Discussion

4.1. Issue of formability

4.1.1. Deformation mechanism

In CAF process, the creep strain rate can be described by [27, 28]:

$$\dot{\varepsilon}_c = A\sigma^n \exp\left(-\frac{Q}{RT}\right) \tag{2}$$

where σ is the instantaneous stress and A, n, Q, R and T are material constant, stress exponent, deformation activation energy, universal gas constant and Kelvin temperature, respectively.

Translating the both sides of Eq. (2) into logarithmic form, the strain rate-stress double logarithmic curves as shown in **Figure 10**. The data of strain rate and stress are taken from the stress relaxation curves in **Figure 3**. Just to be clear, because the creep deformation of T6 and T73 tempers enters into the tertiary stage, these two tempers are not take into account of the discussion for deformation mechanism.

Figure 10(a) shows that each strain rate-stress double logarithmic curve consists of three straight lines corresponding to three different stages in the stress relaxation aging process, namely primary, transition and steady stages. The inflection point between two intersecting straight lines indicates that there is a threshold stress occurred in the transformation of two adjacent stages [25]. Since the threshold stress is the minimum stress required for creep deformation, the different threshold stresses can reflect the differences of CAF formability.

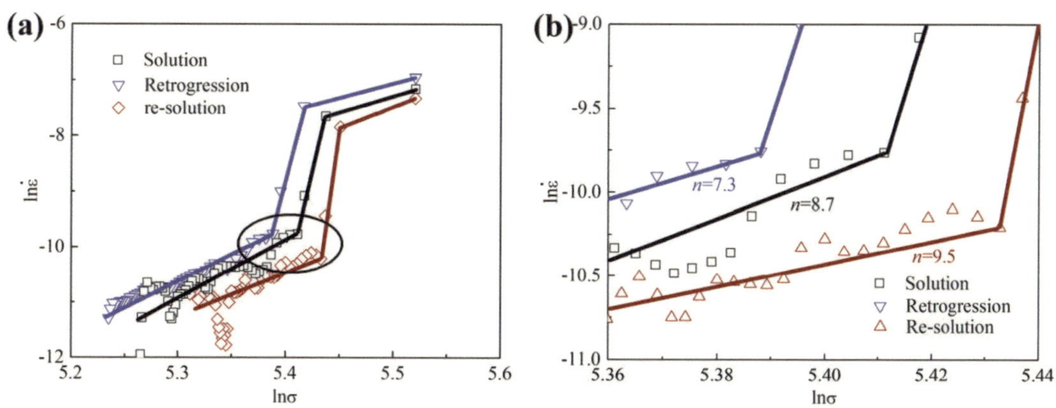

Figure 10. Strain rate versus stress double logarithmic curves of AA7050 with different initial tempers during stress relaxation aging tests: (a) total curves; (b) local enlargement in the circle.

Initial temper	Solution	Retrogression	Re-solution
Upper threshold stresses (MPa)	229.8	225.1	232.9
Lower threshold stresses (MPa)	223.9	218.7	228.8

Table 3. Upper and lower threshold stresses of AA7050 with various initial tempers in CAF at 165°C.

In other words, the smaller threshold stress means the better formability. **Table 3** lists the upper and lower threshold stresses in CAF at 165°C for the specimens with solution, retrogression and re-solution tempers. When the stress drops lower than the upper threshold stress, the strain rate reduces sharply, and the CAF process enters into the transition stage from the primary stage. When the stress falls below the lower threshold stress, the forming process enters into the final steady stage. Whether upper or lower threshold stresses, the order of various initial tempers is re-solution > solution > retrogression. It means that the retrogression temper corresponds to the best CAF formability.

The dominant mechanism of creep deformation can be reflected from the stress exponent n [28], which can be calculated through fitting the slopes of strain rate-stress double logarithmic curves of the steady stage in stress relaxation process. **Figure 10(b)** shows that the values of stress exponent n in the steady stage CAF at 165°C for the AA7050 with solution, retrogression and re-solution tempers are 8.7, 7.3 and 9.5, respectively. The value of n less than 2 indicates that the creep deformation mechanism mainly is diffusion. While the value of n greater than 4 means that the creep deformation is controlled by dislocation movement, including dislocation slip and dislocation climb [29]. It follows that the dominant deformation mechanism of AA7050 in the creep aging process under 165°C is dislocation creep.

4.1.2. Formability and MPts

Since the dislocation movement is affected by the size and distribution of the MPts, there is a direct relationship between the precipitate microstructures and formability of AA7050 during CAF. By comparing the precipitate microstructures of the specimens with different initial tempers after the creep aging tests of 6 h in **Figure 5**, it is shown that the specimen with initial temper of T73 has the largest MPts, and the specimen with initial temper of re-solution has the smallest MPts. It well known that the small-sized precipitates will be cut by dislocation during plastic deformation process of metal, while the big-sized precipitates will be bypassed. Thus, the AA7050 with initial temper of T73 has the best formability because of its big-sized precipitates are bypassed by the moving dislocation, but the alloy with initial temper of re-solution has the worst formability since its small-sized precipitates pin the dislocation [30].

4.2. Issue of performance

4.2.1. Mechanical properties and MPts

It is generally known that the main influencing factor for mechanical properties of the heat-treatable aluminum alloys is the size of MPts. Since its fine MPts similar to the T6 temper, the specimen with initial temper of solution has the highest YS and TS after creep aging tests. Because there is only the coarsening action of MPts, the mechanical strength of the specimens

http://dx.doi.org/10.5772/intechopen.70849

with initial tempers of T6 and T73 has actually lowered. In the creep aged specimen with initial temper of re-solution, the re-precipitation takes place in the creep aging process, thus the fine MPts leads to the higher strength which close to the strength of T6.

4.2.2. Corrosion resistance and GBPs

The corrosion resistance of Al-Zn-Mg-Cu alloys is directly determined by grain boundary structure. The continuously distributed GBPs can create an anodic corrosion channel, where a galvanic reaction will occur between the GBPs as anode and the aluminum matrix as cathode in a corrosive environment [31]. For the solution and retrogression tempers, since their creep aged specimens have continuous GPBs, these two tempers show higher sensitivity of EXCO and SCC. Conversely, the discontinuous GPBs of the creep aged specimens with initial temper of T6, T73 and re-solution cannot form effective anodic corrosion channel, thus these three tempers have better corrosion resistances. The SCC mechanism of the Al-Zn-Mg-Cu alloys is also considered to be the effect of hydrogen embrittlement except the anodic dissolution [21].

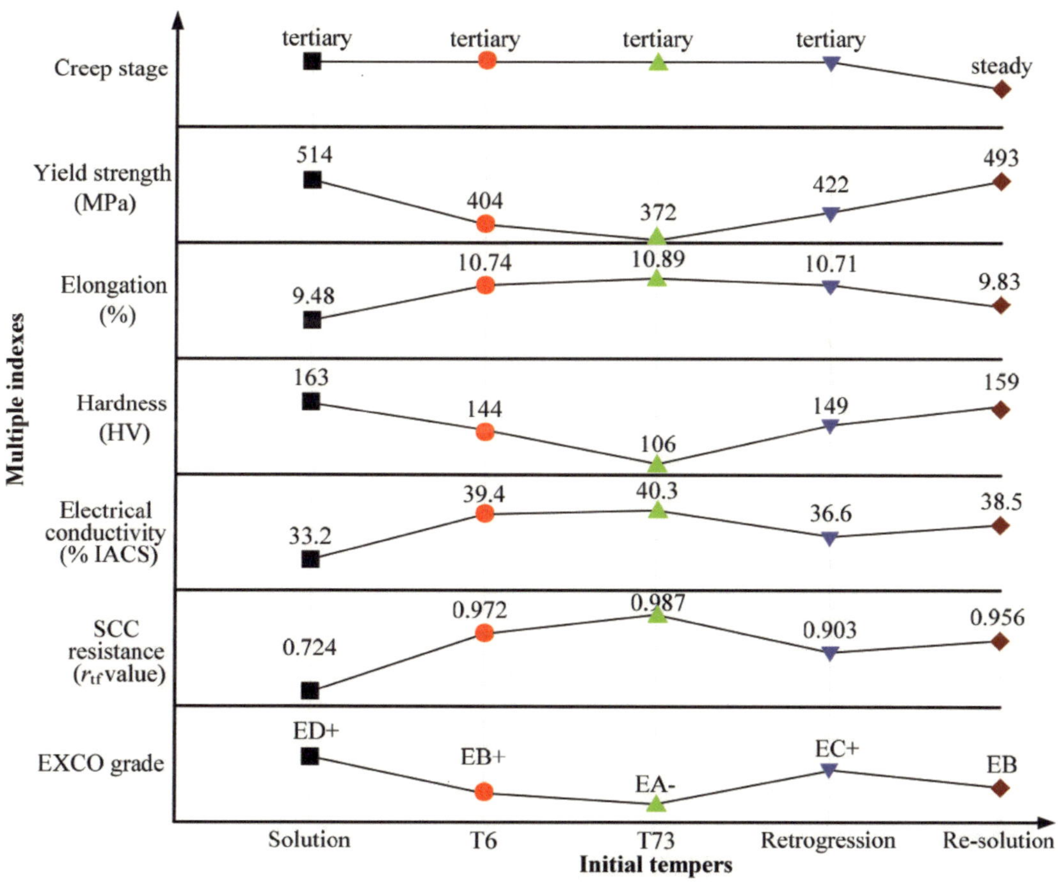

Figure 11. Multiple performance indexes of creep aged AA7050 with different initial tempers.

The small and continuous GBPs in solution and retrogression tempers cannot capture hydrogen atoms, then the SCC easily occurs on the grain boundary where concentrated a bunch of hydrogen atoms. Rather, the large and discontinuous GBPs in T6, T73 and re-solution tempers as the pitfalls of hydrogen atoms avoid the mass concentration of hydrogen atoms in the grain boundary, leading to the low SCC sensitivity.

4.3. Comprehensive evaluation by multiple indexes

Figure 11 displays the various material performances of AA7050 with various initial tempers, which are tested 18 h under creep aging tests conditions of 165°C and 250 MPa. When the tests comes to the end, only the specimen with initial temper of re-solution is absolutely still in the steady creep stage. This shows that it has the potential to sustain a longer CAF period. The mechanical properties of the creep aged specimen with initial temper of re-solution are very close to those of solution temper as the best case, and much more than those of T6, T73 and retrogression. Furthermore, the corrosion resistances of the creep aged specimen with initial temper of re-solution are very close to the best case in T73 temper.

Taking the both requirements of forming precision and material performance into account, a conclusion can be drawn by comparing the multiple indicators shown in **Figure 11**. The re-solution is the best choice for the initial temper of Al-Zn-Mg-Cu alloys in CAF under the given conditions, that is, the temperature of 165°C and the stress of 250 MPa.

5. Conclusions

In terms of the creep deformation and stress relaxation, as well as corresponding properties tests and microstructure observations, the effects of initial tempers on the creep aging formability and performance of an Al-Zn-Mg-Cu alloy (AA7050) have been studied under CAF conditions. The main findings can be summarized as follows:

1. Using various heat treatment methods, five material states of AA7050 are obtained to serve as initial tempers for CAF, namely, solution, T6, T73, retrogression and re-solution. The microstructure analyses show that there is no obvious MPts in the alloy with solution temper. The retrogression temper is transformed from T6 temper through retrogression treatment which makes slight reduction of MPts density and discontinuous distribution of GBPs. The re-solution temper is converted from T73 temper by re-solution treatment which fully dissolves the MPts and makes the GBPs smaller.

2. The T73 temper has the best formability, and the worst presents in the re-solution temper. Through calculating the stress exponents, it can be seen that the CAF of AA7050 under the given thermal-mechanical conditions is mainly controlled by dislocation creep. The TEM observations of the specimens after the creep aging tests of 6 h show that the T73 temper has the biggest sized MPts which can be bypassed by the moving dislocation, thus it has the best formability; the reason of the worst formability appeared in re-solution temper is that the alloy has the smallest sized MPts which can pin down the moving dislocation.

3. The mechanical properties of the creep aged specimens with various initial tempers are different. After creep aging, the specimens with initial tempers of solution and re-solution show almost same high mechanical properties than that of T6, T73 and retrogression. Its reason is that the creep aged specimens with initial tempers of solution and re-solution has fine and homogeneous MPts, while the MPts of the creep aged specimens with initial tempers of T6, T73 and retrogression grow up obviously. The discontinuous GBPs lead to the better EXCO and SCC resistance of the creep aged specimens with initial tempers of T6, T73 and re-solution. Conversely, the continuous GBPs result in the higher corrosion susceptibility of the creep aged specimens with initial tempers of solution and retrogression.

4. In order to achieve the dual objective of precise forming and performance improvement, the comprehensive assessment on multiple performance indexes indicates that the re-solution is the best choice for the initial temper of Al-Zn-Mg-Cu alloys in CAF under the given conditions. Re-solution temper has the potential to withstand long forming period and can bring excellent mechanical properties and corrosion resistances.

Acknowledgements

The authors would like to thank the National Natural Science Foundation of China for Key Program (51235010), the National Science Fund for Excellent Young Scholars (51522509), the Fundamental Research Funds for the Central Universities (3102014KYJD001), the Marie Curie International Research Staff Exchange Scheme (IRSES, MatProFuture, project no: 318968) within the 7th EC Framework Programme (FP7) and the 111 Project (B08040) for their financial supports.

Author details

Heng Li*, Chao Lei and He Yang

*Address all correspondence to: liheng@nwpu.edu.cn

State Key Laboratory of Solidification Processing, School of Materials Science and Engineering, Northwestern Polytechnical University, Xi'an, China

References

[1] Sha G, Cerezo A. Early-stage precipitation in Al-Zn-Mg-Cu alloy (7050). Acta Materialia. 2004;**52**:4503-4516. DOI: 10.1016/j.actamat.2004.06.025

[2] Zhan LH, Lin JG, Dean TA. A review of the development of creep age forming: Experimentation, modelling and applications. International Journal of Machine Tools & Manufacture. 2011;**51**:1-17. DOI: 10.1016/j.ijmachtools.2010.08.007

[3] Lei C, Yang H, Li H, Shi N, Fu J, Zhan LH. Dependence of creep age formability on initial temper of an Al-Zn-Mg-Cu alloy. Chinese Journal of Aeronautics. 2016;**29**:1445-1454. DOI: 10.1016/j.cja.2016.04.022

[4] Arabi Jeshvaghani R, Emami M, Shahverdi HR, Hadavi SMM. Effects of time and temperature on the creep forming of 7075 aluminum alloy: Springback and mechanical properties. Materials Science and Engineering A. 2011;**528**:8795-8799. DOI: 10.1016/j.msea.2011.08.025

[5] Arabi Jeshvaghani R, Shahverdi HR, Hadavi SMM. Investigation of the age hardening and operative deformation mechanism of 7075 aluminum alloy under creep forming. Materials Science and Engineering A. 2012;**552**:172-178. DOI: 10.1016/j.msea.2012.05.027

[6] Guo W, Yang M, Zheng Y, Zhang XS, Li H, Wen XY, Zhang JW. Influence of elastic tensile stress on aging process in an Al-Zn-Mg-Cu alloy. Materials Letters. 2013;**106**:14-17. DOI: 10.1016/j.matlet.2013.04.095

[7] Lin YC, Jiang YQ, Chen XM, Wen DX, Zhou HM. Effect of creep-aging on precipitates of 7075 aluminum alloy. Materials Science and Engineering A. 2013;**588**:347-356. DOI: 10.1016/j.msea.2013.09.045

[8] Chen JF, Zhen L, Jiang JT, Yang L, Shao WZ, Zhang BY. Microstructures and mechanical properties of age-formed 7050 aluminum alloy. Materials Science and Engineering A. 2012;**539**:115-123. DOI: 10.1016/j.msea.2012.01.067

[9] Lin YC, Jiang YQ, Zhang XC, Deng J, Chen XM. Effect of creep-aging processing on corrosion resistance of an Al-Zn-Mg-Cu alloy. Materials and Design. 2014;**61**:228-238. DOI: 10.1016/j.matdes.2014.04.054

[10] Li Y, Shi Z, Lin J, Yang YL, Huang BM, Chung TF, Yang JR. Experimental investigation of tension and compression creep-ageing behaviour of AA2050 with different initial tempers. Materials Science and Engineering A. 2016;**657**:299-308. DOI: 10.1016/j.msea.2016.01.074

[11] Kowalewski ZL, Hayhurst DR, Dyson BF. Mechanisms-based creep constitutive equations for an aluminium alloy. The Journal of Strain Analysis for Engineering Design. 1994;**29**:309-316. DOI: 10.1243/03093247v294309

[12] Lin J, Liu Y, Dean TA. A review on damage mechanisms, models and calibration methods under various deformation conditions. International Journal of Damage Mechanics. 2005;**14**:299-319. DOI: 10.1177/1056789505050357

[13] Ho KC, Lin J, Dean TA. Modelling of springback in creep forming thick aluminum sheets. International Journal of Plasticity. 2004;**20**:733-751. DOI: 10.1016/s0749-6419(03)00078-0

[14] Jeunechamps PP, Ho KC, Lin J, Ponthot JP, Dean TA. A closed form technique to predict springback in creep age-forming. International Journal of Mechanical Sciences. 2006;**48**:621-629. DOI: 10.1016/j.ijmecsci.2006.01.005

[15] Lin J, Ho KC, Dean TA. An integrated process for modelling of precipitation hardening and springback in creep age-forming. International Journal of Machine Tools and Manufacture. 2006;**46**:1266-1270. DOI: 10.1016/j.ijmachtools.2006.01.026

[16] Zhan LH, Lin J, Dean TA, Huang MH. Experimental studies and constitutive modelling of the hardening of aluminium alloy 7055 under creep age forming conditions. International Journal of Mechanical Sciences. 2011;**53**:595-605. DOI: 10.1016/j.ijmecsci.2011.05.006

[17] Xu DK, Rometsch PA, Birbilis N. Improved solution treatment for an as-rolled Al-Zn-Mg-Cu alloy. Part I. Characterisation of constituent particles and overheating. Materials Science and Engineering A. 2012;**534**:234-243. DOI: 10.1016/j.msea.2011.11.065

[18] Xu DK, Rometsch PA, Birbilis N. Improved solution treatment for an as-rolled Al-Zn-Mg-Cu alloy. Part II. Microstructure and mechanical properties. Materials Science and Engineering A. 2012;**534**:244-252. DOI: 10.1016/j.msea.2011.11.073

[19] Marlaud T, Deschamps A, Bley F, Lefebvre W, Baroux B. Evolution of precipitate microstructures during the retrogression and re-ageing heat treatment of an Al-Zn-Mg-Cu alloy. Acta Materialia. 2010;**58**:4814-4826. DOI: 10.1016/j.actamat.2010.05.017

[20] Oliveira AF Jr, de Barros MC, Cardoso KR, Travessa DN. The effect of RRA on the strength and SCC resistance on AA7050 and AA7150 aluminium alloys. Materials Science and Engineering A. 2004;**378**:321-326. DOI: 10.1016/j.msea.2004.02.052

[21] Lin JC, Liao HL, Jehng WD, Chang CH, Lee SL. Effect of heat treatments on the tensile strength and SCC-resistance of AA7050 in an alkaline saline solution. Corrosion Science. 2006;**48**:3139-3156. DOI: 10.1016/j.corsci.2005.11.009

[22] Lei C, Yang H, Li H, Shi N, Zhan LH. Dependences of microstructures and properties on initial tempers of creep aged 7050 aluminum alloy. Journal of Materials Processing Technology. 2017;**239**:125-132. DOI: 10.1016/j.jmatprotec.2016.07.004

[23] Ho KC, Lin J, Dean TA. Constitutive modelling of primary creep for age forming an aluminium alloy. Journal of Materials Processing Technology. 2004;**153-154**:122-127. DOI: 10.1016/j.jmatprotec.2004.04.304

[24] Guyot P, Cottignies L. Precipitation kinetics, mechanical strength and electrical conductivity of AlZnMgCu alloys. Acta Materialia. 1996;**44**:4161-4167. DOI: 10.1016/S1359-6454(96)00033-X

[25] Chen JF, Jiang JT, Zhen L, Shao WZ. Stress relaxation behavior of an Al-Zn-Mg-Cu alloy in simulated age-forming process. Journal of Materials Processing Technology. 2014;**214**:775-783. DOI: 10.1016/j.jmatprotec.2013.08.017

[26] Tsai TC, Chuang TH. Relationship between electrical conductivity and stress corrosion cracking susceptibility of Al 7075 and Al 7475 alloys. Corrosion Science. 1996;**52**:414-416. DOI: 10.5006/1.3292127

[27] Wang H, Wang QD, Yin DD, Yuan J, Ye B. Tensile creep behavior and microstructure evolution of extruded Mg-10Gd-3Y-0.5Zr (wt%) alloy. Materials Science and Engineering A. 2013;**578**:150-159. DOI: 10.1016/j.msea.2013.04.068

[28] Mostafa MM, Al-Ganainy GS, El-Khalek AMA, Nada RH. Steady-state creep and creep recovery during transformation in Al-Zn alloys. Physica B: Condensed Matter. 2003;**336**:402-409. DOI: 10.1016/s0921-4526(03)00318-1

[29] Mahmudi R, Roumina R, Raeisinia B. Investigation of stress exponent in the power-law creep of Pb-Sb alloys. Materials Science and Engineering A. 2004;**382**:15-22. DOI: 10.1016/j.msea.2004.05.078

[30] Guo W, Guo JY, Wang JD, Yang M, Li H, Wen XY, et al. Evolution of precipitate microstructure during stress aging of an Al-Zn-Mg-Cu alloy. Materials Science and Engineering A. 2015;**634**:167-175. DOI: 10.1016/j.msea.2015.03.047

[31] Li JF, Birbilis N, Li CX, Jia ZQ, Cai B, Zheng ZQ. Influence of retrogression temperature and time on the mechanical properties and exfoliation corrosion behavior of aluminium alloy AA7150. Materials Characterization. 2009;**60**:1334-1341. DOI: 10.1016/j.matchar.2009.06.007

Chapter 7

The Effects of Rapid Cooling on the Improved Surface Properties of Aluminium Based Coatings by Direct Laser Deposition

Olawale Samuel Fatoba,
Mamookho Elizabeth Makhatha and
Esther Titilayo Akinlabi

Additional information is available at the end of the chapter

http://dx.doi.org/10.5772/intechopen.71698

Abstract

The deterioration of materials during industrial application poses a serious threat to the materials structural integrity. A material's susceptibility to wear and surface damage can be reduced by alteration of its surface chemistry, morphology and crystal structure. Therefore, modification of surface properties plays an important role in optimizing a material's performance for a given application. Modern industrial applications require materials with special surface properties such as high hardness, wear and corrosion resistance, therefore materials engineers are vital to regularly examine how the microstructure of a material can be altered. Aluminium-based alloys have a wide application in the automotive, domestic and aerospace industries due to their excellent mechanical properties such as good weldability, sound castability and outstanding resistance to corrosion. The purpose of this research is to enhance inherent properties of the materials to create new products or improve on existing ones. The most effective engineering solution to prevent or minimize such surface region of a component is done by fibre lasers. It was concluded that Hypereutectic Al-Si alloys having transition metals are exceptional materials due to their specific properties. The addition of Cu, Fe, Cr, Si, Mg and Ni to Al-based alloys can improve the mechanical properties at both ambient and elevated temperatures.

Keywords: Al-based coatings, modification, mechanical properties, laser, materials, microstructure, laser metal deposition, Ti-6Al-4 V alloy, rapid cooling

1. Introduction

Surface modification of metals has been a useful tool in the improvement of the metal's property. It is a cost-effective method that alters the surface of components rather than the whole structure.

Laser processes are advantageous because it can target a specific area of a component, even if the shape is complex. This book chapter will focus on laser processes, their parameters, and the ternary powder Al-Si to obtain a clearer understanding of the surface treatment. Surface engineering refers to the design and modification of a surface to enhance properties such as corrosion, hardness, and wear. Surface modification has been practiced to enhancing the surface properties of materials in order to better endure wear, friction, and high temperatures through coating and alloying. Although a variety of advanced materials with significant properties have been developed, however, when it concerns a surface engineering application, physical properties of materials are among other factors that need to be considered, which include practicality and cost and time consumption. Surface treatment and coating are the most effective methods to protect surfaces from fatigue and extend the life by reducing damage at contact surfaces [1–3].

1.1. Lasers

Laser is an acronym for light amplification by stimulated emission of radiation. Lasers are mechanisms that give off intense beams of light which are coherent, monochromatic, and highly collimated. The wavelength of the beam is of high purity in comparison to other light sources; hence it is monochromatic. All the photons that constitute to the laser beam have a fixed phase relationship with respect to each other. The light from a laser generally has a very low divergence; it can travel over large distances or can be focused to a very small spot with a brightness which exceeds that of the sun. Due to these properties, lasers have been applied in a wide range of industries [4].

1.2. Laser surface modification

Applications of lasers in materials processing are widely found in part refurbishment and construction of large components from structural industry to small components such as electrochemical systems [5]. Montealegre et al. [6] acknowledged the need in different industrial sectors to improve the performance of material surface for service conditions and reported this need cannot be fulfilled by the conventional surface modification methods and coatings. There are many required component parts with specifications of surface properties when it comes to industrial applications, namely, good corrosion and wear resistance as well as good hardness. Most alloys which contain these specific properties are mostly expensive, and there is a great interest in decreasing the rate of components for fulfilling these requirements. Therefore, laser surface processing has been employed as an effective technique to improve surface properties of materials. To improve surface properties of materials, a laser beam is used to heat and modify the structure and physical characteristics of a material [6, 7]. Surface engineering has been considered as an approach to modify the surface of one metal with another metal that contains higher hardness, a lower coefficient of friction than the substrate [8–10].

Compared with other methods of surface modification, laser surface processing is characterized by the possibility of forming alloys of nonequilibrium compositions, formation of a fine microstructure, development of a strong metallurgical bond between the surface layer and the substrate, a small heat-affected zone, and the combination of a controlled minimal

http://dx.doi.org/10.5772/intechopen.71698

dilution of the substrate by the coating materials [11]. It has major advantages of high productivity, automation worthiness, noncontact processing, elimination of finishing operation, reduced processing cost, improved product quality, and greater material utilization. These characteristics and advantages have led to increasing demand of laser in material processing [12–14].

1.2.1. *Types of laser surface modifications*

1.2.1.1. *Laser transformation hardening*

It is an autogenous heat-treating technique, which involves solid-state transformation without any melting of the material. The material absorbs the laser energy and is controlled by the absorptivity of the surface. The application is limited to alloys that can be heat treated by laser [15, 16] as shown in **Figure 1**.

1.2.1.2. *Laser surface melting*

This technique is performed by heating the alloy's surface with a power density high enough to create a melt pool. A small ration of the top surface material is melted and cooled rapidly; the cooling rate will depend on the thermophysical properties of the metal and the scanning speed of the laser beam as shown in **Figure 2**.

1.2.1.3. *Laser surface alloying*

During laser surface alloying, an additional material is added to the melt pool to improve the properties [17]. The added material mixes with the molten substrate to form a new alloy. It involves melting a thin layer of the substrate with a high-power laser and simultaneously supplying the alloying element in the form of a powder as shown in **Figure 3**.

1.2.1.4. *Laser cladding*

Laser cladding applies the same principle as laser surface alloying, but the difference is the depth of the substrate that was melted. It enables the cladding material to be well bonded to the substrate without much mixing between them, which results in a relatively thick and homogenous overlay of coating material on the substrate as shown in **Figure 4**.

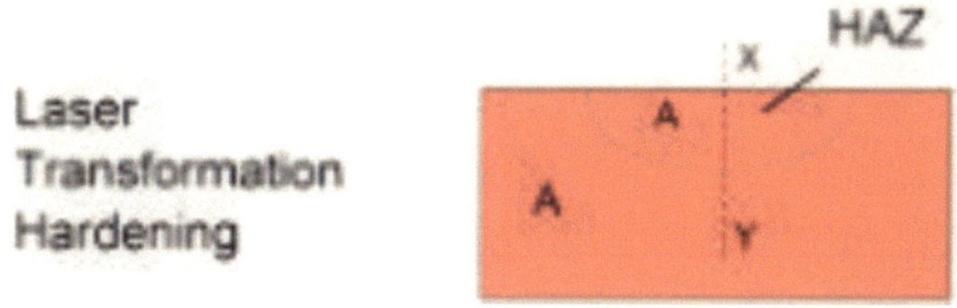

Figure 1. Schematic diagram showing the heat-affected zone of laser transformation hardening [10].

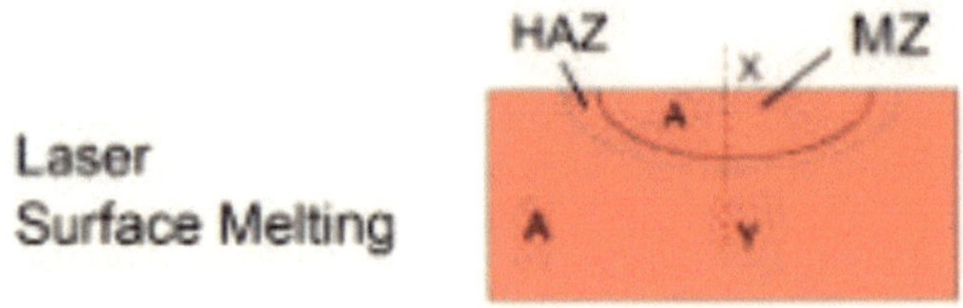

Figure 2. Schematic diagram illustrating the heat-affected zone and melt zone created by laser surface melting [15].

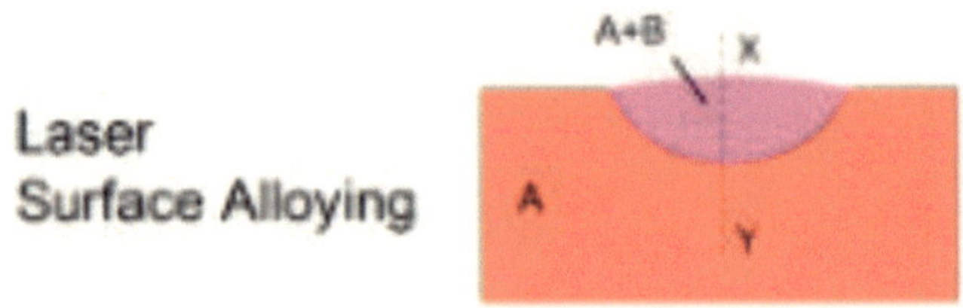

Figure 3. Schematic diagram displaying laser surface alloying [15].

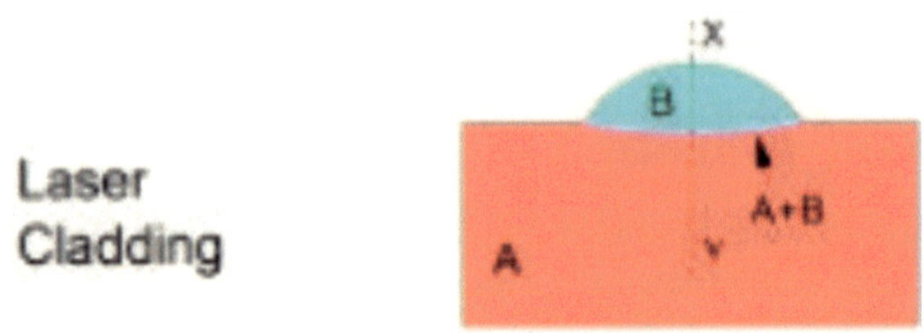

Figure 4. Schematic diagram of laser cladding [15].

1.2.1.5. Laser dispersion

It is used for forming surface composites. This technique injects hard second phase into a melted substrate. To keep the hard particles from melting, decreasing the power or exposure time is necessary. These particles remain solid during processing and after solidification of the melted substrate; the particles are dispersed in a matrix on the substrate as shown in **Figure 5**.

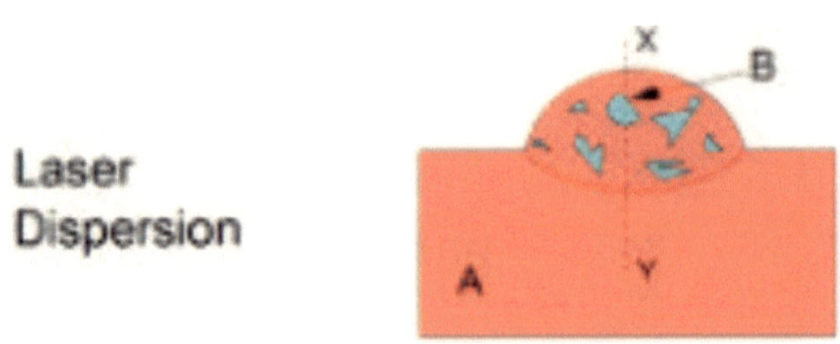

Figure 5. Schematic diagram showing the surface metal matrix formed by laser dispersion [15].

2. Overview on application and improved properties of laser-deposited Al-Si alloy coatings

Grigoriev et al. [18] studied the hypereutectic Al-Si alloys formed by laser surface treatment. Hypereutectic Al-Si alloys are being used in the manufacturing of automotive components and tools. It had been noticed that when the Si content increases, the wear resistance of these alloys increases as well. However, the Si content should not exceed 20 wt.% in standard casting processes because large primary Si particles are formed and they result in the reduction of the alloys' mechanical properties. The alloys that were to be produced had a Si content of 30, 40, and 60% (AlSi30, AlSi40, and AlSi60, respectively). The substrate used was a commercial aluminum alloy (AA6060), which composed of 0.3–0.6 wt.% Si, 0.35–0.6 wt.% Mg, 0.1–0.3% Fe, and the balance being Al. For the coat deposition, a continuous Yb:YAG 1 kW disk laser was used, and argon gas was used as the shielding gas. After processing, the samples were prepared for metallographic observation by cold mounting, polishing, and etching. To study the microstructure, optical microscopy was used, and SEM-ES was used to investigate the chemical composition of the cladding. The phases were analyzed by means of XRD, and the hardness was evaluated. The results showed that the primary Si particles in the AlSi30 were extremely fine and made it impossible to differentiate primary Si from eutectic Si in the cladding as shown in **Figure 6**. It was found that bigger primary Si particle was formed with an increase in the powder feed rate and a decrease in the laser power. The hardness of AlSi30 decreased to $HV_{0.05}$ 80 from $HV_{0.05}$ 190 which was observed from the substrate as shown in **Figure 7**.

Ma et al. [19] studied the phase and microstructure formed in Al-20Si-5Fe-3Cu-1 Mg manufactured by selective laser melting (SLM). The automotive and aerospace industries have encouraged the improvement of advanced materials with low coefficient of thermal expansion, low density, high specific strength, and excellent resistance to wear and corrosion. Hypereutectic Al-Si alloys having transition metals are exceptional materials due to their specific properties. The addition of Cu, Fe, Cr, Mg, and Ni to Al-Si alloys can improve the mechanical properties at

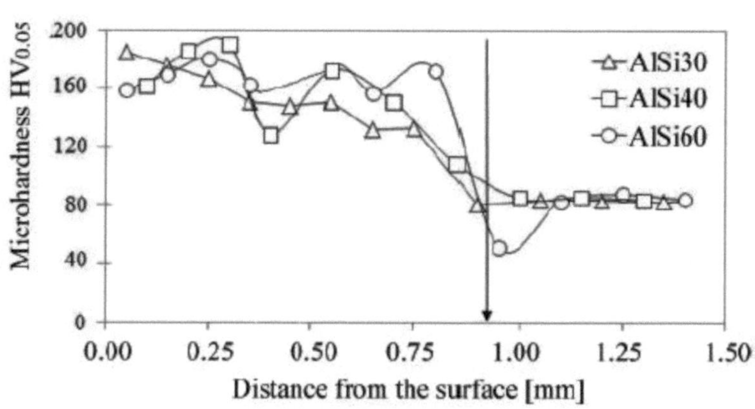

Figure 6. Microstructure of powders with different silicon contents [18].

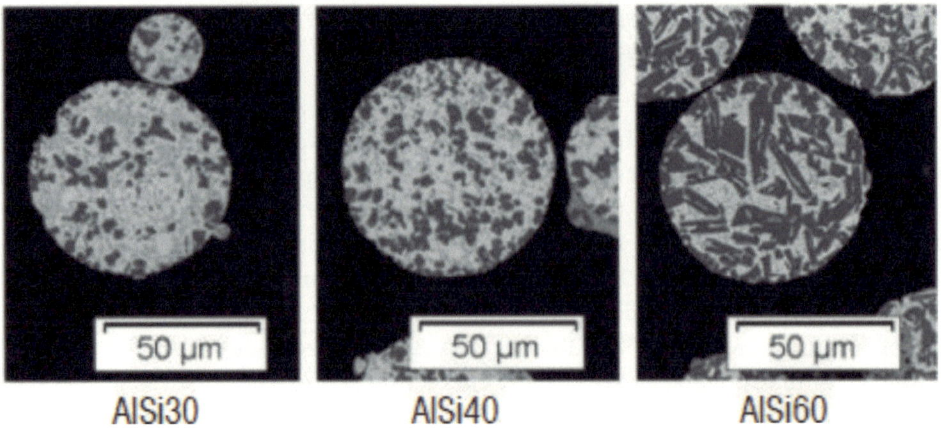

Figure 7. Microhardness distribution of laser cladding produced from alloys with different Si contents [P = 400 W, V = 800 mm/min] [18].

both ambient and elevated temperatures. The evolution of the microstructure and phase transformation of selective laser-melted Al-Si-Fe-Cu-Mg alloy that underwent different heat treatment was investigated. To manufacture the alloy, An SLM device equipped with an Yb:YAG laser was employed. The shielding gas used for the processing was high purity argon. SEM equipped with energy dispersive X-ray spectroscopy (EDS) was used to characterize the microstructure. XRD was used to analyze the present phases after SLM. The results revealed that SLM refined the primary silicon and Al-Si-Fe compound in addition to changing the morphology of eutectic silicon from a rod-like form to a mixture of dendrite and particle. The size of the SLM manufactured Al-Si-Fe compound phase reduced 4.5 times in comparison to the cast alloy (from 10.8 to 2.4 μm). The phase analysis from XRD showed that after SLM, only δ-Al_4FeSi_2 phase was present, but after heat treatment, both δ-Al_4FeSi_2 and β-Al_4FeSi_2 phases were found. It was observed that during cooling, no quasi-peritectic reaction ($L + \delta \rightarrow Si + \beta$) occurred. It can be concluded that selective laser melting of Al-20Si-5Fe-3Cu-1 Mg is better than casting the alloy.

Zhao et al. [20] studied a coating composed of Fe-Al-Si in situ composite synthesized by laser cladding. ASTM A283Gr.D steel is used in the industry as a structural material but has low wear resistance. Dispersion of a hard phase in the matrix of the coating has received interest due to the enhancement of wear resistance and microhardness. To investigate the effect of process parameters on the properties of the steel, laser cladding with Fe-Al-Si powder was done. The powder was preplaced on the steel substrate, and a continuous transverse flow CO_2 laser system was used. The parameters that were changed throughout the work were the composition of the powders, the laser power, as well as the scanning speed. After cladding, the samples were cut, and a Vickers hardness tester was used for the evaluation of the microhardness. The phase and microstructure were studied with XRD and SEM, respectively. A wear tester was used to characterize the wear resistance. The results showed that Fe, SiO_2, and $Al_2Fe_3Si_4$ intermetallic compound were formed on the surface of the steel substrate. It was observed that the increase in the laser power increases the wear resistance up until the power of 1600 kW and beyond this power, the wear resistance decreases. The increase in the scanning speed, until the speed of 400 mm/min, caused the wear resistance of the coating to increase. The hardness of the coating increased with the increased of the power and scanning speed until their peaks were reached. It was seen that the increase in laser power and scanning speed refined the grain size of the coating.

Kempen et al. [21] investigated the mechanical properties of AlSi10Mg manufactured by selective laser melting. Aluminum-silicon alloys have a wide application in the automotive, domestic, and aerospace industries due to their excellent mechanical properties such as good weldability, sound castability, and outstanding resistance to corrosion. Selective laser melting of the alloy can offer a wider range of application, such as structures that are complex involve cavities. To manufacture the alloy, the AlSi10Mg powder was processed using a modified concept laser M1 SLM machine, which is equipped with a 200 W fiber laser. A polarizing microscope and scanning electron microscope were used to observe the microstructures formed. The hardness was determined by means of a Vickers hardness tester, and X-ray patterns were measured with a goniometer. The results revealed that the AlSi10Mg parts produced by selective laser melting exhibited better overall mechanical properties when compared to the casted AlSi10Mg material shown in **Figure 8**. The improvement of these properties was attributed to the formation of $MgSi_2$, the rapid cooling rate resulted in a very fine microstructure, and the distribution of the Si phase was found to be fine as well. The fine microstructure consisted of tiny Al-matrix cells/dendrites (**Figure 9**) decorated with silicon phase. It was therefore concluded that selective laser melting of AlSi10Mg gives rise to mechanical properties than conventional casting.

Lijing et al. [22] examined the microstructure and hardness of selective laser melted of Al-8.5Fe-1.3 V-1.7Si alloy. Al-Fe-V-Si alloys show a great chance in being a competitive alternative to titanium alloys in the aerospace industry; this is due to their ductility, high strength, rapid solidification, and toughness at elevated temperatures. Vacuum induction melting in an argon atmosphere was used to produce the substrate which composed of Al-8.5Fe-1.3 V-1.7Si. A DEYU LM 200 SLM instrument equipped with Yb:YAG fiber laser was for SLM in an argon atmosphere. To observe the effect of scanning speed on the properties, two speeds were used: 200 and 250 mm/s. For microstructural analysis, an electron probe micro-analyzer (EPMA), transmission electron microscopy (TEM) equipped with energy dispersive X-ray spectroscopy (EDS) was used. XRD was used to perform phase analysis, and a Vickers hardness tester was used to establish the microhardness. It was found that SLM-processed samples composed of α-Al,

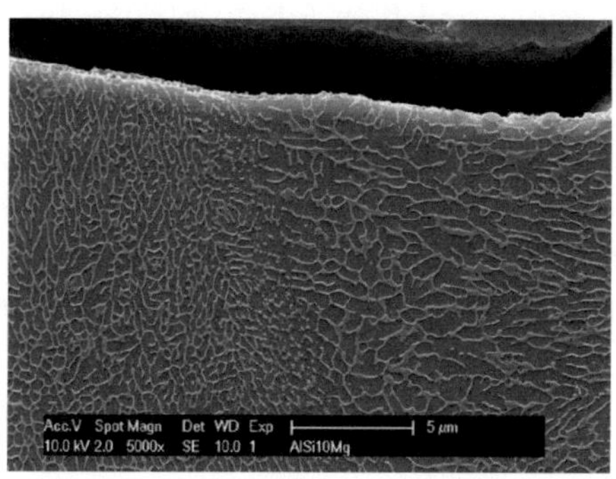

Figure 8. Fracture surface of a SLM-produced AlSi10Mg sample. The border of the broken test sample is shown, where the borderline porosity initiated cracks toward the side of the sample [21].

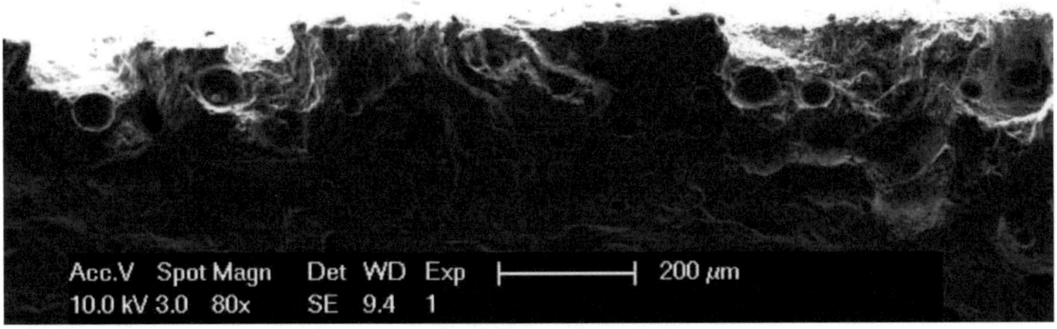

Figure 9. SEM micrograph of a AlSi10Mg SLM part showing the Al-matrix cells decorated with Si phase. A cross section perpendicular to the layers and the scanning direction [21].

$Al_{12}(Fe,V)_3Si$, and θ-$Al_{13}Fe_4$ phases. The laser-melted zone displayed a region of a mixture zone of α-Al and nanoscale $Al_{12}(Fe,V)_3Si$ or a cellular-dendritic structure, the melt pool border was found to have submicron θ-$Al_{13}Fe_4$ particles, and the heat-affected zone had an identical mixture zone of $Al_{12}(Fe,V)_3Si$ and α-Al. The hardness of the SLM-processed samples was much higher than that of the cast alloy. The hardness of the sample processed with a scanning speed of 200 mm/s was found to be in the range of 145–175 HV and 135–170 HV was the range of hardness for the scanning speed of 250 mm/s, and the hardness of the cast alloy was 40–45 HV. It can be concluded that SLM gives better surface properties than casting.

Anandkumar et al. [23] analyzed the sliding wear resistance and microstructure of an Al-12 wt. % Si/TiC laser clad coating. Al-alloy matrixes (Al-MMCs) have excellent mechanical properties, good wear resistance, and low density and have found application in automotive or aerospace industries. SiC has been extensively used for the reinforcement of Al-MMCs, but it dissolves significantly in molten aluminum; this results in the precipitation of Al_4C_3 at temperatures of 670–1350°C. The coating produced was composed of 40 wt.% TiC powder, and the balance was Al-12wt.% Si powder. The powders were mixed with a Turbula powder mixer. For the laser cladding, a continuous wave Nd:YAG laser system was used to clad the powder on Al–7 wt.% Si cast alloy plates. The laser power used was 1.8 kW, and the scanning speed was 12 mm/s. After laser cladding, the samples were prepared for analysis by standard metallographic polishing methods and then followed by etching with Keller's reagent. To characterize the microstructure, optical microscopy, scanning electron microscopy, energy dispersive X-ray spectroscopy, and X-ray diffraction were used. A Vickers hardness tester was used to determine the microhardness, and the dry sliding wear tests were performed with a ball-cratering apparatus. The results showed that the microstructure of Al-12wt.%Si/TiC coatings had TiC particles distributed in a matrix that consisted of 35 vol.% primary α-Al dendrites and interdendritic α-Al + Si eutectic. Because of partial dissolution of TiC in liquid aluminum, 3% of Ti_3SiC_2 was found to be present in the coating. The hardness of the coating was determined to have increased to 165 HV from the hardness of 115 HV exhibited by the substrate. The wear coefficient of the Al-Si/TiC composite was lower than that of the Al-12wt.% Si alloy and the substrate. The 2.2 mm^3/Nm was found to be the wear coefficient of the composite, 8.2 mm^3/Nm was determined for the alloy, and the substrate presented a wear coefficient of 4.5 mm^3/Nm. Hence, it was concluded that the coating produced by laser cladding had superior mechanical properties.

Zhao et al. [24] studied ultrafine Al-Si hypereutectic alloy produced by direct metal deposition. Hypereutectic Al-Si alloys are being applied in the automobile, aeronautical, and military industries because of their good corrosion resistance, low density, low thermal expansion coefficient, good casting ability, and high wear resistance. The large proportions of coarse primary Si present in Al-Si alloys formed during conventional casting processes can decrease their anti-wear characteristic and ductility; to overcome this, the primary Si of the alloy should be refined. The substrate used was 6061 aluminum alloy which was prepared by sandblasting, and Al-Si and silicon carbide powders were used. A direct metal deposition machine was used to deposit 10 successive layers by scanning the laser back and forth over the previous layer. A coaxial nozzle was used to inject the powders at a feed rate of 4.9 g/min. The samples were cut along the transverse section and were prepared by metallographic polishing and etching in Keller's reagent. SEM-EDS was used to characterize the microstructure of the layer, XRD was used to determine the phases present, and an automatic microhardness tester was used to evaluate the microhardness. The results showed that an in situ hypereutectic Al-Si alloy containing primary Si is ultrafine. It was observed that an increase in the scanning speed and laser power led to an increase in the volume fraction of primary Si, as well as the size. The size of eutectic Si grain decreased with the increase in scanning speed and reached the minimum value at a laser power of 850 W. the microhardness tests displayed that an increase in scanning speed resulted in an increase in the microhardness. It was concluded that direct metal deposition refined the microstructure and, therefore, resulted in improved properties.

Li et al. [25] examined the effect heat treatment had on AlSi10Mg alloy manufactured by selective laser melting. AlSi10Mg is extensively used in automotive and aerospace industries and heat exchanger products. The wide range of application is due to the high mechanical properties, low thermal expansion, reduced recycling costs, and light weight of this alloy. It has been revealed that in a tensile environment, coarse and acicular eutectic silicon phases instigate cracks, which in turn brings about the deterioration of mechanical properties. The modification of this coarse phase is important in improving the mechanical properties of AlSi10Mg alloy. The alloy was produced with a gas-atomized powder processed with a selective melting machine where the parameters were a scanning speed of 1140 mm/s, a laser power of 350 W, scan spacing of 170 μm, and a powder layer thickness of 50 μm. During selective laser melting, argon gas was used as a shielding gas. After SLM, the samples were subjected to solution heat treatment at different temperatures (450, 500, and 550°C) for 2 h and water quenched. Half of the samples were exposed to artificial aging at 180°C for 12 h. Prior to characterization, the samples were ground, polished, and etched in Keller's reagent. SEM was used to examine the microstructure, and XRD was used to analyze the phases present. A high-precision electronic universal testing machine was used for tensile tests, and a Vickers hardness tester was used to get the microhardness. The results showed that the solubility of Si atoms in Al matrix rapidly decreased with the increase in the heat treatment temperature, and after aging the solubility decreased even further. An ultrafine eutectic microstructure was formed as a result of the high cooling rate and thermal fluctuation of selective laser melting;

this led to significantly improve tensile properties and hardness. The size of Si particles was initially small after SLM but coarsened with heat treatment and become coarser with aging.

Gu and Van Gelder [26] characterized the coating layer of laser-deposited Al-Si over laser weld bead as shown in **Figures 10** and **11**. Steel sheets in the automotive industry should have excellent corrosion resistance as it is the primary protection of these sheets. The weld bead was prepared by applying a localized coating by means of well-defined spot heating and concurrently spraying coating material powder (Al-Si) on the heated spot. A Trump disk was used to perform laser welding and laser-localized coating. A steel sheet was used as the substrate. After processing, the samples were water quenched. Scanning electron microscopy coupled with energy dispersive spectroscopy was used to characterize the morphology of the coating layer. The results showed that the thickness of the layer varied from 100 to 140 μm depending on the feed rate, laser power, and processing speed. A thin layer was produced by a faster speed and lower power. EDS revealed that the layer consisted of Fe-rich intermetallic compound as also reported by [27, 28]. For a higher laser power, Fe-Al and Fe_3Al phases were present, and for a lower energy and fast process, Fe-Al, Fe_3Al, and α-Fe intermetallic phases were identified. The coating was found to be a stable Fe-Al compound enriched with Fe which would provide improved corrosion resistance for the steel material underneath it.

Liang et al. [29] studied the corrosion resistance and high strength of an Al-Si-based casting alloy. Because Al-Si-based casting alloys exhibit good castability, great strength-to-weight ratio, and reduced manufacturing costs, they have a variety of applications in the automotive industries. Although these properties are desirable, the alloys were stated to have a high corrosion rate in environments that contained salt. Alloying was a technique employed to improve the properties. The two reference metals used composed of silicon, iron, copper, manganese, magnesium, zinc, and aluminum of different compositions, and the new alloy has the same

Figure 10. Top view of a laser-generated coating on weld bead [26].

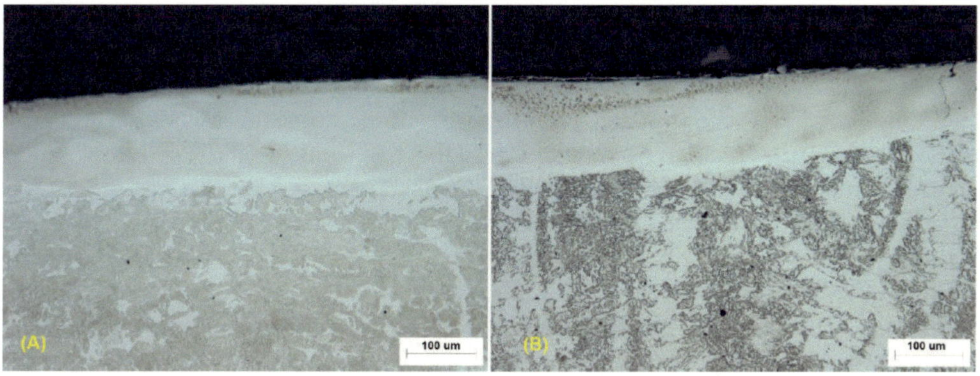

Figure 11. Optical micrographs of the coating before (A) and after (B) heat treatment [26].

metals with an addition of strontium. The metals were melted in clay graphite crucibles heated at 740°C and poured into a steel mold, and then the ingots were aged at 180°C for 4 h before they were allowed to cool be undergo analysis. Tensile tests were done, the microstructures were assessed using optical and scanning electron microscopy, and the corrosion behavior was evaluated by means of immersion tests and polarization. XRD was used to observe the surface of the samples before and after immersion tests were done in a 3.5 wt. % NaCl solution for 240 h. The microstructural tests revealed that all three alloys had an intermetallic that contained eutectic silicon, α-Al dendrites, and needle-shaped Fe. The addition of strontium made the α-Al dendrites of the new alloy more uniform and columnar than those of the reference alloys. The tensile tests revealed that the new alloy had a higher strength and could be elongated; this was due to the eutectic silicon being more spherical and much finer. The corrosion tests revealed that all three alloys had the same corrosion mechanism, but the new alloy had a much higher corrosion resistance. The XRD performed after immersion tests showed that the new alloy had fewer sites of pit corrosion. The authors therefore conclude that the new alloy has superior microstructural and mechanical properties (**Figure 12**).

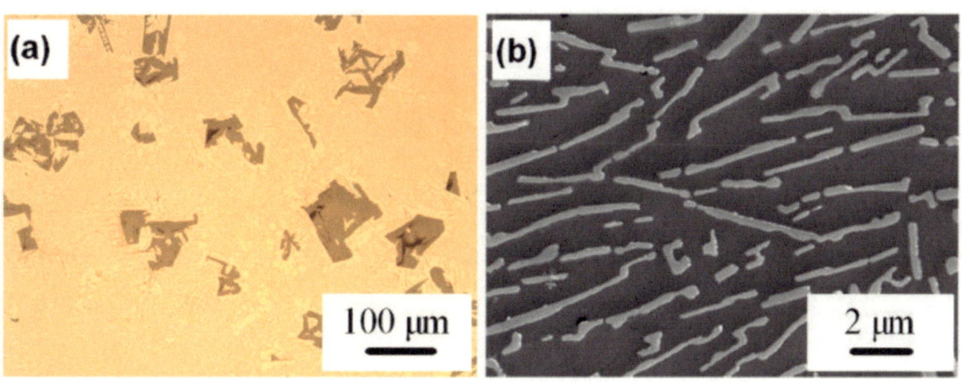

Figure 12. (a) Microstructure of as cast Al-20Si alloy and (b) high magnification micrograph showing the eutectic Al-Si [24].

El-Salam et al. [30] analyzed how the mechanical and structural properties of Al-Si alloy were affected by the Sn content. Al-Si cast alloying is mainly used for applications such as automobile components and instrument panels in the transportation industries. Tin has outstanding anti-welding properties with iron and low strength and modulus; this makes it a crucial component in bearing. Adding Sn to Al-Si or adding Si to Al-Sn produces the possibility of meeting the requirements needed for maintaining a balance of strength and soft surface characteristics. Different amounts of Sn (0.5–2 wt.%) were added to Al-3 wt.% Si alloy to observe the effect. The samples were made by melting 99.99% pure Al, Sn, and Si in a graphite crucible at different compositions, and the ingots were normalized at 550°C for 53 h. The hardness and stress-strain were tested as well as the microstructure of the alloys by means of a transmission electron microscope. The stress-strain curves obtained for the alloys show that the hardness of the alloys decreases when the temperature increases and increases when the Sn content is raised as also reported by Makhatha et al. [2017]. After homogenization, it was found that the spheroidized Si particles were encased in β-Sn layers that formed a peritectic structure. The ternary alloy, Al-Sn-Si, had overall better mechanical properties in comparison to the initial binary alloy Al-Si. The authors therefore conclude that the addition of Sn to Al-Si alloys enhances the mechanical properties of the alloys and a higher Sn content leads to an alloy with superior alloys.

Kang et al. [31] studied the microstructure of hypereutectic Al-Si alloys produced by means of selective laser melting (SLM), as well as its wear behavior. Aluminum alloys have numerous applications in the automobile and aircraft industries due to their excellent properties such as corrosion resistance, formability, and high stiffness-to-weight ratio. Al-Si alloys have a low thermal expansion coefficient and a high strength to weight ratio and resist wear; these properties allow the alloy to be used as a material for pistons and engine blocks. It has been observed before that the increase of silicon content until eutectic constitution will increase the resistance to wear of the alloys. An investigation was done on the impact that laser power had on the microstructure and mechanical properties of the alloy. The powders were mixed, and a commercial SLM machine was used to fabricate the alloy on an aluminum base plate at laser powers that ranged from 120 to 225 W. The microhardness of the samples was obtained by employing a Vickers pyramid, the microstructure was observed using optical microscopy and SEM-EDS, and dry ball-on-disk wear tests were done on the samples. After the tests were conducted, it was found that relatively packed hypereutectic Al-SI alloys were achieved in situ from a combination of elemental powders using a SLM procedure under argon conditions. The microhardness and relative density increased with the increase of power until 210 W; beyond this power, the mechanical properties reduced drastically. Analyzing the microstructure revealed that the increase of laser power caused the porosity of the samples decrease and the form of pores altered from an irregular shape to a spherical one. The authors therefore concluded that a laser power of 210 W was the optimum to use as it presented the highest relative density and microhardness and exhibited the lowest wear rate of all the processed specimens.

Zhao et al. [32] studied in situ composite coating Fe-Al-Si that was fabricated by laser cladding. Steel has many applications, but its poor wear resistance limits the extent of which it can be used. The intermetallic compound Fe-Al has good corrosion resistance and reasonable wear resistance and can be fabricated by laser processing to improve surface properties. The effect of powder consumption, scanning speed, and laser power on the wear resistance, microhardness, and microstructure was examined. The substrate used was ASTM A283Gr.D steel, and the coating applied was a composite powder consisting of Fe-Al-Si. The powder was preplaced onto the substrate, and the laser equipment used was a continuous CO_2 laser. The parameters that were changed were the ratios of the powders, the laser power, and the scanning speed of the laser beam. Once the laser processing was complete, the samples were cut, and the hardness was obtained by using a Vickers microhardness tester, the wear resistance was assessed by means of a wear tester, and the microstructure and phases were examined using SEM and XRD, respectively. The intermetallic compound found in the in situ coating was $Al_2Fe_3Si_4$, and it existed with Fe and SiO_2. The grain size of the coating was revealed to decrease with the increase of the scanning speed and laser power of the beam used. When the scanning speed and power of the laser increased, it caused the microhardness to increase; it increased to three times that of the substrate. The wear tests showed that the increase of the laser parameters increased the wear resistance of the coating; the wear resistance of the coating was 3.5 better than the substrate. It can be concluded that the Fe-Al-Si in situ composite coating had enhanced properties in comparison the ASTM A283Gr.D steel.

Ma et al. [33] investigated how the mechanical properties of Al-20Si synthesized by selective laser melting were influenced by annealing. Aluminum-based alloys are applied as lightweight structural materials due to their low density and high strength. Hypereutectic Al-Si alloys have a variety of applications in the aerospace and automobile industries as a result of their high resistance to wear and corrosion, low density, good machinability and thermal conductivity, and high thermal stability. Research on SLM processing of Al-Si alloys mainly focus on the hypoeutectic regime and not enough on the hypereutectic. A selective laser melting device equipped with Nd-YAG laser was used to prepare Al-20wt.% Si samples in a form of cylindrical tensile bars. The powders used were gas atomized. During the laser processing, argon gas was used as a shielding gas. An Al-20Si alloy was casted by graphite mold casting for comparison. The samples were then subjected to heat treatment by annealing. Optical microscopy and scanning electron microscopy equipped with energy dispersive spectroscopy were used for the microstructural characterization of both the SLM and cast alloys. The phases were analyzed by XRD, and testing device was used to test the strain of the samples as part of the tensile tests. The results displayed that the size of the eutectic and particulate Si was refined when SLM processing was employed. With the increase in the annealing temperature, the Si particles became coarser. The strength of the SLM Al-20Si was higher than those produced by different techniques (**Table 1**).

Viswanathan et al. [34] analyzed $TiC\text{-}Al1_3Fe_4$ composite layer formed on Al-Si alloy by laser processing. Cylinder blocks, pistons, valve lifters, and cylinder heads are often made of Al-Si alloys due to their good corrosion resistance, great formability, high thermal conductivity, high specific strength, and low density. The limitation of being used in other industrial applications is because of the poor wear resistance and low hardness. The substrate used was Al-12% Si with a composition of 12 wt.% Si, 1 wt.% Fe, 1.5 wt.% Cu, 1.5 wt.% Mg, 1.5 wt.% Ni, 0.2 wt.%Ti, and the remainder being Al. Two different coatings were applied on the substrate: 75–25% TiC-Fe (Coating A) and 25–75% TiC-Fe (Coating B). The laser processing was done using a continuous wave CO_2 laser where argon was used as shielding gas. After coating, the samples were sectioned and prepared for metallographic examination. The microstructure

Preparation methods	YS (MPa)	UTS (MPa)	Ductility (%)
As cast-1	105	162	4.6
As cast-2	95	120	0.37
As cast-3	—	91.5	0.49
Ultrasonic treated	80	130	1.2
Extruded-1	155	190	3.6
Extruded-2	—	280	>30
Extruded-2 + ECAP	—	350	>30
RS + wrought	131	210	19.7
As cast-1 + wrought	118	177	7.7

Table 1. Tensile properties of Al-20Si alloys processed by different methods (YS = yield strength; UTS = ultimate tensile strength) [24].

was assessed using an optical microscope, the hardness was obtained using a microhardness tester with a load of 100 g, the phases were evaluated by means of XRD, and the wear was tested using a block-on-ring dry sliding wear tribometer. The tests revealed that a different type of iron-based aluminum phase, $Al_{13}Fe_4$, was formed during laser processing. The sample from Coating A exhibited good metallurgical bonding between the coating and the substrate, whereas Coating B did not form a layer, but it dissolved into the substrate. Coating A had an even hardness distribution all over the composite layer and had a hardness that was about six times higher than that of the substrate. The TiC reinforced with $Al_{13}Fe_4$ led to a reduced wear rate of the composite layer. It can be concluded that the higher TiC content will enhance the tribological properties of Al-Si alloys.

Chen et al. [35] investigated laser cladding of Al-Si powders on a magnesium-based Mg-Gd-Y-Zr alloy. The effect of scanning speed on the properties of the surface was also examined. Alloys based on magnesium are extensively used in industries such as aerospace and automobile due to their good castability and machinability, high strength, and low density. The range of application of magnesium alloys is limited by their low corrosion resistance and poor resistance to wear. The substrate was a magnesium-based Mg-Gd-Y-Zr alloy that was casted and homogenized for 12 h at a temperature of 500°C; it was then sectioned, polished with alumina sand papers, and rinsed with pure ethanol. The powders were mixed at a weight ratio of 3:1 (Al to Si) and were placed on the substrate homogenously and were pressed to expel the gas between the particles of the powder. For laser cladding, a continuous carbon dioxide gas laser was used, and the shielding gas applied was high purity argon. The clad samples were polished and etched; then SEM was employed to analyze the microstructure of the cross section of the sample. The phase constituents of the clad layers were studied using XRD and the microhardness on the cross section of the samples was done with a microhardness tester, and a load of 50 g was used. The potentiodynamic polarization was measured by means of a three-electrode cell in an aqueous solution of 3.5 wt.% NaCl. Before the corrosion tests were done, the samples were polished and cleaned with ethanol. The clad layer was found to be made up of Mg, $Mg_{17}Al_{12}$ phases, Mg_2Si, and isolated Al_2 (Gd, Y) particles. It was concluded that these intermetallic composites were responsible for the strengthening of the clad layer. The highest microhardness was achieved with the slowest scanning speed because it had a deeper hardened layer. The corrosion resistance of the laser clad layer was much higher than that of the substrate.

3. Conclusion

The main aim of this study is to show the improved properties of Al-Si alloy by laser metal deposition by developing high wear and corrosion-resistant coatings using various combinations of laser metal deposition powders as reinforcements. From the literature reviewed, all authors agreed that the Ti6Al4V alloy has poor wear resistance as well as low hardness property, thus having the tendency to gall. From the reviewed literature, it can be concluded that laser surface modification techniques have the ability to improve the microstructure, mechanical, and tribological properties of Ti6Al4V alloy.

The reviewed literature indicates that laser cladding can be employed as a technique that will improve the surface properties. It refines the microstructure of the alloy, therefore improving mechanical properties. The corrosion resistance can be bettered by adding other elements to the surface of the material.

Author details

Olawale Samuel Fatoba[1]*, Mamookho Elizabeth Makhatha[1] and Esther Titilayo Akinlabi[2]

*Address all correspondence to: drfatobasameni@gmil.com

1 Department of Mechanical Engineering Science, Faculty of Engineering and the Built Environment, University of Johannesburg, South Africa

2 Department of Metallurgy, Faculty of Engineering and the Built Environment, University of Johannesburg, South Africa

References

[1] Ching HA, Choudhury D, Nine MJ, Abu Osman NA. Effects of surface coating on reducing friction and wear of orthopaedic implants. Science and Technology of Advanced Materials. 2014;**15**(1):1-7

[2] Bacci T, Pradelli G, Tesi B, Gianoglio C, Badini C. Surface engineering and chemical characterization in ion-nitrided titanium and titanium alloys. Journal of Materials Science. 1990; **25**:4309-4314

[3] Dressler S. Plasma treatments. In: Sudarshan TS, editor. Surface Modification Technologies: An Engineer's Guide. New York: Marcel Dekker; 1989. pp. 317-419

[4] Toma M. Laser fusing of alloy powders for applying hard facing layers onto mild steel substrate. Damascus University Journal. 2005;**21**(2):95-117

[5] Bao R, Yu H, Chen C, Qi B, Zhang L. Development of laser cladding wear resistant coating on titanium alloys. Surface Review and Letters. 2006;**13**:645-654

[6] Montealegre MA, Castro G, Rey P, Arias JL, Vanquez P, Gouzalez M. Surface treatments by laser technology. Contemporary Materials. 2010;**I-1**:19-30

[7] Fatoba OS, Popoola API, Aigbodion VS. Experimental study of hardness values and corrosion behaviour of laser alloyed Zn-Sn-Ti coatings of UNS G10150 mild steel. Journal of Alloys and Compounds. 2016a;**658**:248-254

[8] Baker TN. (2010) Laser surface modification of Ti alloys. In: Surface Engineering of Light Alloys - Aluminium, Magnesium and Titanium Alloys. Cambridge, UK: Woodhead Publications; 2010. pp. 398-443. ISBN 1-84569-537-2. https://strathprints.strath.ac.uk/8071/

[9] Popoola API, Fatoba OS, Aigbodion VS, Popoola OM. Tribological evaluation of mild steel with ternary alloy of Zn-Al-Sn by laser deposition. International Journal of Advanced Manufacturing Technology. 2016a:1-7. DOI 10.1007/s00170-016-9170-7

[10] Fatoba OS, Popoola API, Pityana SL, Adesina OS. Computational dynamics of anti-corrosion performance of laser alloyed metallic materials. Paul M. ed. Fiber Laser; 2016: InTech. DOI: 10.5772/62334. Available from: http://www.intechopen.com/books/fiber-laser/computational-dynamics-of-anti-corrosion-performance-of-laser-alloyed-metallic-materials

[11] Fatoba OS, Popoola API, Fedotova T, Pityana SL. Electrochemical studies on the corrosion behaviour of laser alloyed Zn-Sn coatings on UNS G10150 steel in 1M HCl solution. Silicon. 2015a;**7**(4):357-369

[12] Lo KH, Cheng FT, Kwok CT, Man HC. Improvement of cavitation erosion resistance of AISI 316 stainless steel by laser surface alloying using fine WC powder. Surface and Coatings Technology. 2003;**165**:258-267

[13] Oberlander BC, Lugscheider E. Comparison of properties of coatings produced by laser cladding and conventional methods. Materials Science and Technology. 1992;**8**:657-665

[14] Adebiyi DI, Fatoba OS, Pityana SL, Popoola API. Parameters optimization, microstructure and microhardness of silicon carbide laser deposited on titanium alloy. In: Proceedings of International Conference on Surface Modification Technologies, 29th June-1st July, 2016, Milan, Italy; 2016. pp.1-6

[15] Kwok CT, Cheng FT, Man HC, Ding WH. Corrosion characteristics of nanostructured layer on 316L stainless steel fabricated by cavitation-annealing. Journal of Materials Letters. 2006;**60**:2419-2422

[16] Popoola API, Fatoba OS, Popoola OM, Pityana SL. The influence of heat treatment and process parameter optimization on hardness and corrosion properties of laser alloyed X12CrNiMo steel. Silicon. 2016;**8**(4):579-589

[17] Fatoba OS, Popoola API, Fedotova T. Characterization and corrosion behaviour of Zn-Sn binary alloy coatings in 0.5M H_2SO_4 solution. Journal of Electrochemical Science and Technology. 2015;**6**(4):65-74

[18] Grigoriev SN, Tarasova TV, Gvozdeva GO, Nowotny S. Structure formation of hypereutectic al-Si-alloys produced by laser surface treatment. Journal of Mechanical Engineering. 2014;**60**:389-394

[19] Ma P, Jia Y, Prashanth KG, Scudino S, Yu Z, Eckert J. Microstructure and phase formation in al-20Si-5Fe-3Cu-1Mg synthesized by selective laser melting. Journal of Alloys and Compounds. 2016;**657**:430-435

[20] Zhao L-Z, Zhao M-J, Li D-Y, Zhang J, Xiong G-Y. Study on Fe–al–Si in situ composite coating fabricated by laser cladding. Applied Surface Science. 2012;**258**:3368-3372

[21] Kempen K, Thijs L, Van Humbeeck J, Kruth J-P. Mechanical properties of AlSi10Mg produced by selective laser melting. Physics Procedia. 2012;**39**:439-446

[22] Lijing Z, Yingying L, Shaobo, Hu Z. Selective laser melting of al–8.5Fe–1.3V–1.7Si alloy: Investigation on the resultant microstructure and hardness. Chinese Journal of Aeronautics. 2015;**28**:564-569

[23] Anandkumar R, Almeida A, Vilar R. Microstructure and sliding wear resistance of an al–12 wt.% Si/TiC laser clad coating. Wear. 2012;**282-283**:31-39

[24] Zhao LZ, Zhao MJ, Song LJ, Mazumder J. Ultra-fine al–Si hypereutectic alloy fabricated by direct metal deposition. Materials and Design. 2014;**56**:542-548

[25] Li W, Li S, Liu J, Zhang A, Zhou Y, Wei Q, Yan C, Shi Y. Effect of heat treatment on AlSi10Mg alloy fabricated by selective laser melting: Microstructure evolution, mechanical properties and fracture mechanism. Materials Science & Engineering A. 2016;**663**:116-125

[26] Gu H, Van Gelder A. Coating layer characterization of laser deposited al-Si coating over laser weld bead. Physics Procedia. 2016;**83**:597-605

[27] Makhatha ME, Fatoba OS, Akinlabi ET. Effects of rapid solidification on the microstructure and surface analyses of laser-deposited Al-Sn coatings on AISI 1015 steel. International Journal of Advanced Manufacturing Technology. 2017. 1-15. https://doi.org/10.1007/s00170-017-0876-y

[28] Aigbodion VS, Popoola API, Fatoba OS. Evaluation of hardness values and corrosion behaviour of laser alloyed 20Al-20Sn-60Ti coatings of UNS G10150 mild steel. International Journal of Advanced Manufacturing Technology. 2015;**81**:1-11. DOI: 10.1007/s00170-015-8111-1

[29] Liang ZX, Ye B, Zhang L, Wang QG, Yang WY, Wang QD. A new high-strength and corrosion-resistant al–Si based casting alloy. Materials Letters. 2013;**97**:104-107

[30] Abd El-Salam F, Abd El-Khalek AM, Nada RH, Wahab LA, Zahran HY. Effect of Sn content on the structural and mechanical properties of al–Si alloy. Materials Science and Engineering A. 2010;**527**:1223-1229

[31] Kang N, Coddet P, Liao H, Baur T, Coddet C. Wear behavior and microstructure of hypereutectic al-Si alloys prepared by selective laser melting. Applied Surface Science. 2016; **378**:142-149

[32] Zhao LZ, Zhao MJ, Li DY, Zhang J, Xiong GY. Study on Fe–al–Si in situ composite coating fabricated by laser cladding. Applied Surface Science. 2011;**258**:3368-3372

[33] Ma P, Prashanth KG, Scudino S, Jia Y, Wang H, Zou C, Wei Z, Eckert J. Influence of annealing on mechanical properties of al-20Si processed by selective laser melting. Metals. 2014;**4**:28-36

[34] Viswanathan A, Sastikumar D, Kumar H, Nath AK. Laser processed TiC–$Al_{13}Fe_4$ composite layer formation on al–Si alloy. Optics and Lasers in Engineering. 2012;**50**:1321-1329

[35] Chen E, Zhang K, Zou J. Laser cladding of a mg based mg–Gd–Y–Zr alloy with al–Si powders. Applied Surface Science. 2016;**367**:11-18

Chapter 8

Effect of the Processing Conditions on the Microstructural Features and Mechanical Behavior of Aluminum Alloys

Tomasz Tański and Przemysław Snopiński

Additional information is available at the end of the chapter

http://dx.doi.org/10.5772/intechopen.70682

Abstract

Aluminum and aluminum alloys are widely used for aircraft structures, where they are subjected to demanding conditions and where is an increased demand for weight reduction and fuel savings. Aluminum comprises 8% of the earth's crust and is, therefore, the most abundant structural metal. Its production since 1965 has surpassed that of copper and now comes next to iron. This increased use of aluminum alloys leads to a need for deeper understanding of their mechanical properties and the impacts of processing parameters. The mechanical properties can determine by controlling the microstructures of the alloys. For example, precipitation hardening is the main strengthening mechanism improving the tensile and yield strength. Solute atoms, precipitates and dispersoids influences to the yield strength, since they act as distributed pinning points for mobile dislocations, thus increasing the shear stress required to move the dislocations. Another approach is the manipulation of a grain size that can be performed by alloying or plastic deformation processes. Therefore, the precise understanding of each mechanism that can influence the properties of aluminum and its alloys is very important. The aim of this chapter is to shed light on the influence of the processing history on the microstructure and mechanical properties.

Keywords: aluminum, heat treatment, structure, properties, severe plastic deformation

1. Aluminum—introduction

Aluminum is the third most abundant element in the earth's crust and the most abundant metallic element. For the last five decades, it has been second only to iron in an industrial use. It is worth to point out that the potential of aluminum as engineering material was found well before it became an industrial material. It was supposed that the most useful field for this metal will be in its alloys. At the beginning, the application of aluminum was limited to small- or high-valuable items due to the very high cost of this material. Nowadays, pure aluminum and its alloys play a fundamental role in engineering. Aluminum is the most heavily consumed non-ferrous metal in the world, with concurrent annual production at 130 million metric tonnes. About 50% of this total alumina production ~65 million tonnes is "primary aluminum". The process of primary aluminum production is divided into three separate stages [1–3].

- Mining of the raw material (bauxite and a variety of ores);
- Preparation of an aluminum oxides from ores;
- Production of primary aluminum.

The total world production of primary aluminum has increased from 13 million metric tones in 1974 to about 65 million metric tones in 2016 (**Figure 1**).

At present, consumers and engineers are demanding energy efficiency, thus aluminum can play a fundamental role in driving this change. Due to the fact that by replacing the steel parts with those made from aluminum, a significant decrease in weight can be achieved, many car

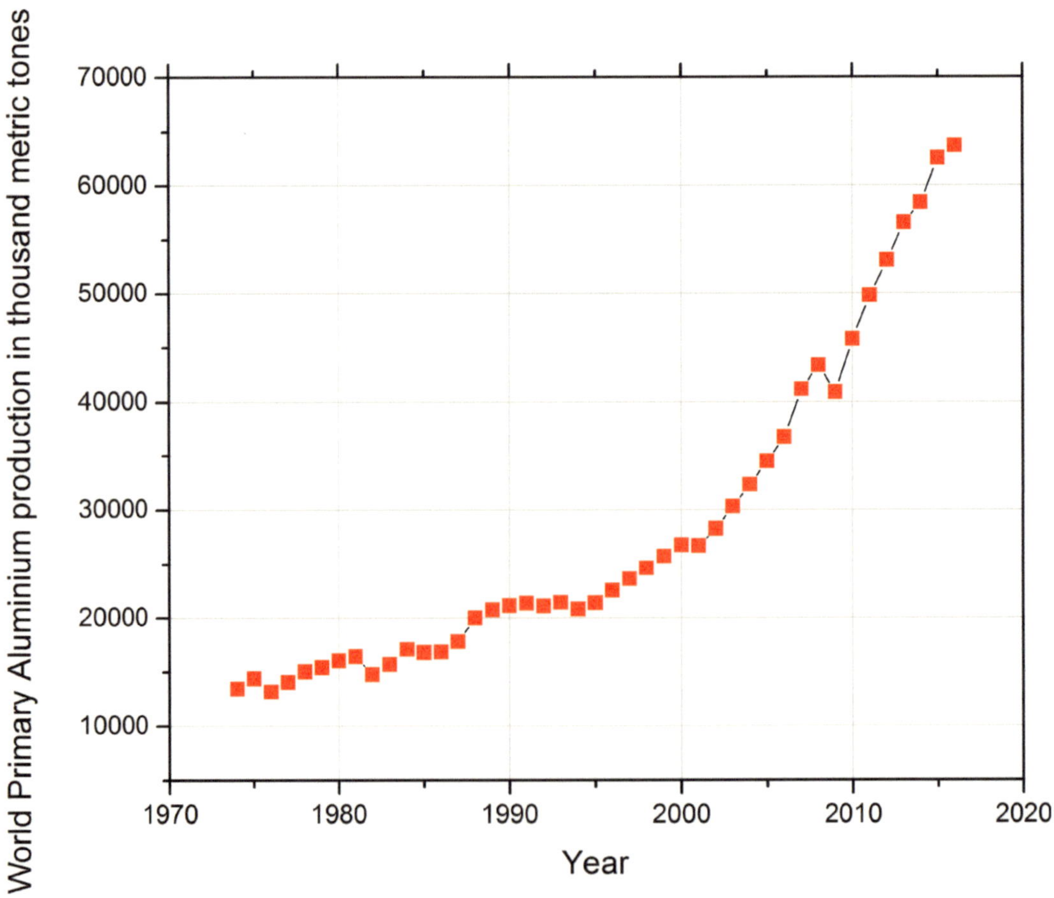

Figure 1. World primary aluminum production [4].

http://dx.doi.org/10.5772/intechopen.70682

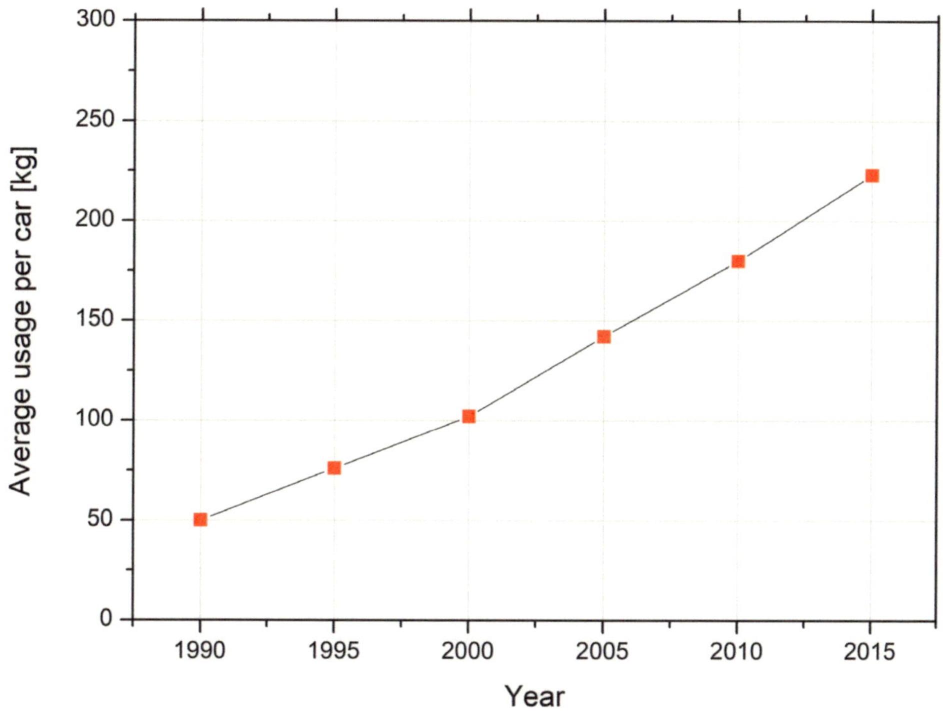

Figure 2. Average use of aluminum per car—data in Western Europe [5].

companies are now moving to aluminum to achieve this goal. Moreover, its usage in the automotive industry is accelerating (**Figure 2**) because it offers the fastest, safest and environmentally friendly way to increase performance and fuel economy. Aluminum and its alloys are mainly used in an automotive, aerospace and construction engineering due to their unique properties such as corrosion resistance and high specific strength. Its usage in automotive applications has increased by more than 80% in the past 5 years. It is predicted that a total amount of about 50 kg of aluminum content per car produced in Europe in 1990 will increase to about 250 kg in the 2020 year. However, to meet the engineers' demand, the properties of Al and its alloys have to be increased or modified, which can be obtained through the microalloying, heat treatment, plastic deformation or the combination of this treatment [2, 3, 6].

2. Strengthening of aluminum

Like all known pure metals, in comparison to its alloys, pure aluminum has a low strength. Therefore, many elements are added to solid solution of aluminum. All known alloying elements that are used for the production of Al alloys can be classified into three principal groups: basic, ancillary additions and impurities. Depending upon the nature of an alloy, the same elements could play different roles. In a great majority of aluminum alloys, four kinds of alloying elements are used: magnesium, zinc, copper and silicon. These additions can be classified as "basic" or "principal", due the fact that they are introduced into Al solid solution in (relatively) large amounts and create their microstructure mechanical and physical properties. This introduction of relatively large amounts of alloying elements can be done because they are characterized by considerable solubility in Al. Due this fact this introduced atoms are obstacles for dislocation movement and enhances strength. The solid solution strengthening can take place through three basic mechanisms:

- Lattice strain field interactions between dislocations and alloying atoms result in a decrease of dislocation movement.
- Alloying elements that are in solid solution impose lattice strains on surrounding host atoms.
- Alloying atoms tend to diffuse and segregate around the dislocation to find atomic sites more suited to their radii. This decreases the entire strain energy and "anchor" or "pin" the dislocation.

The final properties of such alloys depend on a complex interaction of chemical composition, solidification sequence of main phases during crystallization process history [1, 6–10].

The most important function of the alloying of Al solid solution is to enhance alloy mechanical properties. A strengthening effect proceeds when lattice strain field interactions occur between alloying elements and dislocations. These dislocations are imperfections in the atomic structure of the material. This single atom can substitute the aluminum atoms in the lattice. In addition, they may fit in the atomic lattice being substitutional atoms (**Figure 3**). In contrast to

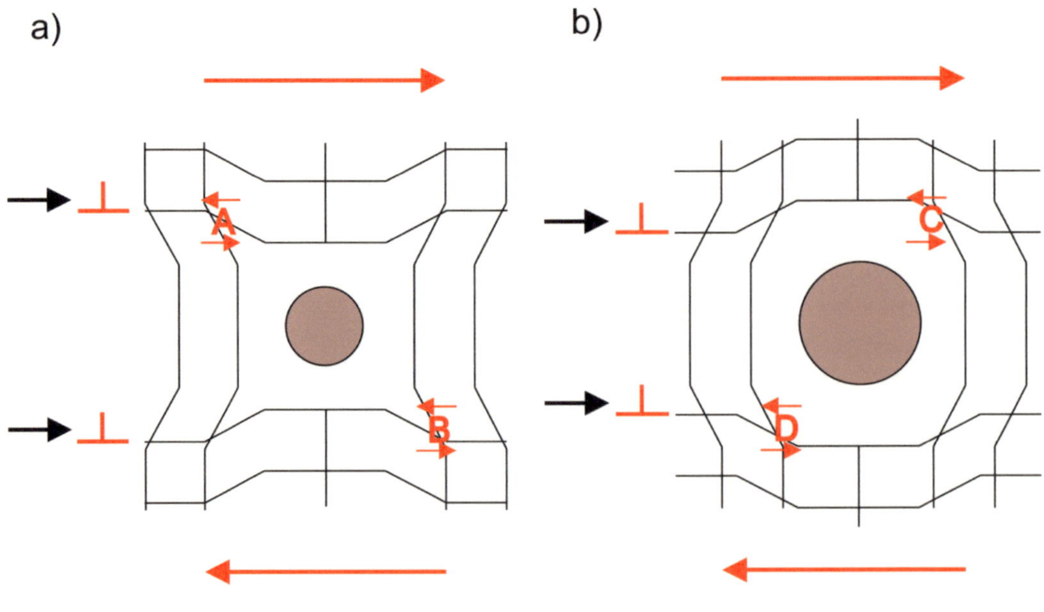

Figure 3. The principle of solid solution strengthening mechanism due atomic radii mismatch, (a) smaller substitutional alloying element generates a local shear at A and B that opposes motion of dislocation to the right, (b) element generates a local shear at C and D that opposes motion of dislocation to the right [9].

substitutional atoms, the interstitial atoms have considerably smaller atomic radii than aluminum. Due to the fact that aluminum has a small atomic radii, the interstitial atoms do not frequently form substitutional atoms. Atoms introduced to the aluminum lattice can cause a distortion in crystal lattice and these distorted regions impede the dislocation movement which causes a strengthening effect. The difference of the Al atomic radii and that used in alloying element ((RAl − R2)/RAl) × 100% reach the maximal value, for instance for magnesium (11.7%) and copper (10.5%). These alloying elements provide the greatest solid solution strengthening effect ($\Delta\sigma_b$/1 at.% = 30–40 MPa). The addition of alloys in aluminum may also influence other important properties such as castability. This property, to a very significant extent, will define whether an alloy could be used in industry or not [6–10]. For most engineering parts made from aluminum two types of alloys are used:

- Non-heat-treatable or work-hardening that are solid solution (and eventually strain) hardened, showing a good combination of strength and formability.
- The heat-treatable alloys that obtain their required strength through the heat treatment—(precipitation treatment).

The mechanical properties of Al alloys can also be enhanced by the formation of fine uniformly dispersed in Al matrix particles of the second phase within the original phase matrix. This process is known as precipitation (age) hardening. The fundamental demand for an alloy to be strengthened through this process is that solid solubility decrease with decreasing temperature. The age hardening of Al alloys is accomplished by two individual heat treatments. At the first stage, a material is subjected to a solution heat treatment in which all of the β phases dissolves and forms a single-phase solid solution. Moreover, in a great majority of Al alloys, the diffusion rates are very slow for this reason the solution treatment has to be conducted for relatively long periods. The solution treatment is followed by rapid quenching. At this stage of heat treatment, a non-equilibrium situation exists in which α phase solid solution with some atoms of an alloying element is present. The alloy in this state is weak and ductile. To achieve the strengthening effect, the second type of treatment must be used—aging. At this stage, the supersaturated solid solution is heated to an intermediate temperature, at which diffusion rates become appreciable. The secondary β phase precipitate start to form as fine dispersed particles in a specified precipitation sequence that may consist of the coherent, semicoherent and non-coherent precipitates. The character, coherence and subsequently the strength depend on the temperature and time of artificial aging and a lattice misfit between a strengthening precipitate and Al matrix. The strength enhancement is reached because dislocations interact with these precipitates. Depending on the character of a precipitate and crystallographic orientation in relation with an aluminum matrix, different interactions can occur. Precipitates can be impenetrable or penetrable by dislocations. In the first case, a dislocation is forced by the applied stress bow around the precipitate and bypasses it, leaving a dislocation loop around the particle which is called Orowan loop (**Figure 4**). In the second case, the precipitate can be sheared by the dislocations and moves through the crystal. This phenomenon can occur when the precipitate is coherent with the Al matrix. Generally, coherent particles can be penetrable or not, while large particles are usually incoherent thus impenetrable. For some Al alloys such as Al-Cu, precipitation strengthening can occur spontaneously at room

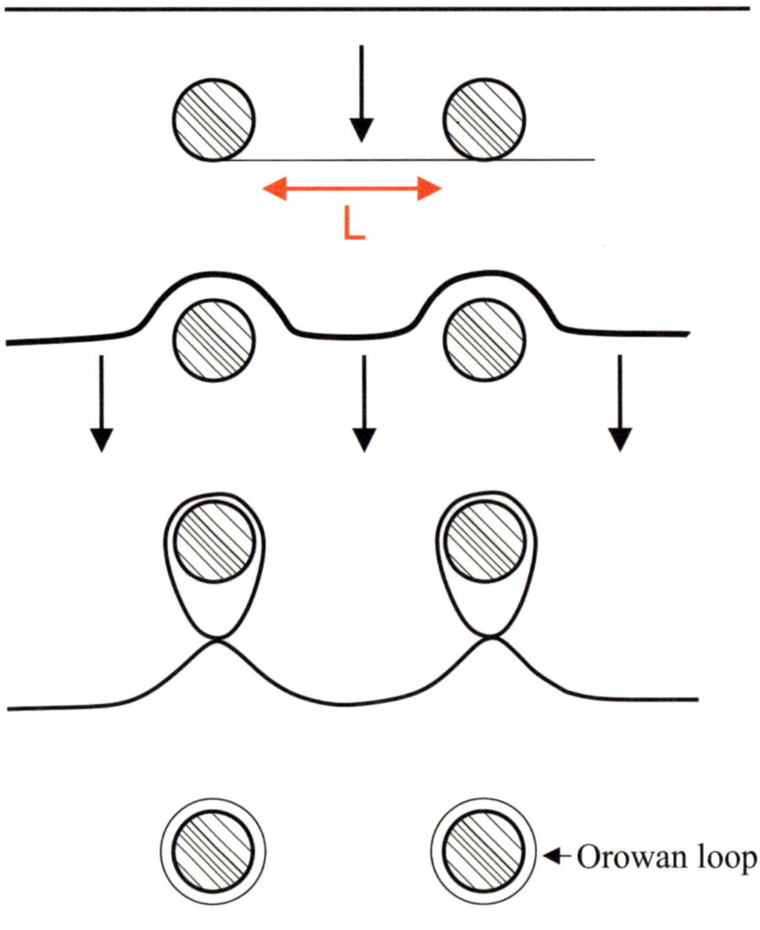

Figure 4. Dislocation bypass by the Orowan bowing mechanism [8].

temperature over long time periods. The growth of mechanical properties during natural aging is continuous or becomes stable. Whereas in artificially aged alloy with an increase in aging time, the strength of the material reaches a maximum value, and after that diminishes. This decrease of mechanical strength is known as overaging and usually can be related to a growth of precipitates and coherency loss [7–13].

Another principal tool used for strength enhancement of many materials is the manipulation of their grain size. When the material is deformed and when the resistance to plastic flow is governed by dislocation glide and diffusion-controlled processes are not an issue, a decrease in the grain size causes the strengthening effect [14–17]. Thus, the mechanical strength is related to the grain size, d, through the Hall-Petch equation which states that the yield stress, σ_y, is given by:

$$\sigma_y = \sigma_0 + k_y d^{-1/2} \quad (1)$$

To transform a coarse-grained microstructure into an ultrafine, it is required to impose an exceptionally high strain into a sample to introduce a high density of dislocations and to allow these dislocations to re-arrange to form an array of grain boundaries—usually the low-angle grain boundaries are transformed into a high-angle grain. The conventional metal-working procedures, such as drawing, extrusion, rolling or forging, are restricted in their ability to produce ultrafine grains for two important reasons. First, there is a limitation on the overall strain that may be imposed using these procedures because the processing techniques incorporate corresponding reductions in the cross-sectional dimensions of the work-pieces. The second reason is that the strains imposed in conventional processing are insufficient to convert coarsely grained structures into this ultra fine-grained (UFG) because most of the industrially used alloys exhibit low workability at ambient and low temperatures. As a consequence of the limitations mentioned above, attention has been paid to develop an alternative metal-working procedure, based on the application of severe plastic deformation (SPD) techniques, where a sample is subjected to the extremely high strains which are imposed at relatively low temperatures without changing cross-sectional dimensions of the samples. Many different SPD techniques are now available and summaries of these various procedures published in works of many scientists are in this several reviews [14–18]. Nevertheless, major emphasis has been placed to date on the two techniques of equal channel angular pressing (ECAP) and high pressure torsion (HPT) and, accordingly, one of these procedures—ECAP will be used in this report.

3. Principles and processing by ECAP

There are numerous studies showing the behavior of aluminum alloys samples subjected to a severe plastic strain [12–18], including metal processing through the standard industrial methods of rolling or extrusion, but all of these processes necessitate a change in the dimensions of the work samples. In contrast to these methods, ECAP processing differs from them. The general principle of ECAP procedure is shown schematically in **Figure 5**.

Processing by ECAP uses a specially designed die consisting of two channels that are bent through a sharp angle near the die center. The sample is usually pre-machined to fit tight the channel, and then is pressed through die using a plunger. The ECAP die is defined by two angles: the channel angle Φ that represents the intersection angle of two parts of the channel, and second is the curvature angle Ψ that represents the angle at the outer arc of curvature where the two parts of the channel intersect. The cross-sectional dimensions of the work sample are not changed during processing thus the process can be repeated to obtain high strain accumulation. Additionally, it is possible to initiate different slip systems by sample rotation between consecutive passes. This processing routes are termed in nomenclature as route A where the sample is pressed repetitively without rotation, routes B_A and B_C where the sample is rotated by 90° along the longitudinal axis in alternate direction and same direction, respectively, and route C where the sample is rotated by 180° between each passes. The equivalent strain imposed in one pass of ECAP is dependent primarily upon the die channel angle Φ and, to a lesser extent, on the angle Ψ. It can be shown from first principles that the shear strain ε_N is given by a relationship of the form [14–20].

$$\varepsilon_N = \frac{N}{\sqrt{3}}\left[2\cot\left(\frac{\Phi}{2}+\frac{\Psi}{2}\right)+\Psi\cos\left(\frac{\Phi}{2}+\frac{\Psi}{2}\right)\right] \tag{2}$$

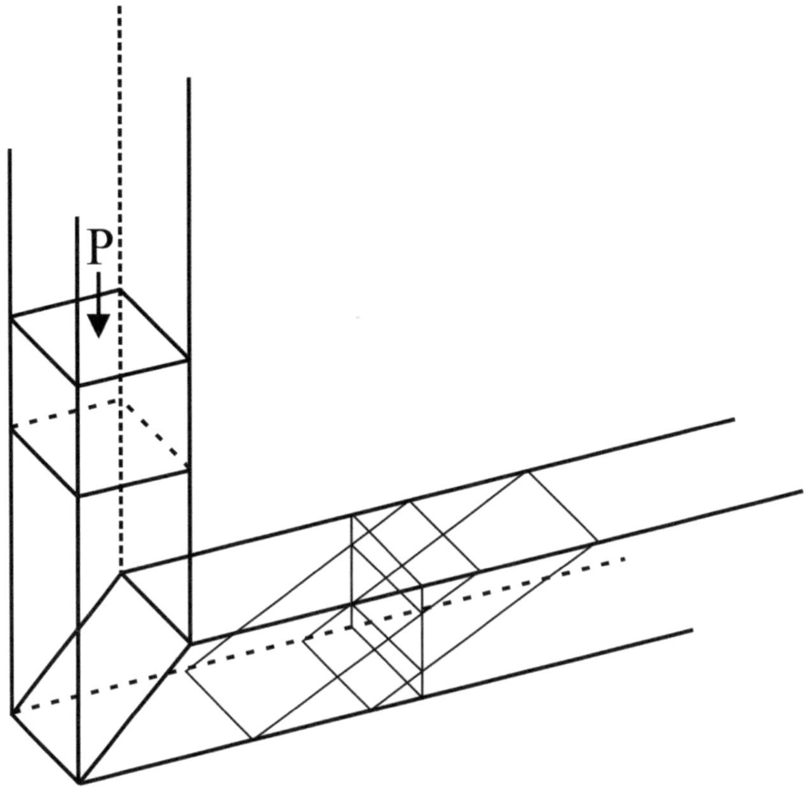

Figure 5. The general principle of ECAP processing [14].

where N is the number of ECAP passes. In conventional ECAP, it is generally assumed that the billet fills the corner of the die at the intersection of the two parts of the channel, and this produces a uniform microstructure throughout the billet.

4. Aluminum-magnesium alloys

Al-Mg alloys are commonly used in transport and for structural components in the automotive industry due to their combination of good properties that are excellent corrosion and high specific strength. As of now, mechanical properties and corrosion resistance of Al-Mg alloys are required to be improved to broaden the application area in a new industrial field. It is well known that the great mechanical properties of Al-Mg alloys come from magnesium solution strengthening. The solubility of magnesium in aluminum is very high ~14.9% at 450°C (**Figure 6**) thus alloying even with small quantity of Mg, results in a significant increase in strength due the solution strengthening [21–26].

With an increase in Mg content, Al-Mg alloys become susceptible to intergranular corrosion thus their application field is limited. This susceptibility to corrosion is due to the formation of anodic β (Al_3Mg_2) phase at grain boundaries. Due the fact that the β phase is strongly anodic

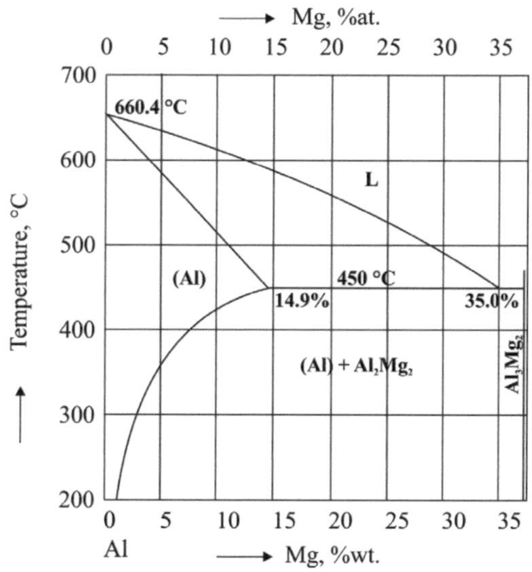

Figure 6. Aluminum-magnesium binary diagram [11].

in relation to the Al matrix, it corrodes and dissolves very fast when alloy is exposed to a conductive medium such as seawater. This is because of the formation of galvanic coupling between the Al matrix and the β phase. The dissolution of continuous grain boundary β phase results in the decrease in the bond strength of grain boundaries, and leads to elements failure. The size, morphology and its distribution were reported to be affected by a heat treatment conditions. It was found that precipitation of β phase at the grain boundaries increases with time of heat treatment at an elevated temperature. The precipitation velocity of β-Al_3Mg_2 phase can be controlled by temperature. Precipitation of β phase can be continuous at 160°C, but at higher temperature of annealing (220°C) becomes discontinuous. The formation of β′ and β phase is associated with the change in mechanical properties during the heat treatment. Recently, more details of the precipitation process in Al-Mg alloys, such as the lattice structure, limited formation temperatures and shape of different precipitates were obtained from many investigation results [21–26].

Based on the data published in the nomenclature, the precipitation sequence [27–31] in Al-Mg alloys can be described as follows:

$$\text{Supersaturated solid solution-GP zone} - \beta'' - \beta' - \beta \text{ phase} \quad (3)$$

where GP (Guinier-Preston) zones (short-range ordered Al_3Mg) have a modulated structure and β″ phase (long-range ordered Al_3Mg, a = 0.408 nm) has an L12 structure in which Al and Mg atoms are alternatively aligned along the [100] directions. β′ phase, Al_3Mg_2, is reported to have a hexagonal (a = 1.002 nm, c = 1.636 nm) structure and semi-coherent with the matrix. The equilibrium β phase, Al_3Mg_2, is of fcc structure (a = 2.824 nm) with a unit cell containing 1186 atoms. However, the precipitation sequence can be slightly changed if aging temperature or Mg content is different.

5. Material and experimental procedure

The chemical composition of the aluminum-magnesium alloy used in this investigation is given in **Table 1**.

Mg	Fe	Si	Cu	Ti	Al
2.86	0.07	0.07	0.01	0.01	Rest

Table 1. Chemical composition of AlMg3 aluminum alloys/wt.%.

The first stage of the experiment was to investigate the aging behavior of the solution-treated samples. The process was conducted in a resistance furnace followed by quenching and artificial aging. To characterize mechanical properties and investigate the aging response, hardness measurements were performed using Rockwell hardness tester ZWICK ZHR 4150. For selected samples, the static tensile tests were also performed. In the second stage, samples were subjected to the ECAP process. Two different dies were used in this investigation. First $\Phi = 120°$, providing an equivalent strain equal to ~0.6 in each pass and second $\Phi = 90°$ with an additional twist angle at the outer channel providing an equivalent strain ~1. To decrease the friction MoS_2 was used as a coating lubricant. For a metallographic study, samples were prepared according to Struers standards. To reveal structure constituents, samples were etched using Keller's, Weck's and Barker's reagent. The microstructure was characterized with an Axio Observer Image Analyzer light microscope under bright field and polarized light. To obtain information about the chemical composition of the precipitates, scanning electron microscope equipped with an energy-dispersive X-ray spectroscopy (EDS) detector was used. The examinations of the thin foils microstructure and phase identification were made on the high-resolution transmission electron microscope JEM 3010UHR from JEOL, at an accelerating voltage of 200 kV.

6. Results

6.1. Structure

The representative structures of the Al-Mg alloy are presented in **Figure 7a–c**. It can be visible that as-cast microstructure can be characterized as fine dendritic. Moreover, in the interdendritic regions, β-Al_3Mg_2 phase exists and are visible insoluble inclusions of the Si- and Fe-rich phases. Precipitation treatment led to the disappearance of the fine dendritic microstructure and has no significant impact on the grain size. An EDS chemical composition microanalysis presented in **Table 2** and **Figure 8** allow one to confirm the presence of the main structure constituents. However, due to the fact that expected size of the strengthening precipitates is >200 nm, the EDS analysis can be overestimated. Therefore, to identify the morphology and crystal structure of precipitates, transmission electron microscopy (TEM) was used (**Figure 9a–e**). It is clear, that during the precipitation treatment process from supersaturated solid solution, the hardening secondary phases β'-Al_3Mg_2 precipitates [32]. This secondary precipitate, of which average diameter was measured to be approximately

Figure 7. Microstructures of AlMg3 alloy (a) as-cast state (Weck's reagent), (b) precipitation-treated state (Barker's reagent)—polarized light, (c) precipitation-treated (Keller's reagent).

Point/phase	Element/line	The average mass concentration of elements (%)	
		Weight (%)	Atomic (%)
1/Mg_2Si	MgK	56.46	59.84
	AlK	5.74	5.48
	SiK	37.81	34.68
2/Al_3Fe	AlK	54.45	71.22
	FeK	45.55	28.78

Table 2. Results of pointwise EDS chemical composition microanalysis.

~100 nm (**Figure 9a**) in the nomenclature is known as metastable with a hexagonal crystal structure (a = 1.002 nm, c = 1.636 nm), semi-coherent to the aluminum matrix. The β′ precipitates are formed mainly through nucleation and growth on the structural defects of the matrix. Moreover, it can be also observed that after precipitation treatment, Al_3Fe and Mg_2Si phases are still present which confirms the presented TEM study. Al_3Fe impurities precipitates are usually formed after annealing at 550°C and can pin down the dislocation motion while at lower temperature Al_6Fe forms. Next, to the small Al_3Fe phase, larger—about 2 μm in length —Mg_2Si precipitate with many dislocations around can be observed (**Figure 9b**).

In this study, Rockwell hardness measurements were used as an initial assessment of the influence heat treatment conditions on the mechanical properties. The experimental results permitted a correlation between hardness, microstructure and heat treatment conditions to be established. The mechanical test results are summarized and listed in **Tables 3** and **4**. Solution treatment temperature was selected to be just below the solidus temperature (based on the Al-Mg binary diagram—**Figure 6**). In this study, the temperature of artificial aging was selected to be 160 and 180°C, respectively. Based on the data listed in **Tables 3** and **4**, it can be concluded that the AlMg3 alloy exhibits a high aging potential. It can be observed that there is a 50% increase in hardness of heat-treated samples. Taking into account the energy costs, the most beneficial conditions seem to be solution treatment and artificial aging for 8 h while the temperature has minor influence on the strength. The observed increase in hardness is a result

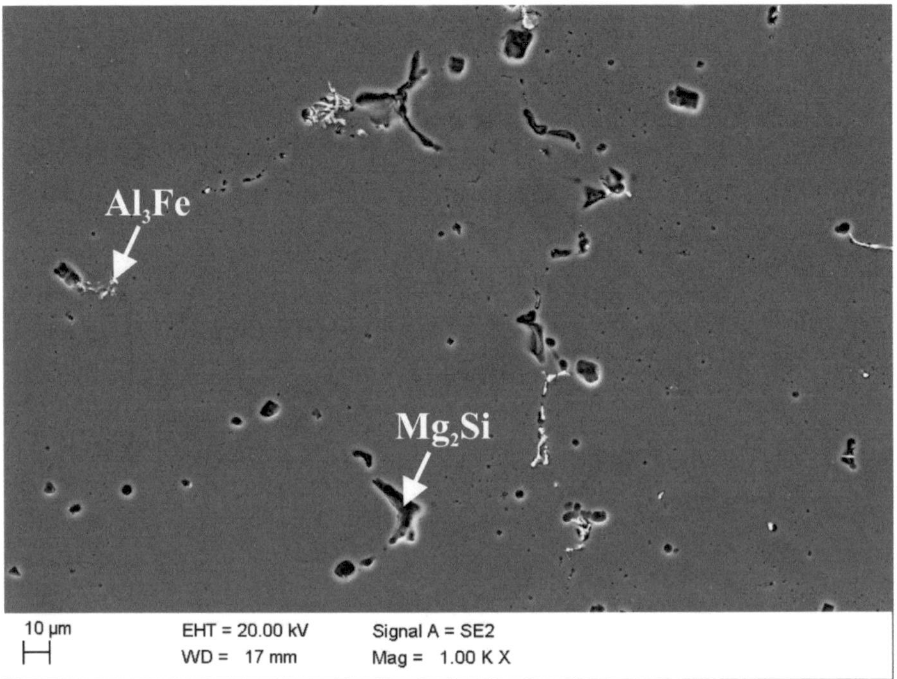

Figure 8. The SEM microstructure of AlMg3 alloy in precipitation-treated state.

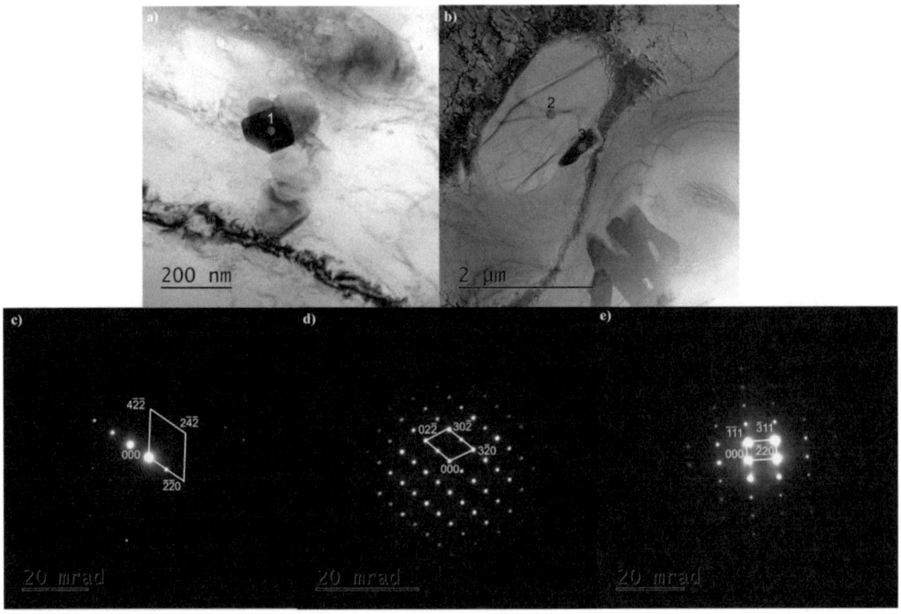

Figure 9. TEM microstructures of the AlMg3 alloy in precipitation-treated state showing size and morpholohgy of (a) β′-Al_3Mg_2 hardenning phase, (b) Mg_2Si and Al_3Fe phases, (c) diffraction spot from point 1 (Al_3Mg_2), (d) diffraction spot from point 2 (Mg_2Si) and (e) diffraction spot from point 3 (Al_3Fe).

Solution treatment conditions		Artificial aging time (h) (160°C)											
Temp. (°C)	Time (h)	0	4	8	12	0	4	8	12	0	4	8	12
		Hardness (HRF)				Tensile strength (MPa)				Elongation at failure (%)			
580	8	45	65	63	66	198	224	232	236	29	30	25	26
	12	46	68	69	67	197	214	225	233	29	29	28	27

Table 3. Summary of the mechanical properties of AlMg3 alloy aged at 160°C.

Solution treatment conditions		Artificial aging time (h) (160°C)											
Temp. (°C)	Time (h)	0	4	8	12	0	4	8	12	0	4	8	12
		Hardness (HRF)				Tensile strength (MPa)				Elongation at failure (%)			
580	8	45	65	63	66	198	222	236	232	29	28	26	23
	12	46	68	69	67	197	234	238	225	29	25	24	23

Table 4. Summary of the mechanical properties of AlMg3 alloy aged at 180 °C.

of the precipitation of the semicoherent secondary $\beta' - Al_3Mg_2$ phase from the supersaturated solid solution sssα [33, 34].

To examine the influence of the selected heat treatment conditions on the tensile properties, static tensile tests were carried out. It can be concluded from the presented data that in both investigated cases an increase in tensile strength is quite similar. **Tables 3** and **4** show the relationship between the tensile strength and heat treatment conditions. The values of the tensile strength slightly increase with aging time. Aging at higher temperature increases the tensile strength more rapidly. Tensile strength reaches the maximum after 12 h of artificial aging (at 160°C) and 8 hours (at 180°C). However, when the sample is aged at 180°C for 12 h, a decrease in ultimate tensile strength is observed—sample is overaged. Time of solution treatment has a minor influence on the mechanical properties. Additionally, in contrast to the tensile strength, the values of elongation at failure decreases with time. When the temperature of artificial aging is lower (160°C), the values of elongation decreases slightly from 29% (solution-treated sample) to about 25%. While aging at 180°C decreases elongation to about 23%. This is due to the fact that higher aging temperature increases the rate of the precipitation process which enhances the mechanical properties but on the other hand decreases the ductility.

6.2. Severe plastic deformation

Figure 10a illustrates the evolution of the microstructure of the Al-3%Mg alloy subjected to six ECAP passes followed by a precipitation treatment. Based on the metallographic analysis, it can be concluded that the individual grains cannot be clearly distinguished. The grains become elongated due to the imposed shear strain. Slip, shear and micro-shear bands create the band-like microstructure. These bands also refine microstructure of the Al-3%Mg alloy. Moreover, it can be observed that deformation bands have a preferred crystallographic orientation. In addition, each single shear or micro-shear band causes lattice rotations lying within the

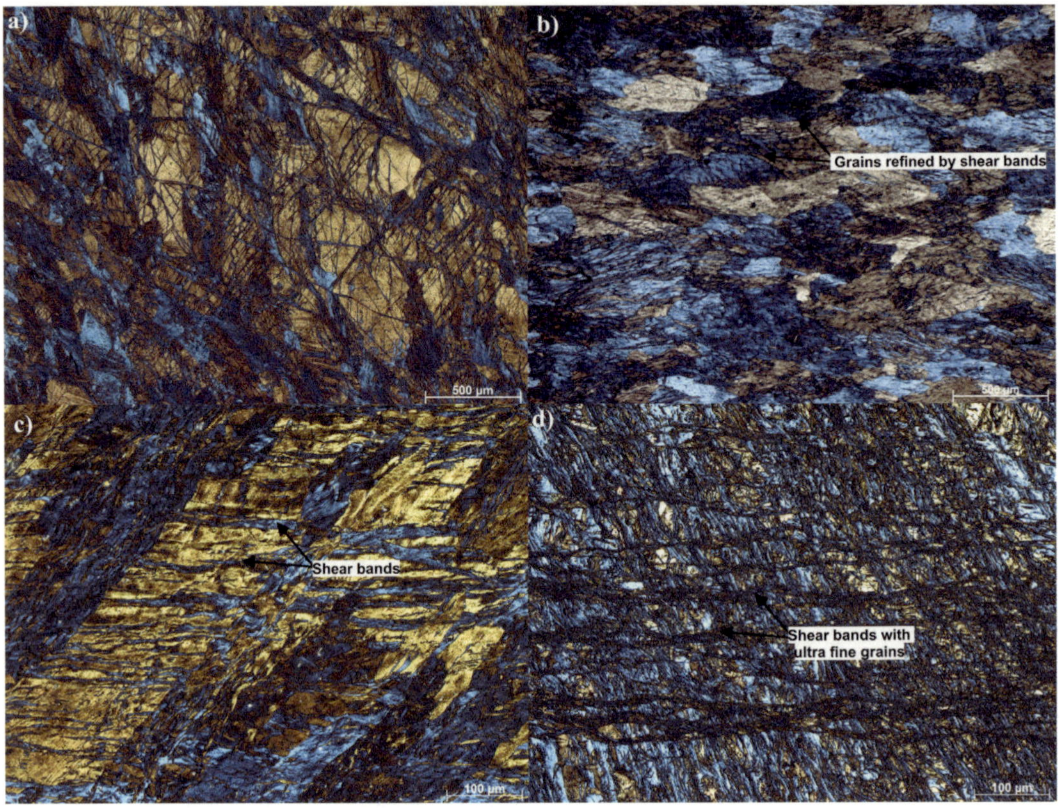

Figure 10. Optical microscope microstructures of the AlMg3 alloy (a) precipitation treated +6 ECAP passes (120° die, route B_C), (b) initial state +1 ECAP pass (90° die with 30° twist, route A), (c) initial state +2 ECAP passes (90° die with 30° twist, route A), (d) initial state +4 ECAP passes (90° die with 30° twist, route A).

low-to-moderate grain/(sub)grain boundary misorientation and thus their interactions can result in a violent appearance of the deformation bands having high-angle misorientations. Moreover, it can also be visible that the microstructure consists of the regions with the higher density of deformation bands and those less affected. This is due to the fact that, during shear deformation in the ECAP process, individual grains in the work samples cannot be deformed uniformly because of the orientation difference. **Figure 10b–d** illustrates the microstructure evolution of AlMg3 aluminum alloy subjected to the ECAP with a modified die. It can be observed that the obtained microstructures differ from that obtained using the conventional equipment. The grains in the microstructure after one ECAP pass are elongated, parallel to the transverse direction (TD), while when the number of ECAP passes increases, the obtained microstructures becomes more complex. It can be seen in **Figure 6c** (two ECAP passes) that additional shear is introduced by the twist angle—shear bands intersect at an angle of about 60° after two ECAP passes (there is an extra shear after each pass at an angle of 30°). When a number of accumulated strain increases to about ~4 (**Figure 10d**)—four ECAP passes, the microstructure looks as completely refined. There is no possibility to distinguish the individual grains. Due to this fact, to study the changes in the microstructure at higher magnification, TEM study was used.

The microstructures of the as deformed (FCC) metals are very complex. The size of individual grains differs thus grains with diversified size may coexist. Different structure constituents, such as dislocation-free grains, non-equilibrium grain boundaries, dislocation cell and (sub) grain structures, low-angle GBs (LAGBs), high-angle grain boundaries (HAGBs), stacking faults (SFs) and nanotwins, can also be identified [15–22, 33–36]. **Figure 11a** shows the typical microstructure of the as deformed material where a dislocation forest forms dense dislocation walls (DDWs). These DDWs forming cell blocks are free from dislocations. The size of cell blocks is about 1 μm. One of the most substantial characteristics of the microstructure after

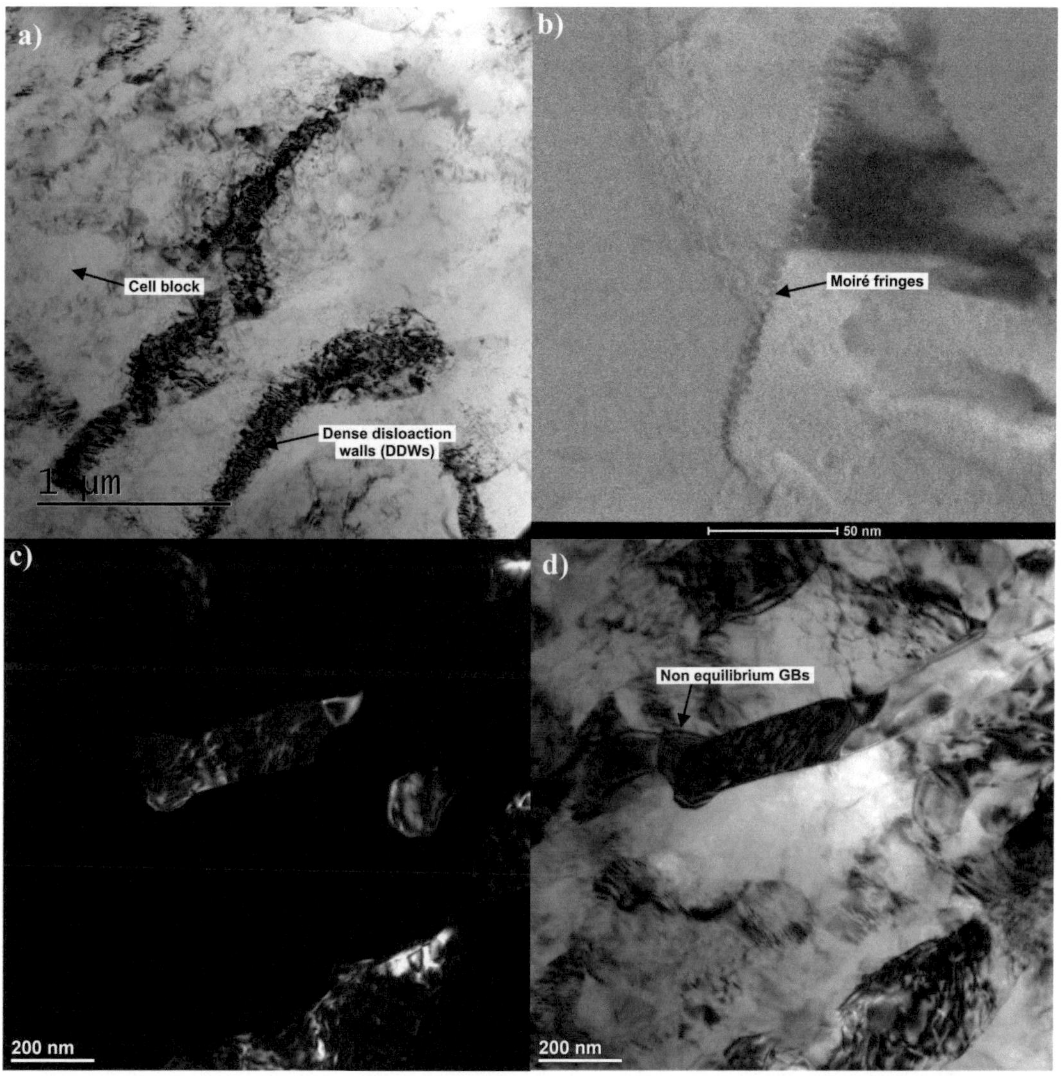

Figure 11. Bright field TEM images showing general microstructure of AlMg3 alloy (a) and (b) bright filed TEM images showing microstructure of AlMg3 alloy—precipitation treated and six ECAP passes, (c) dark field TEM image of AlMg3 alloy—four ECAP passes (90° die with 30° twist, route A), (d) bright field TEM image from the same area.

SPD processing is that some grain boundaries may give the appearance of extinction contours. The presence of diffuse grain boundaries is due the introduction of non-equilibrium grain boundaries having an excess extrinsic dislocations. This indicates the presence of high internal stresses and elastic distortions in the crystal lattice. It has been also suggested that these non-equilibrium GBs may have a strong effect on grain boundary processes such as sliding or diffusion. One can also see dislocation tangles in the grain interiors which can be considered as early stages of dislocation cell structure formation. This is related to an increase of strain accumulation in the material after the consecutive passes of ECAP process. Moreover, the presented TEM investigation results indicate that some grain interiors may also be free from dislocations and the dislocation density varies from crystallite to crystallite. Some grains also have sharp boundaries and are completely free of dislocations and no (sub)grains or dislocation cells can be observed in these grains. Such straight and narrow grain boundaries are believed to be in an equilibrium state and are high-angle grain boundaries (HAGBs). TEM investigation has also revealed the presence of Moiré fringes (**Figure 11b**). Early formation of very low-angle boundaries—typically for $\varphi < 2°$, is usually reported in the nomenclature by an existence of Moiré fringes. This also indicates that lamellar dislocation and cell boundaries continuously form during ECAP process. **Figure 11c** and **d** shows dark field and bright field images of AlMg3 alloy in as-cast state subjected to four passes using a modified die. It is clearly visible that with an increase in the number of ECAP passes, the grain/subgrain size decreases while the dislocation accumulation rise.

To evaluate the influence of the severe plastic deformation process using ECAP method of samples subjected to different strain paths, hardness measurements and tensile properties were determined. Due to the small sample size obtained from modified ECAP equipment, determination of the tensile properties was not possible. The effects of a number of passes and used ECAP die on mechanical properties are summarized in **Table 5**.

It is obvious that the plastic deformation processes result in increase strength, however, processing by SPD allows to impose much greater amounts of plastic strain in comparison to the conventional processes. Based on the data analysis presented in **Table 5**, it can be observed that there is a significant increase in the tensile properties of the 6× ECAPed sample (about 50% in comparison to a precipitation-treated state). This growth of mechanical properties may be

	Hardness ($Hv_{0.3}$)	**Ultimate tensile strength (MPa)**	**Elongation at failure (%)**
As-cast	45	198	31
Precipitation treated	65	232	25
Precipitation treated + 6 ECAP passes (120° die)	126	384	2.4
As-cast + 1 ECAP pass (90° die, 30° twist)	95	—	—
As-cast + 2 ECAP passes (90° die, 30° twist)	107	—	—
As-cast + 3 ECAP passes (90° die, 30° twist)	112	—	—
As-cast + 4 ECAP passes (90° die, 30° twist)	120	—	—

Table 5. Summary of the mechanical properties of AlMg3 alloy in different states.

http://dx.doi.org/10.5772/intechopen.70682

attributed to the accumulation of many excess dislocations generated through dislocation multiplication during ECAP by a volume fraction of the grain boundaries in the fine-grained microstructure. However, it can also be seen that the material after SPD processing loses its ductility rapidly. The processing using a modified ECAP die also results in a significant increase in hardness which grows gradually with an increase in ECAP cycles. This considerable strength increase may be attributed to three strengthening mechanisms, that is, solution strengthening, dislocation strengthening and grain size strengthening. Solution strengthening originates from the elastic distortions, however, the dislocation strengthening and grain refinement strengthening are the main contributors to the significant strength increase of AlMg3 alloy.

7. Conclusions

In this study, we investigate the effect of an ECAP die, its modification and combination of this process with a heat treatment with the aim to estimate the influence of process parameters on the structural evolution and mechanical properties of the AlMg3 aluminum alloy. Based on the analysis of the obtained results, the following conclusions can be stated.

The microstructure of the Al-3%Mg alloy in a precipitation-treated state consists of the α-Al matrix, large precipitates of Mg_2Si, Al_3Fe phases and β'-Al_3Mg_2 fine precipitates with hexagonal structure. The presence of a β'-Al_3Mg_2 phase causes a strengthening effect.

The mechanism of the microstructure refinement after the ECAP process consists of the creation of dislocation cell structures, non-equilibrium grain boundaries, (sub)grain boundaries and HAGBs. (Sub)grains develop from dislocation cells. The refined microstructure consists of the areas with deformation bands (shear bands) separated by HAGBs and non-refined regions with LAGBs and very low-angle boundaries (showing Moiré fringes on TEM). It is believed that the grain boundaries first transform into LAGBs and finally HAGBs with an increase in plastic strain accumulation.

The evolution of hardness of AlMg3 alloy when processed by ECAP with modified die shows a significant increase after the first pass followed by a more gradual increase with subsequent pressings until a saturated value is achieved after ~four ECAP passes.

The improved mechanical properties of the AlMg3 aluminum alloy obtained through the combination of ECAP process with heat treatment were a consequence of the solid solution strengthening, precipitation hardening and grain refinement.

Author details

Tomasz Tański* and Przemysław Snopiński

*Address all correspondence to: tomasz.tanski@polsl.pl

Silesian University of Technology, Gliwice, Poland

References

[1] Bloeck M. 5 - Aluminium sheet for automotive applications, In Advanced Materials in Automotive Engineering, edited by Jason Rowe, Woodhead Publishing, Sawston, Cambridge UK. 2012, P. 85-108, ISBN 9781845695613

[2] De Matteis G, Brando G, Mazzolani FM. Pure aluminium: An innovative material for structural applications in seismic engineering. Construction and Building Materials. 2012; **26**:677-686

[3] Miller WS, Zhuang L, Bottema J, Wittebrood AJ, De Smet P, Haszler A, Vieregge A. Recent development in aluminium alloys for the automotive industry. Materials Science and Engineering A. 2000;**280**:37-49

[4] http://www.world-aluminium.org/statistics/alumina-production/

[5] http://www.totalmateria.com/page.aspx?ID=CheckArticle&site=ktn&NM=137

[6] shby MF, Jones DRH. Engineering Materials 1 - An Introduction to their Properties & Applications. Linacre House, UK/Woburn, MA, USA: Butterworth-Heinemann An imprint of Elsevier Science; 1996. p. 1-322

[7] Zolotorevsky VS, Belov NA, Glazoff MV. Chapter two—Structure and microstructure of aluminum alloys in As-cast state. In: Casting Aluminum Alloys. Amsterdam: Elsevier; 2007. p. 95-182 ISBN 9780080453705

[8] Callister WD. Materials science and engineering: An introduction (2nd edition). Materials and Design. 1991;**12**:59

[9] Votano J, Parham M, Hall L. Hanbook of Aluminum: Volume 2: Alloy Production and Material Manufacturing; John Wiley & Sons, Inc., United States. 2004. p. 1-731

[10] Baldwin W. Metallography and Microstructures. USA: ASM International; 2004. p. 2733

[11] Mackenzie DS. Handbook of Aluminum; Marcel Dekker, Inc., New York. 2003. DOI: 10.1201/9780203912607

[12] Zolotorevsky VS, Belov NA, Glazoff MV. Structure and Microstructure of Aluminum Alloys in As-Cast State, Amsterdam, The Netherlands: Elsevier, 2007

[13] Langdon TG. Twenty-five years of ultrafine-grained materials: Achieving exceptional properties through grain refinement. Acta Materialia. 2013;**61**:7035-7059

[14] Fukuda Y, Oh-ishi K, Furukawa M, Horita Z, Langdon TG. The application of equal-channel angular pressing to an aluminum single crystal. Acta Materialia. 2004;**52**:1387-1395

[15] Valiev RZ, Estrin Y, Horita Z, Langdon TG, Zehetbauer MJ, Zhu Y. Producing bulk ultrafine-grained materials by severe plastic deformation: Ten years later. JOM. 2016; **68**:1216-1226. DOI: 10.1007/s11837-016-1820-6

[16] Furukawa M, Horita Z, Nemoto M, Langdon TG. Review: Processing of metals by equal-channel angular pressing. Journal of Materials Science. 2001;**36**:2835-2843

[17] Liu MP, Roven HJ, Murashkin MY, Valiev RZ, Kilmametov A, Zhang Z, Yu Y. Structure and mechanical properties of nanostructured Al-Mg alloys processed by severe plastic deformation. Journal of Materials Science. 2013;**48**:4681-4688. DOI: 10.1007/s10853-012-7133-4

[18] Furukawa M, Horita Z, Nemoto M, Langdon TG. The use of severe plastic deformation for microstructural control. Materials Science and Engineering A. 2002;**324**:82-89

[19] Sabirov I, Murashkin MY, Valiev RZ. Nanostructured aluminium alloys produced by severe plastic deformation: New horizons in development. Materials Science and Engineering A. 2013;**560**:1-24

[20] Kawasaki M, Ahn B, Kumar P, Il Jang J, Langdon TG. Nano- and micro-mechanical properties of ultrafine-grained materials processed by severe plastic deformation techniques. Advanced Engineering Materials. 2017;**19**:1-17

[21] Lee BH, Kim SH, Park JH, Kim HW, Lee JC. Role of Mg in simultaneously improving the strength and ductility of Al-Mg alloys. Materials Science and Engineering A. 2016;**657**: 115-122

[22] Vinogradov A, Nagasaki S, Patlan V, Kitagawa K, Kawazoe M. Fatigue properties of 5056 Al-Mg alloy produced by equal-channel angular pressing. Nanostructured Materials. 1999;**11**:925-934

[23] Meng C, Zhang D, Cui H, Zhuang L, Zhang J. Mechanical properties, intergranular corrosion behavior and microstructure of Zn modified Al-Mg alloys. Journal of Alloys and Compounds. 2014;**617**:925-932

[24] Nakayama K, Tsuruta H, Koyama Y. Formation of giant atomic clusters in the β-Samson (β-Al_3Mg_2) phase of the Al-Mg alloy system. Acta Materialia. 2017;**128**:249-257

[25] Sato T, Kojima Y, Takahashi T. Modulated structures and GP zones in Al-Mg alloys. Metallurgical Transactions A. 1982;**13**:1373-1378

[26] Bouchear M, Hamana D, Laoui T. GP zones and precipitatemorphology in aged Al-Mg alloys. Philosophical Magazine A. 1996;**73**:1733-1740

[27] Starink MJ, Zahra AM. β′ and β precipitation in an Al-Mg alloy studied by DSC and TEM. Acta Materialia. 1998;**46**:3381-3397

[28] Starink MJ, Zahra A. The kinetics of isothermal β precipitation in Al-Mg alloys. Journal of Materials Science. 1999;**34**:1117-1127

[29] Hamana D, Bouchear M, Betrouche M, Derafa A, Rokhmanov NY. Comparative study of formation and transformation of transition phases in Al-12 wt.%Mg alloy. Journal of Alloys and Compounds. 2001;**320**:93-102

[30] Goswami R, Spanos G, Pao PS, Holtz RL. Precipitation behavior of the β phase in Al-5083. Materials Science and Engineering A. 2010;**527**:1089-1095

[31] Yi G, Littrell KC, Poplawsky JD, Cullen DA, Sundberg E, Free ML. Characterization of the effects of different tempers and aging temperatures on the precipitation behavior of Al-Mg (5.25 at.%)-Mn alloys. Materials and Design. 2016;**118** Submitted

[32] Król M, Tański T, Snopiński P, Tomiczek B. Structure and properties of aluminium–magnesium casting alloys after heat treatment. Journal of Thermal Analysis and Calorimetry. 2017;**127**:299-308. DOI: 10.1007/s10973-016-5845-4

[33] Tański T, Snopiński P, Borek W. Strength and structure of AlMg3 alloy after ECAP and post-ECAP processing Materials and Manufacturing Processes. 2017;**32**(12)

[34] Tański T, Snopiński P, Borek W. Strength and structure of AlMg3 alloy after ECAP and post-ECAP processing. Materials and Manufacturing Processes. n.d.:1-7

[35] Tański T, Snopiński P, Pakieła W. Structure and properties of ultra fine grained aluminium alloys after laser surface treatment. Materialwissenschaft und Werkstofftechnik. 2016;**47**:419-427. DOI: 10.1002/mawe.201600517

[36] Tański T, Snopiński P, Hilser O. Microstructure and mechanical properties of two binary Al-Mg alloys deformed using equal channel angular pressing. Materialwissenschaft und Werkstofftechnik. 2017;**48**:439-446. DOI: 10.1002/mawe.201700020

Chapter 9

Gluconates as Corrosion Inhibitor of Aluminum in Various Corrosive Media

Omotayo Sanni and Abimbola Patricia Popoola

Additional information is available at the end of the chapter

http://dx.doi.org/10.5772/intechopen.71467

Abstract

Corrosion processes are responsible for huge losses in industry. Though organic, inorganic and mixed material inhibitors were used for a long time to combat corrosion, the environmental toxicity of organic corrosion inhibitors has prompted the search for inorganic corrosion inhibitors. The effect of gluconates as novel corrosion inhibitors on the corrosion of aluminum alloy in acidic and saline media was investigated by electrochemical and weight loss techniques. The effect of inhibitor concentration was also investigated. High resolution scanning electron microscopy equipped with energy dispersive spectroscopy (HR-SEM/EDS) was used to characterize the surface morphology of the metal before and after corrosion. Experimental results revealed that gluconates in the studied solution decreased the corrosion rate at the different concentrations studied. The experimental results obtained from potentiodynamic polarization method showed that the presence of the gluconates in 3.5% NaCl and 0.5 M H_2SO_4 solutions decreases the corrosion current densities (icorr) and corrosion rates (CR), and increases the polarization resistance (Rp). It was observed that the inhibitor efficiency depends on the corrosive media, concentration of the inhibitor and the substrate material. The adsorption characteristics of the gluconates were also described. Good correlation exists between the results obtained from both methods.

Keywords: aluminum alloy, corrosion, inhibition, gluconates, potentiodynamic polarization, weight loss, SEM/EDS

1. Introduction

1.1. Aluminum alloy

Aluminum compounds occur in aluminum types of clay, but the most useful ore for producing pure aluminum is bauxite. It consists of 45.60% aluminum oxide, along with various impurities such as iron, sand and other metals [1]. Aluminum is durable, relatively soft, light weight, malleable and ductile metal with physical appearance ranging from silvery to dull gray depending on the roughness of the surface. Aluminum has almost one – third of the stiffness and density of steel. It is easily drawn, machined, cast and extruded. Aluminum is also a good electrical and thermal conductor. Aluminum and its alloys are very important material because of their wide use in many industries. They are used in the aviation, aerospace and automotive industries, in the production of military hardware, in ship building and household appliances [2–4]. Their usefulness is derived from their very good mechanical and physical properties such as their weight-to-high strength ratio, good machining properties, recyclability, and their outstanding corrosion resistance. The corrosion resistance of aluminum and its alloys is attributed to the formation of a stable protective thin film of aluminum oxide when exposed to the atmosphere. Aluminum alloys are therefore used widely as materials for electricity cables, bottle tops, food and beverage containers, and roofing sheets [5]. Despite the huge benefit of aluminum when compared to other metals, it is not always resistant to corrosion completely. When the environment is highly basic or acidic, the protective layer breakdown can occur and its instant renewal may not be fast enough to prevent corrosion [6–9]. Corrosion of aluminum leads to passivation of active cathode material, the electrical resistance of the solid product is increased and the electrolyte is contaminated by its soluble products and increases the self-discharge rate. In efforts to reduce aluminum corrosion, the main approach is to separate the aluminum metal from corrosive environments and this can be accomplished by means of corrosion inhibitors. Therefore, the ability of aluminum to resist corrosion attack in various environments is inadequate and a continuous search for methods for controlling this behavior remains important in the field of corrosion control.

1.2. Corrosion of metals

The corrosion of metals is a serious problem in most industries. Corrosion phenomena, control and prevention are unavoidable major scientific issues that must be addressed daily as far as there are increasing needs of metallic materials in all facets of technological development [10]. Technological progresses require the use of materials in aggressive environments, such as corrosive atmospheres [11]. Corrosion is a degradation of metal due to environmental reactions dissolution; it is an electrochemical process and usually begins at the surface [12, 13]. As a result of daily encounters with corrosion problems such as; waste of valuable resources, costly maintenance, plant shutdowns, expensive overdesign, loss or contamination of products and reduction in efficiency; technological developments are impeded and safety endangered [14, 15]. The multifacet aspect of corrosion problems coupled with responsibilities associated with them only increase the complexity of the subject [16, 17]. This problem has caused the stimulation of many corrosion control research in various oil production facilities around the world [18]. The cost due to corrosion does not only increase rapidly, the quality of the products and the efficiency of the plants is also reduced. In the United States this amount to over 276 billion USD per year. The highest segments of the cost of corrosion are associated with utilities, transportation and infrastructure. Millions of dollars are lost each year because of corrosion including the cost of prevention and maintenance. NACE studies recently estimates global cost of corrosion at $2.5 trillion annually at CORROSION 2016 conference held at Vancouver, B.C., which examined the economics of corrosion and the role of corrosion management in establishing industry best practices [19]. Corrosion can be localized by forming a crack or pit, or expand across a wide area by uniformly corroding the surface [20]. Pitting corrosion takes place in materials with surface films protection like corrosion products, the metal exposed releases electrons easily and tiny pits is initiated by the reactions [21]. Galvanic corrosion usually occurs on the surface of metal which have more than one phase, in other for local cells to be set up on heterogeneous surface with different potentials. The preferential attack of grain boundaries without substantial attack of the grains themselves is called intergranular corrosion; it is often caused by the precipitation of fine intermetallics at grain boundaries [22]. Localized corrosion may possibly result in structure failure of metallic materials if it is allowed to reach critical levels [23]. Corrosion resistance of alloys has become a continuous research due to its severe consequences [24]. One of the most useful techniques in protecting and controlling metals against corrosion is the use of inhibitors, especially in saline and acidic media.

2. Corrosion control and monitoring

The practice carried out to assess and predict the corrosion behavior in equipment and operational plant is known as corrosion monitoring. The objectives of corrosion monitoring are:

- To provide information based on the operational state of the equipment with the intent of avoiding unplanned shut-downs occurring from unforeseen deterioration of the plant.
- To provide information on the operating variables and interrelation between corrosion processes to allow more efficient use of the plant.
- To provide information that can be used to prevent potential disasters and safety failures by plant inspection departments.
- To assess process fluids contamination levels.

The principle behind corrosion protection is to influence the reactions which occur during corrosion, it being possible to control the corrosive agent and the package contents, and also the reaction itself in such a manner that corrosion is minimized. Corrosion protection is subdivided into three subtopics: passive, permanent and temporary corrosion protection.

2.1. Passive corrosion protection

Damage is prevented in passive corrosion protection by mechanically isolating the package contents from the aggressive corrosive agents, for example by using films or other coatings and protective layers. This type of corrosion protection changes neither the aggressiveness of the corrosive agent nor the general ability of the package contents to corrode; this is why this approach is known as passive corrosion protection. Corrosion may occur within a very short time if the protective film or layer is destroyed at any point in time.

2.2. Anti-corrosion coating protection

The main purpose of permanent corrosion protection methods is to provide protection at the place of use. Machines are usually located in factory sheds and are protected from extreme variations in temperature, which frequently causes condensation. The stresses presented by biotic, chemical and climatic factors are relatively slight in this situation. Examples of permanent corrosion protection methods are: galvanization, tin plating, copper plating, coating and enameling.

2.3. Temporary corrosion protection

The stresses occurring during handling, storage and transport are greater than those that occur at the place of use. Stresses manifested are; extreme variations in temperature, which result in a risk of condensation. In maritime transport, the elevated salt content of the air and water in the sea salt aerosols may cause damage, due to the strongly corrosion-promoting action of salts. The main temporary corrosion protection methods are as follows: desiccant method, protective coating method and corrosion Inhibitor methods.

Corrosion scientists are relentless in search of more and better efficient and cheap ways of mitigating the corrosion of metals and alloys. One of the ways of combating corrosion is the addition of inhibitors to the corroding environment. The various types of inhibitors can be classified as organic and inorganic. Corrosion inhibitors are used widely in industries to reduce or prevent corrosion rate of metals in acidic, alkaline media, and industrial processes such as acid descaling, acid pickling and cleaning of refinery equipment and oil well acidizing.

3. Corrosion inhibitors

Certain substances which, when added to a corrosive system, modify the surface of the material to reduce the corrosion rate, are called inhibitors [25]. The use of chemical inhibitors to mitigate the rate of corrosion is quite varied. In the processing industries and oil extraction, inhibitors have always been considered to be one of the first ways against corrosion. Corrosion inhibitor performance has historically been described by two concepts namely: corrosion inhibitor availability and corrosion inhibitor efficiency. This essentially examines the ratio of the uninhibited and inhibited corrosion rates, and expresses this as a percentage. An inhibitor that reduced corrosion rates by a factor of 10 would be 90% efficient. This leads to the development of the concept of "availability of corrosion inhibitor." In this concept, performance is determined based on the summation of total metal loss over a period of time, assuming periods of uninhibited corrosion and inhibited corrosion. Corrosion availability concept is currently replacing the inhibitor efficiency concept with respect to use in corrosion inhibition concepts for systems design [26]. A great number of scientific researchers have responded to this need and it has generated increased research studies into the use of inhibitors. Compounds containing nitrogen as corrosion inhibitors have been investigated thoroughly by many authors; imidazole and benzimidazole derivatives [27–29], bipyrazole [30], quaternary ammonium salts [31], substituted aniline-N-salicylidenes [32], cationic surfactants [33], phosphorous containing inhibitors [34, 35]. The research on corrosion inhibition of aluminum alloy is ongoing due to this fact our research work focuses on the effectiveness of gluconates as corrosion inhibitors for aluminum alloy in 0.5 M H_2SO_4 and 3.5% NaCl solutions. Attempt at making novel contribution to this growing research area has necessitated the present investigation. Weight loss and potentiodynamic polarization techniques were used to examine the rate of deterioration. The effect of zinc gluconate and ferrous gluconate on the aluminum alloy surface in the test media was studied using high resolution scanning electron microscope equipped with energy dispersive spectroscopy (HR-SEM/EDS). Studies on surface characterization, adsorption mechanisms and electrochemical analysis of corroded materials were also carried out.

3.1. Industrial applications of corrosion inhibitors

3.1.1. Petroleum production

Petrochemical and refineries industries use different film-forming inhibitors to control wet corrosion. Water-soluble and water-soluble-oil-dispersible type inhibitors are continuously injected or oil-soluble and oil-soluble-water-dispersible type inhibitors are intermittently applied to control corrosion.

3.1.2. Internal corrosion of steel pipelines

Gathering pipelines, operating between oil and gas wells and processing plants, have corrosion problems similar to those in refineries and petrochemical plants. The flow regimes of multiphase fluids in pipelines influence the corrosion rate. At high flow rates, flow-induced corrosion and erosion-corrosion may occur, whereas at low flow rates, pitting corrosion is more common. Internal corrosion of pipelines is controlled by cleaning the pipeline and by adding continuous and/or batch inhibitors.

3.1.3. Water

Potable water is frequently saturated with dissolved oxygen and is corrosive unless a protective film, or deposit, is formed. Cathodic inhibitors, such as calcium carbonate, silicates, polyphosphates, and zinc salts, are used to control potable water corrosion. Water is used in the cooling system in many industries.

3.1.4. Acids

Acids are broadly used in pickling, cleaning of oil refinery equipment, heat exchangers and in oil well acidizing. Mixed inhibitors are widely used to control acid corrosion.

3.1.5. Automobile

Inhibitors are used in an automobile for two reasons which are; to reduce the corrosivity of fluid systems (internal corrosion), and to protect the metal surfaces exposed to the atmosphere (external corrosion). Internal corrosion is influenced by the coolants, flow, aeration, temperature, pressure, water impurities, corrosion products, operating conditions and maintenance of the system.

3.1.6. Paints

Finely divided inhibiting pigments are frequently incorporated in primers. These polar compounds displace water and orient themselves in such a way that the hydrophobic ends face the environment, thereby augmenting the bonding of the coatings on the surface.

4. Corrosion prevention of aluminum alloy

Gluconic acids are environmentally suitable non-toxic compounds. It is abundant in plants, fruits and other foodstuffs. Gluconic acid and its derivatives are used in the formulation of pharmaceuticals, foods, and cosmetics as mineral supplements to prevent the deficiency and as buffer salts. They are also used as ingredients in various hygienic products. In industrial applications, they are used for scale removal in metal cleanings, industrial and household cleaning compounds including mouth washer, metal finishing, water treatments, and as paper and textile auxiliaries. For these reasons, it is very useful to study the corrosion behavior of aluminum alloy in acidic and saline media using gluconates as inhibitor. Gluconate is a large hydrocarbon oxyanion [36], aluminum would be expected to anodize in the presence of this kind of ion. The aim of this research was to evaluate the effect of ferrous gluconate and zinc gluconate concentration on corrosion of aluminum alloy in different environments and to estimate optimal concentrations of ferrous gluconate and zinc gluconate required for satisfactory inhibition. Diverse researches carried out by different authors regarding the improvement of aluminum alloy using different inhibitors in various media have been attempted by different authors. Popoola et al. [37] confirm that in aggressive environment, the protective film on the surface of aluminum usually breakdown leading to initiation and further propagation of pits. The authors studied the corrosion inhibition of aluminum alloy in 2 M HCl and HNO_3 with *Arachis hypogaea* natural oil as an inhibitor using potentiodynamic polarization and gravimetric techniques at 25°C. The obtained results showed that *Arachis hypogaea* oil in 2 M HCl and HNO_3-solutions decreased the corrosion rate of aluminum alloy at different concentrations of inhibitor considered. Umoren and Solomon [38] researched on the influence of bromide and iodide ions on the inhibitive effect of polyacrylamide (PA) on aluminum corrosion in HCl solution show that the halide additives synergistically increased the inhibition efficiency of polyacrylamide and the increase in inhibition efficiency was found to be more obvious in iodide than bromide ions. Obot and Obi-Egbedi [39] confirmed fluconazole (FLC) as a corrosion inhibitor for aluminum in hydrochloric acid solution. Garcia et al. [40] worked on the effect of solution pH on the anti-corrosion performance of different inhibitors $K_2Cr_2O_7$, $CeCl_3$, and $Ce(dbp)_3$ on aluminum alloy (AA2024-T3) using the multi electrode and conventional potentiodynamic polarization methods. The results showed that the $K_2Cr_2O_7$ at 10^{-4} M after 30 min of exposure time behaves as the best inhibitor across the studied pH range. The efficiency of the $Ce(dbp)_3$ and $CeCl_3$ at 10^{-4} M was highly affected by the pH of the solution, where both cerium-based inhibitors gave similar protection to the K_2Cr_2. Onen et al. [5] analyzed the use of titanium (iv) oxide (TiO_2) as an inhibitor for the corrosion of aluminum in H_2SO_4 solution. Various researchers have come to the same conclusion that some compounds that exist naturally can be successfully used as corrosion inhibitors. Avwiri and Igho [41] studied the effect of *Vernonia amygdalina* on the acidic corrosion of 3RS and 2S alloys using weight loss test. The results showed that the corrosion rate of the aluminum coupons in the HNO_3 and HCl solutions decreased with an increase in the exposure time due to passivation of the surface by Al_2O_3 and $AlCl_3$ respectively. The authors concluded that the studied inhibitor is an excellent inhibitor in HNO_3 and HCl solution at 0.2 and 0.4 g/l concentration at 29°C. Umoren et al. [42] studied the corrosion inhibition of aluminum in hydrochloric acid medium using exudate gum from *Raphia hookeri* as inhibitor at 30–60°C by thermometric and weight loss methods. The experimental results showed that exudate gum is a good inhibitor of aluminum in hydrochloric acid medium and the efficiency of the inhibitor increases with an increase in concentration of the inhibitor but decreases with a rise in temperature. The authors concluded that *Raphia hookeri* performs as an inhibitor for aluminum corrosion in 0.1 M hydrochloric acid medium. Umoren and Solomon [38] investigated the influence of potassium iodide additives on the inhibition efficiency of *Aningeria robusta* extract for aluminum in 2 M hydrochloric medium using hydrogen evolution method at 30 and 60°C. Comparison between the corrosion rates data with and without additives was used to determine the inhibition efficiency by the authors. The result of the investigation showed that the *Aningeria robusta* extract is an inhibitor for acid corrosion of aluminum. *Lavandula angustifolia* was used as an inhibitor by [43]. The authors investigated the use of natural oil extracted from *Lavandula angustifolia* as an inhibitor for aluminum alloy in 3% sodium chloride medium using polarization measurements and weight loss techniques. The results showed that *L. angustifolia* oil acts as an effective inhibitor for the Al-3 Mg alloy corrosion in 3% NaCl medium at the temperature range studied. It was concluded that *Lavandula angustifolia* oil can be used to prevent pitting corrosion of aluminum. The combination of *Arachis hypogaea* natural oil was successfully used in the work of [37] as an inhibitor 2 M HNO_3 and 2 M HCl acids solutions. Bark extract have been successfully studied as corrosion inhibitor for aluminum in 1 M NaOH by [44]. Comparative study of different plant extract has been carried out by [45]. In order to contribute to the ongoing research, the current research work is focused on analyzing the effectiveness of novel corrosion inhibitors ferrous gluconate (FG) and zinc gluconate (ZG) for the corrosion of aluminum alloy in acidic (0.5 M H_2SO_4) and saline (3.5% NaCl) media.

5. Gluconate as corrosion inhibition for aluminum alloy

5.1. Behavior of aluminum alloy substrate in sodium chloride solution

The potentiodynamic polarization behavior of the received aluminum sample in saline solution is shown in **Figure 1**, where the polarization curve can be seen and the data obtained can be seen in **Table 1**.

In the sodium chloride solution, the aluminum sample gives a corrosion potential of −0.75893 V. The active corrosion path was formed as a result of the dissolution of the aluminum matrix. The surface of the tested aluminum coupon at different areas shows different dark shades, and this was examined by SEM/EDS. It is most likely that the darker areas are where the aluminum has not suffered severe corrosion from the sodium chloride, at those

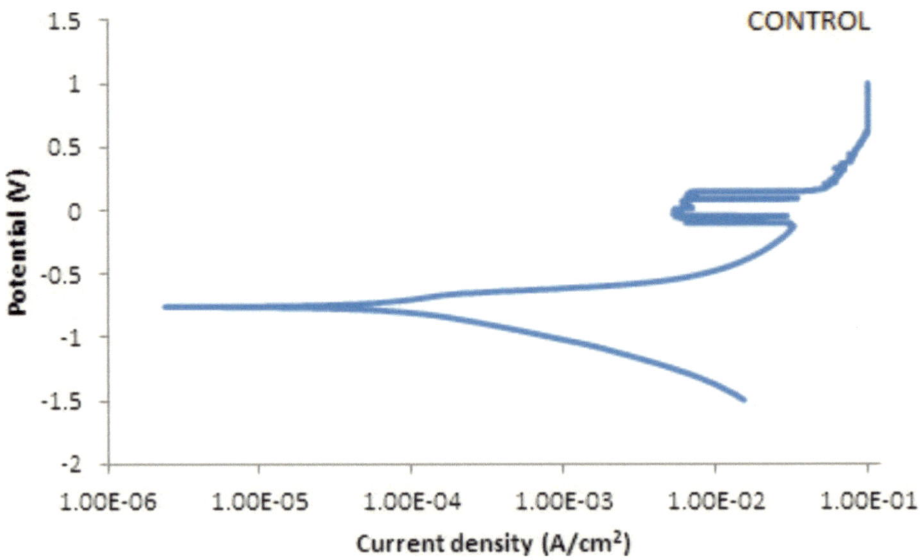

Figure 1. Linear polarization curve for aluminum alloy in sodium chloride solution [46].

areas surface oxides layers have been formed that impeded corrosion. At the whitish areas aluminum dissolution must have occurred the chloride ion has penetrated the matrix of the aluminum surface. This was confirmed by the EDS analysis on the different areas which revealed that the darker areas had a high Al peak while the whitish areas consist of O, Na, S, Fe and C. In the border between the whitish and the darker areas there was a band of hole which seems to be associated to the dissolution of the Al alloy as a result of the aggressiveness of the sodium chloride. This hole did not exist on the as received sample [46] (**Figure 2**).

The result show the adsorption of the oxide film, formation of basic hydrochloric aluminum salt which separates from the lattice and goes into the system:

$$Al^{3+} + 4\,Cl^{-} \rightarrow AlCl_4^{-} \tag{1}$$

$$Al^{3+} + 2\,Cl^{-} + 2OH \rightarrow Al(OH)_2Cl \tag{2}$$

The oxide film is thinned to the extent that aluminum ion can pass from the metal to the sodium chloride interface. The chloride ion gets entry by penetrating through the oxide film or diffusion of halide ion through the oxide film and attack the alloy.

S/N	C (% g/v)	Icorr (A/cm²)	βc (V/dec)	βa (V/dec)	LPR Rp (Ωcm²)	–Ecorr (V)	CR (mm/yr)
1	0.0	8.59E-06	0.050867	0.080038	1572.3	0.75893	0.277870

Table 1. Polarization data for aluminum alloy in sodium chloride solution.

http://dx.doi.org/10.5772/intechopen.71467

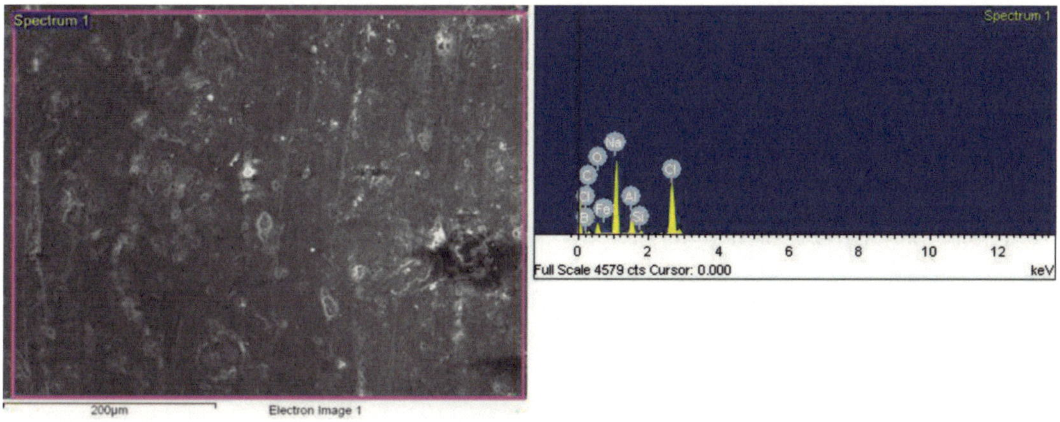

Figure 2. SEM/EDS spectra of corroded aluminum alloy in saline solution [46].

5.2. Behavior of aluminum alloy substrate in H_2SO_4 acid medium

The potentiodynamic polarization behavior of the as-received aluminum alloy sample in acid medium is shown in **Figure 3** where the polarization curve can be seen, the data obtained can be seen in **Table 2**.

The linear polarization curve determines the active or passive characterization of aluminum alloy in acid solution. In the acid solution, the aluminum sample gives a corrosion potential of −0.55960 V. The active corrosion path was formed as a result of the dissolution of the aluminum matrix.

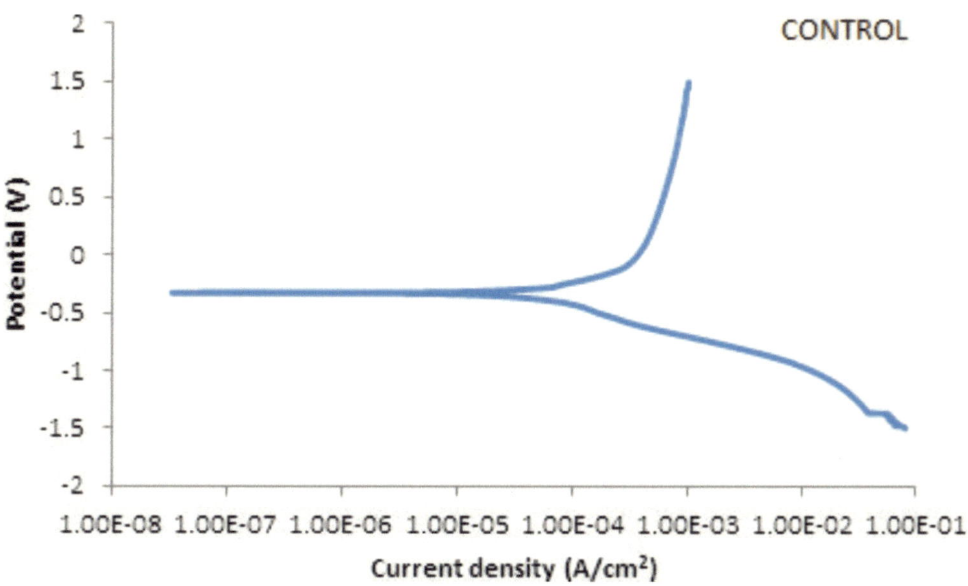

Figure 3. Tafel polarization curve for aluminum alloy in H_2SO_4 acid solution [46].

S/N	C (% g/v)	Icorr (A/cm^2)	βc (V/dec)	βa (V/dec)	LPR Rp (Ωcm^2)	–Ecorr (V)	CR (mm/yr)
1	0.0	1.73E-05	0.025551	0.12493	532.51	0.33054	0.55960

Table 2. Polarization data for aluminum alloy in 0.5 M H_2SO_4 solution.

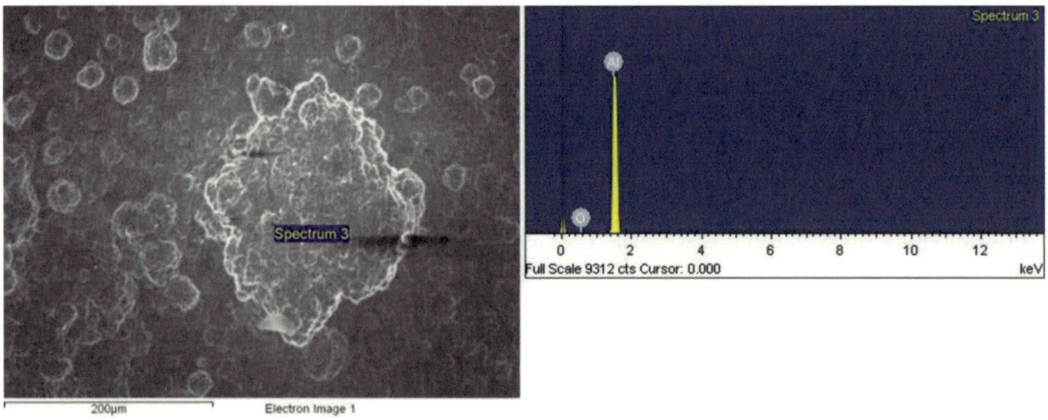

Figure 4. SEM/EDS spectra of corroded aluminum alloy in H_2SO_4 acid solution [46].

A morphological study of the surface of aluminum specimen in acid solution was carried out by SEM after immersion in the test solution for 28 days at 28°C [46]. **Figure 4** shows the SEM images of the aluminum surface after 28 days immersion in 0.5 M H_2SO_4 acid. A severely corroded surface was observed after immersion in the uninhibited system, due to corrosive attack of the acid solution. The corrosion product layer on the metal surface in uninhibited 0.5 M H_2SO_4 acid and the corrosion damage is clearly visible on the metal surface. The oxide film is thinned to the extent that aluminum ions can pass from the aluminum alloy to the solution interface.

$$Al_2O_3 + 3H_2SO_4 \rightarrow Al_2(SO_4)_3 + 3H_2O \quad (3)$$

The sulfuric ion gets entry by penetration through the oxide film and attack the metal.

6. Characterization of inhibited aluminum alloy in saline (3.5% NaCl) medium

The obtained results from weight loss and potentiodynamic polarization methods at different concentrations of ferrous gluconate (FG), zinc gluconate (ZG) and the synergistic effect of the two inhibitors in saline solution at 28°C are represented in Figures below.

6.1. Aluminum alloy in saline medium with ferrous gluconate (FG) as inhibitor

6.1.1. Weight loss method

Figure 5 is the plot of weight loss against exposure time for aluminum alloy coupons in saline environment in the absence and presence of different concentrations of FG at 28°C.

In the absence of FG, the weight loss increased with an increase in exposure time in 3.5% sodium chloride solution. The weight loss was 0.001 gm at the beginning of the analysis but increased with an increase in exposure time which gave a value of 0.011 gm after 28 days of exposure. In the presence of FG, the weight loss value was reduced given the highest value of 0.007 gm at the end of 28 days exposure time at all the concentrations of inhibitor studied. This clearly indicates a 36% reduction in weight loss value in the presence of FG when compared to the value in the absence of FG [47, 49].

Figure 6 shows the graph of corrosion rate of aluminum alloy coupons in the absence and presence of varied concentrations of FG in 3.5% NaCl solution at 28°C.

Observation made indicates that corrosion rate of aluminum alloy reduced in the presence of all the different concentrations of FG considered. In the presence of 0.5% g/v concentration of FG, the corrosion rate value was 0.051 mm/yr at the end of 28 days exposure time compared with the absence of FG which gave 0.110 mm/yr at the end of 28 days of exposure [47, 49]. This shows 54% reduction in corrosion rate value compared with the absence of FG. It can be deduced that FG inhibits the corrosion of aluminum alloy in saline solution. Also shown in **Figure 7** is the graph of percentage inhibition efficiency (%IE) with exposure time in the presence of different concentrations of FG.

A reduction in IE of FG from 100% depending on the concentration of FG added to the corrosive environment was observed. The inhibition efficiency at 0.5% g/v concentration

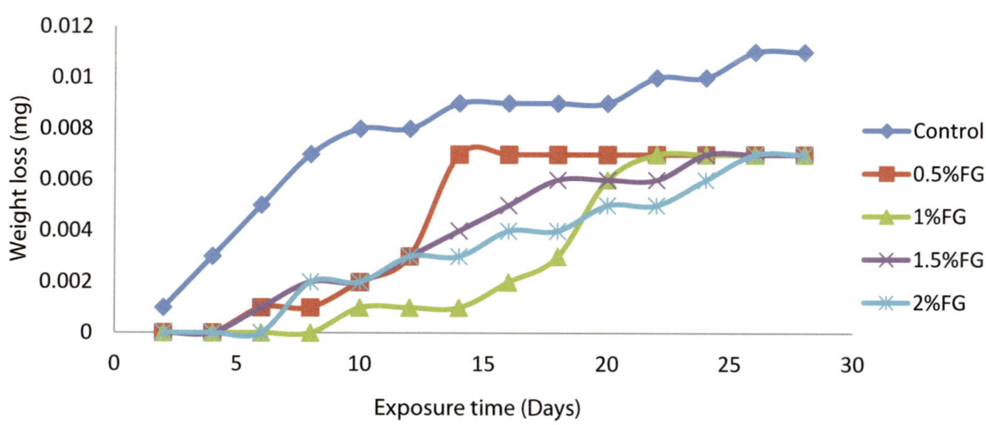

Figure 5. Weight loss versus exposure time for aluminum alloy immersed in 3.5% NaCl solution with varied ferrous gluconate addition [47].

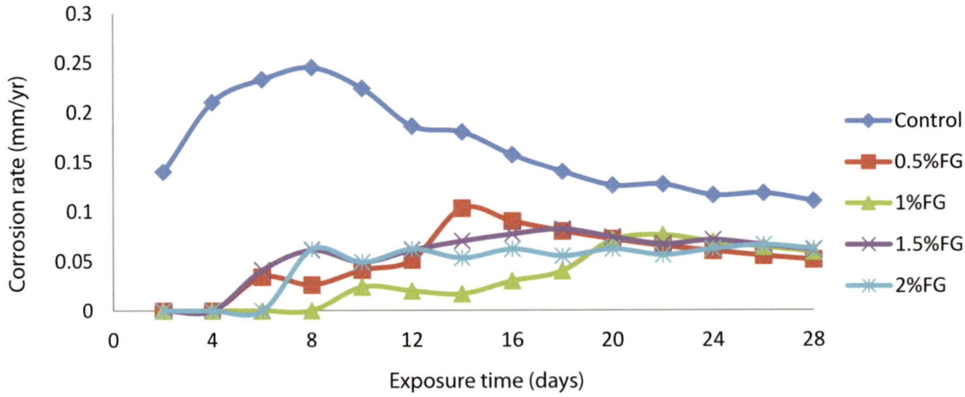

Figure 6. Corrosion rate versus exposure time for aluminum alloy immersed in 3.5% NaCl solution with varied ferrous gluconate addition [47].

of FG reduced to 54%, 1.0% g/v reduced to 46%, 1.5% g/v reduced to 45% and 2.0% g/v reduced to 44% at the end of 28 days exposure time to the corrosive environment. The optimum inhibition efficiency was observed at 1.0% g/v concentration of FG in 3.5% NaCl solution at 28°C.

6.1.2. Adsorption studies

The Langmuir adsorption isotherm was found to best describe the adsorption behavior of FG as inhibitor. Plot C/against C yield straight lines with regression coefficient, R^2 equals 1. The results plotted in **Figure 8** suggest that FG in 3.5% NaCl solution at 28°C obeyed Langmuir adsorption isotherm.

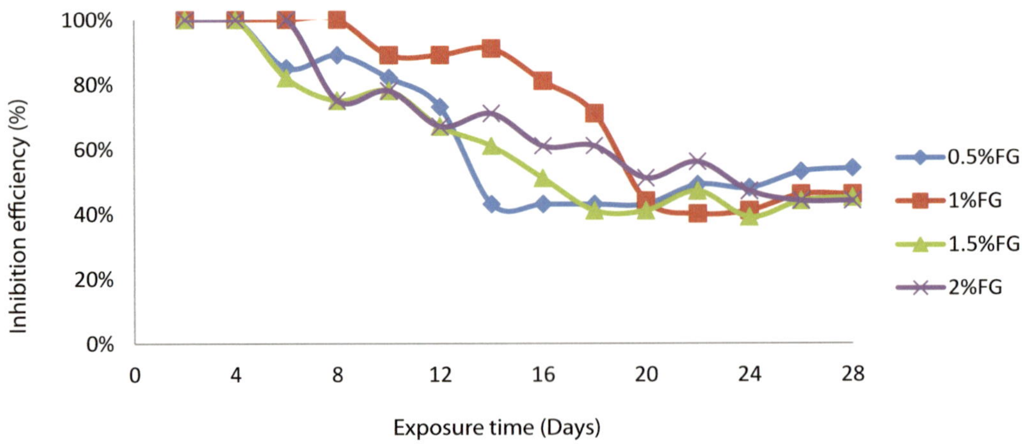

Figure 7. Inhibition efficiency versus exposure time for aluminum alloy immersed in 3.5% NaCl solution with varied ferrous gluconate addition [47].

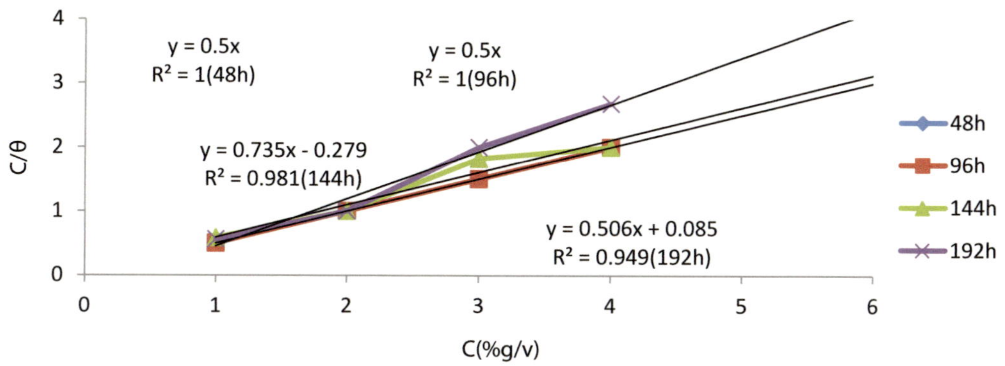

Figure 8. Langmuir isotherm for the adsorption of different concentrations of FG on aluminum alloy surface in 3.5% NaCl solution obtained at 28°C [47].

6.1.3. Microstructural characterization of FG inhibited aluminum alloy in saline medium

The morphology of the test sample showing the inhibitive action of ferrous gluconate in saline solution with 1.0% g/v concentration of FG after weight loss experiment is shown in **Figure 9**. The severity of the damage is much less when compared with the morphology of the sample in the absence of FG. The effective corrosion inhibition of FG could be attributed to the film formed on the aluminum alloy surface which acts as a barrier between the aluminum alloy and the corrosive environment boundary, thereby preventing further corrosion reaction. From the EDS in **Figure 9**, the presence of oxygen and other elements is as a result of ferrous gluconate constituent [47].

In the presence of 2.0% g/v concentration of FG, the corrosion is slightly reduced with little corrosion product on the aluminum alloy surface when compared with the morphology of the aluminum alloy in the absence of FG. This clearly revealed that FG is a corrosion inhibitor for the aluminum alloy in acidic solution. The effective corrosion inhibition of FG could

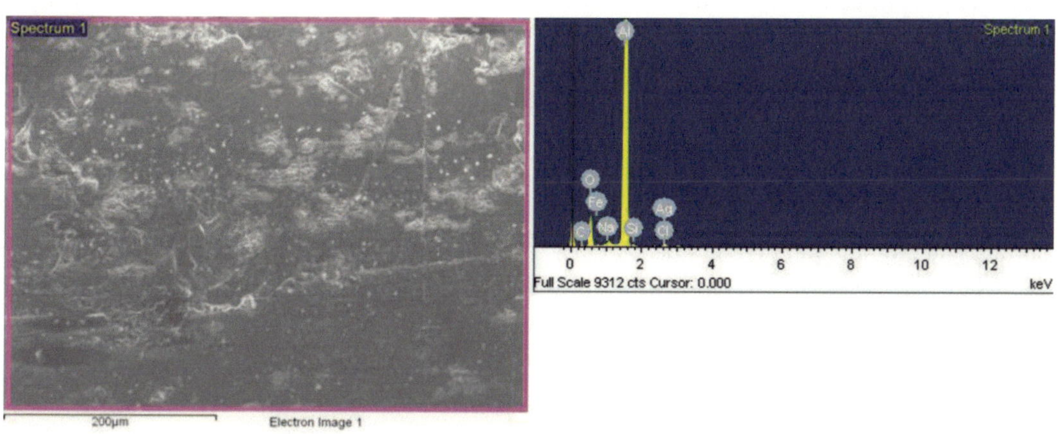

Figure 9. SEM/EDS spectra of corroded aluminum alloy in the presence of FG [47].

be attributed to the film formed on the surface of the sample. The EDS in **Figure 9** clearly shows the presence of carbon, oxygen indicating the formation of oxide films as confirmed by Raman spectroscopy. Other elements are also present which is as a result of ferrous gluconate constituent.

6.1.4. Potentiodynamic polarization method

Table 3 shows the electrochemical corrosion parameters that is, potentiodynamic polarization-corrosion potential (Ecorr), linear polarization resistance (Rp), anodic and cathodic Tafel slopes (βa, βc), corrosion rate (CR) and potentiodynamic polarization corrosion-current density (icorr) obtained by extrapolation of the Tafel lines.

Figure 10 gives the potentiodynamic polarization curves for the inhibition of aluminum alloy in the absence and presence of different concentrations of ZG in 0.5 molar sulfuric acid solutions at 28°C. From **Table 3**, it can be concluded that the corrosion current density (icorr) values reduced in the presence of all the concentrations of ZG studied. The anodic and cathodic values were observed to change in the presence of all the different concentrations of ZG studied in 0.5 M H_2SO_4 solution. The polarization resistance (Rp) values increased and corrosion rate (CR) values decreased in the presence of inhibitor. Also, no definite trend was observed in the corrosion potential (Ecorr) values in the presence of inhibitor [47]. This result showed the influence of ZG on aluminum alloy in 0.5 M H_2SO_4 solution and the ZG acts as a mixed type inhibitor.

6.2. Synergetic effect of the two inhibitors (FG+ZG) on aluminum alloy in H_2SO_4 acid medium

6.2.1. Weight loss method

Figure 11 shows the variation of weight loss with exposure time for aluminum alloy coupons in 0.5 molar sulfuric acid solution in the absence and presence of different concentrations of FG and ZG in the same proportion at 28°C. The Figure shows an increase in weight loss value in the absence of inhibitors with an increase in exposure time. In the presence of inhibitors, there was a decrease in weight loss of the aluminum alloy coupons with the following values after 28 days exposure time to the corrosive medium; 0.027 gm, 0.025 gm, 0.022 gm and 0.020 gm

S/N	C (% g/v)	Icorr (A/cm²)	βc (V/dec)	βa (V/dec)	LPR Rp (Ωcm²)	–Ecorr (V)	CR (mm/yr)
1	0.0	1.73E-05	0.025551	0.124930	5.33E+02	0.33054	0.559600
2	0.5	6.46E-06	0.054932	0.032904	1.38E+03	0.33054	0.209090
3	1.0	5.43E-06	0.042048	0.035871	1.55E+03	0.33054	0.175710
4	1.5	6.30E-09	0.892180	1.727600	4.06E+07	0.34318	0.000206
5	2.0	3.78E-06	0.172780	0.072640	5.87E+03	0.35751	0.123690

Table 3. Polarization data for ZG inhibited aluminum alloy in 0.5 M H_2SO_4 solution.

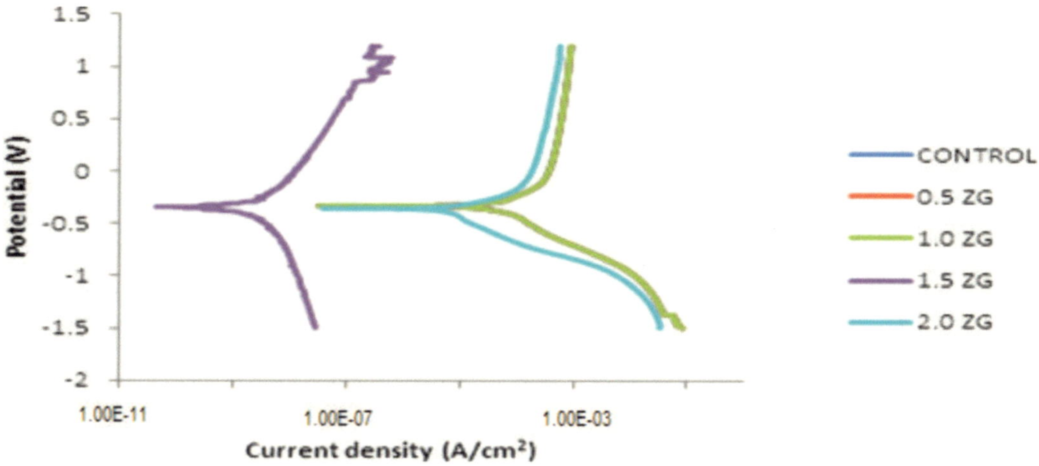

Figure 10. Tafel polarization curves for aluminum alloy in 0.5 M H_2SO_4 solution in the absence and presence of different concentrations of ZG at 28°C [48].

at 0.5:0.5, 1.0:1.0, 1.5:1.5 and 2.0:2.0 g/v concentrations of inhibitors respectively. These values translated to 27% for 0.5:0.5 g/v, 32% for 1.0:1.0 g/v, 41% for 1.5:1.5 g/v and 46% for 2.0:2.0 g/v when the weight loss values in the presence of inhibitors were compared with the values in the absence of inhibitors. The Figure also denotes that increase in concentration of the inhibitor leads to a decrease in the weight loss of the aluminum alloy coupon.

The variation of corrosion rates of aluminum alloy coupons exposed to 0.5 M H_2SO_4 solution at 28°C in the absence and presence of different concentrations of FG and ZG are presented in **Figure 11**. The Figure shows that in the absence of inhibitors, the corrosion rate of aluminum alloy decreased from 0.395 to 0.346 mm/yr after 28 days exposure time to the corrosive environment (**Figure 12**).

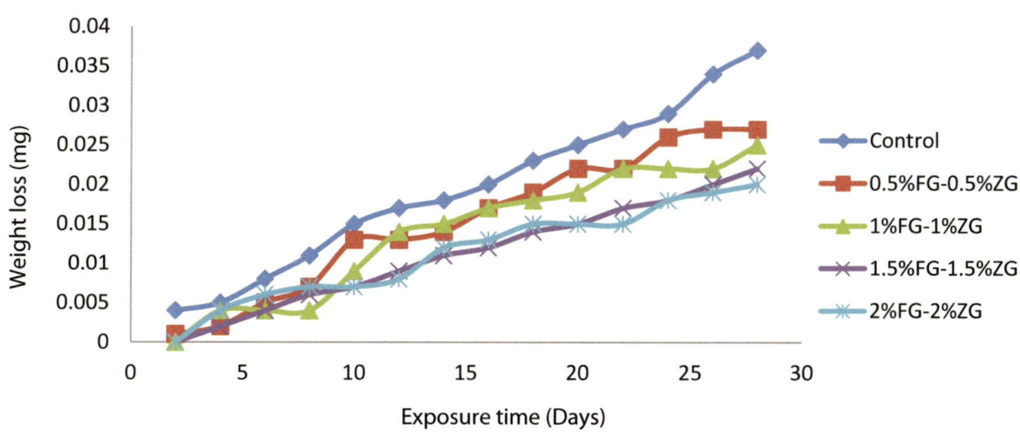

Figure 11. Weight loss versus exposure time for aluminum alloy immersed in 0.5 M H_2SO_4 solution with varied ferrous gluconate and zinc gluconate additions [46].

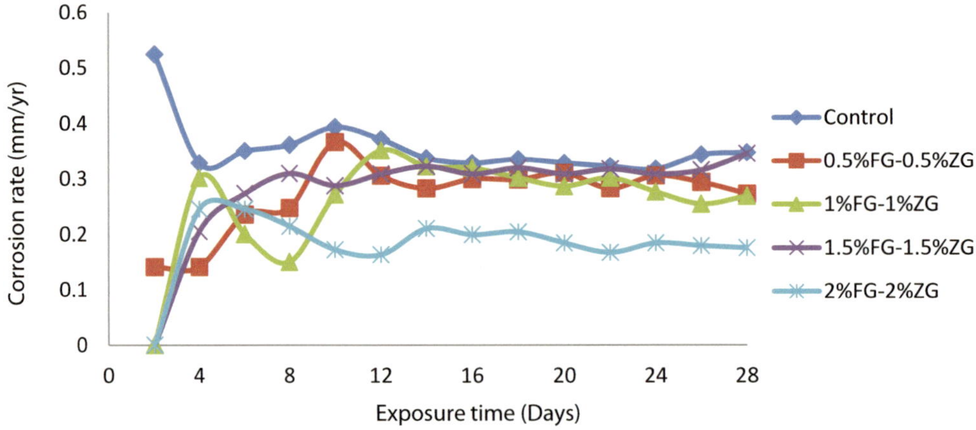

Figure 12. Corrosion rate versus exposure time for aluminum alloy immersed in 0.5 M H_2SO_4 solution with varied ferrous gluconate and zinc gluconate additions [46, 50–51].

The graph shows a sinusoidal trend. In the presence of inhibitors, the corrosion rate values was reduced from 0.346 to 0.272 mm/yr at 0.5:0.5 g/v concentration of FG-ZG, 0.268 mm/yr at 1.0:1.0 g/v, 0.345 mm/yr at 1.5:1.5 g/v and 0.175 mm/yr at 2.0:2.0 g/v. The percentage reduction in the corrosion rate value of aluminum alloy after 28 days of exposure in the presence of different concentrations of inhibitors are 21% for 0.5:0.5 g/v, 23% for 1.0:1.0 g/v, 0.3% for 1.5:1.5 g/v and 49% at 2.0:2.0 g/v.

Figure 13 shows the variation of percentage inhibition efficiency (%IE) against exposure time with different concentrations of inhibitors at 28°C. From the results, it was observed that the inhibition efficiency increased with an increase in the concentration of the inhibitors. The optimum inhibition efficiency was observed at 0.5 g/v concentration of the inhibitors.

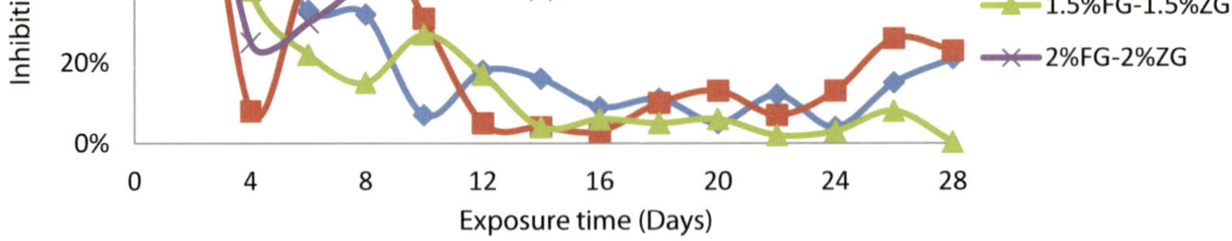

Figure 13. Inhibition efficiency versus exposure time for aluminum alloy immersed in 0.5 M H_2SO_4 solution with varied ferrous gluconate and zinc gluconate additions [46].

http://dx.doi.org/10.5772/intechopen.71467

6.2.2. Adsorption studies

The Langmuir adsorption isotherm provides the best description for the adsorption behavior of the synergistic effect of FG and ZG as inhibitors. Plot C/against C yield a straight line with regression coefficient, R^2 close to 1. The result plotted in **Figure 14** suggests that the inhibitors in 0.5 M H_2SO_4 solution at 28°C obeyed Langmuir adsorption isotherm.

6.2.3. Potentiodynamic polarization method

Table 4 shows the electrochemical corrosion kinetics parameters: polarization resistance (Rp), potentiodynamic polarization-corrosion potential (Ecorr), potentiodynamic polarization corrosion-current density (icorr), corrosion rate (CR), anodic and cathodic Tafel slopes (βa and βc) obtained by extrapolation of the Tafel lines. **Figure 15** shows the cathodic and anodic polarization curves obtained for aluminum alloy in 0.5 molar sulfuric acid solution in the absence and presence of different concentrations of FG and ZG at 28°C.

The results showed a decrease in corrosion current density (icorr) and corrosion rate (CR) values in the presence of different concentrations of inhibitors studied. The corrosion potential (Ecorr) and polarization resistance (Rp) values increased in the presence of inhibitors. Shifts in anodic and cathodic region suggest the inhibitor is a mixed type corrosion inhibitors for aluminum alloy in 0.5 M H_2SO_4 solution at 28°C.

6.3. Correlation between experimental results and analyses (aluminum)

6.3.1. Weight loss and electrochemical corrosion test methods (aluminum)

The percentage inhibition efficiency (%IE) of aluminum alloy in 3.5% sodium chloride solution and 0.5 molar sulfuric acid environments in the presence of different concentrations of FG, ZG and the synergistic effect of the two inhibitors was obtained from weight loss and potentiodynamic

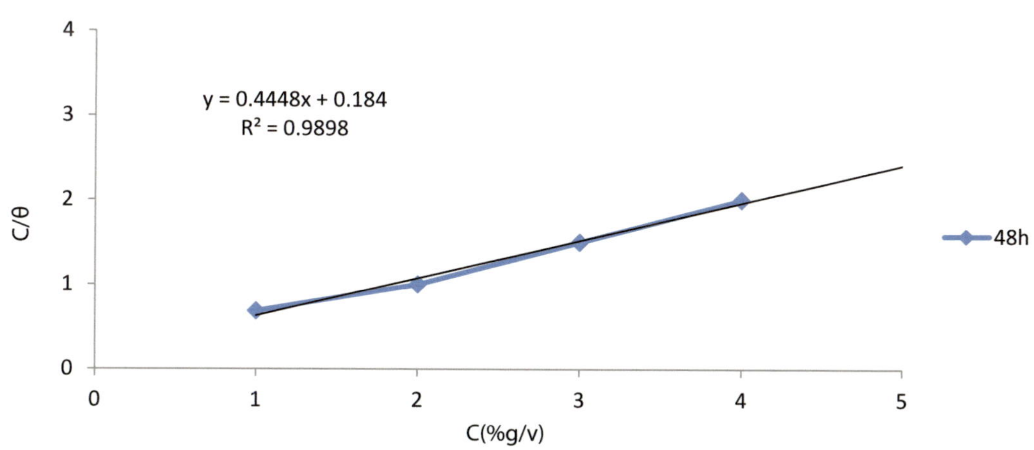

Figure 14. Langmuir isotherm for the adsorption of different concentrations of FG and ZG on aluminum alloy surface in 0.5 M H_2SO_4 solution obtained at 28°C [46].

S/N	C (% g/v)	Icorr (A/cm²)	βc (V/dec)	βa (V/dec)	LPR Rp (Ωcm²)	–Ecorr (V)	CR (mm/yr)
1	0.0	1.73E-05	0.025551	0.124930	5.33E+02	0.33054	0.559600
2	0.5	4.29E-06	0.033853	0.025551	1.47E+03	0.33054	0.138900
3	1.0	2.71E-06	0.020610	0.059142	2.45E+03	0.11486	0.087732
4	1.5	6.30E-09	0.892180	1.727600	4.06E+07	0.34318	0.000206
5	2.0	2.57E-07	0.043434	0.263020	6.30E+04	0.99295	0.008324

Table 4. Polarization data for FG and ZG inhibited aluminum alloy in 0.5 M H_2SO_4 solution.

polarization methods. The data computed for the %IE using linear polarization resistance (LPR), potentiodynamic polarization-corrosion rate (PP-CR), potentiodynamic polarization-corrosion current density (PP-icorr) and weight loss method (WLM) are presented in the Figures above at different concentrations of inhibitors studied. From the given graphs, it is evident that obtained data from the different methods are in good correlation at all the concentrations of inhibitors studied.

6.3.2. Surfaces analyses (aluminum alloy)

From the SEM/EDS micrographs, aluminum alloy in acidic medium was observed to be more corrosive than aluminum alloy in sodium chloride medium after the immersion tests in both environments in the absence of inhibitors. Significant difference in the morphology of the as-received and the corroded samples of aluminum alloy in the absence and presence of inhibitors revealed that the inhibitors was able to hinder the dissolution of aluminum alloy in all the environments investigated. More uniform films were observed on the surfaces of

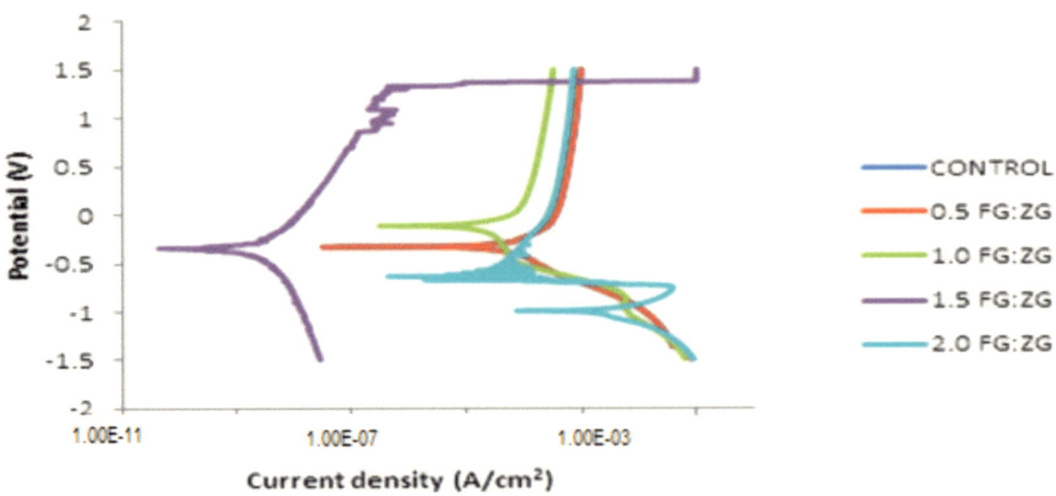

Figure 15. Tafel polarization curves for aluminum alloy in 0.5 M H_2SO_4 medium in the absence and presence of different concentrations of FG and ZG at 28°C [46].

the aluminum alloy in sulfuric acid solution compared to sodium chloride solution, which implied that the corrosion rate in acidic medium is higher when compared to corrosion rate in sodium chloride medium in the presence and absence of inhibitors.

7. Conclusions

The following conclusions have been drawn from this research;

- The aluminum alloy exhibited similar corrosion behavior in the presence of different concentrations of inhibitors in the solution studied, with different corrosion resistances in the different solutions which was confirmed from the similar trend from corrosion rates and inhibition efficiency of the materials from weight loss and potentiodynamic polarization curves.
- In saline environment in the presence of different concentration of inhibitors, the corrosion resistant of aluminum alloy was significantly different, which was evident by their corrosion rates and inhibition efficiency values. FG performed best at 1.0% g/v concentration in 3.5% NaCl solution which gave inhibition efficiency value of 100% from the beginning of the analysis to the 8th day of exposure time to the corrosive medium. ZG performed best at 2.0% g/v concentration which gives inhibition efficiency value of 100% from the beginning of the analysis to the 14th day of exposure time to the corrosive solution and the inhibition efficiency of the synergism of the two inhibitors was 100% from the beginning of the analysis to the 10th day of exposure time to the corrosive medium.
- It could be concluded that ZG performed best in 3.5% NaCl solution. The inhibitors performed fairly well in 0.5 M H_2SO_4 solution. The corrosion rates and inhibition efficiency values obtained from the weight loss and potentiodynamic polarization methods show good agreement.
- The main task of this research work has been accomplished. The optimum concentrations for the application of the inhibitors which should be useful guide for corrosion engineers have been obtained.
- It can be seen from above that FG, ZG and FG+ZG acts as a good corrosion inhibitor for aluminum alloy in saline and acid environments with the following final deduction: Aluminum alloy: in both test environments, the inhibitive performances of the inhibitors are excellent. Finally, the inhibitors have been established to be practically reliable and effective as demonstrated by the performance characteristics of the inhibitors that were used in this study.

Author details

Omotayo Sanni* and Abimbola Patricia Popoola

*Address all correspondence to: tayo.sanni@yahoo.com

Department of Chemical, Metallurgical and Materials Engineering, Tshwane University of Technology, Pretoria, South Africa

References

[1] Owoeye FT, Olokode OS, Aiyedun PO, Anyanwu BU. Comparative analysis of corrosion of aluminum plates in different concentrations of lime solution. Pacific Journal of Science and Technology. 2012;**13**:105-111

[2] Caporali S, Fossati A, Lavacchi A, Perissi I, Tolstogouzov A, Bardi U. Aluminium electroplated from ionic liquids as protective coating against steel corrosion. Corrosion Science. 2008;**50**:534-539

[3] Fang H, Chen K, Zhang Z, Zhu C. Effect of Yb addition on microstructure and properties of 7A60 aluminium alloy. Transactions of the Nonferrous Metals Society of China. 2008;**18**:28-32

[4] Rosaliza R, Wan WB, Senin HB. The effect of inhibitor on the corrosion of aluminium alloys in acidic solutions. Materials Chemistry and Physics. 2008;**107**:281-288

[5] Onen AI, Nwufo BT, Ebenso EE, Hlophe RM. Titanium (IV) oxide as corrosion inhibitor for aluminium and mild steel in acidic medium. Journal of the Electrochemical Society. 2010;**5**:1563-1573

[6] Hurlen T, Lian H, Odegard O, Valand T. Corrosion and passive behaviour of aluminium in weakly acid solution. Electrochimica Acta. 1984;**29**:579-585

[7] Oguzie EE, Onuoha GN, Ejike EN. Effect of *Gongronema latifolium* extract on aluminium corrosion in acidic and alkaline media. Pigment & Resin Technology. 2007;**36**:44-49

[8] Oguzie E. Corrosion inhibition of aluminium in acidic and alkaline media by *Sansevieria trifasciata* extract. Corrosion Science. 2007;**49**:1527-1539

[9] Nnanna LA, Nwadiuko OC, Ekekwe ND, Ukpabi CF, Udensi SC, Okeoma KB, Onwuagba BN, Mejeha IM. Adsorption and inhibitive properties of leaf extract of *Newbouldia leavis* as a green inhibitor for aluminium alloy in H_2SO_4. American Journal of Materials Science. 2011;**1**:143-148

[10] Loto CA, Popoola API. Effect of *Cola acuminata* and *Camellia sinensis* mixed extracts on the corrosion inhibition of mild steel in 0.5 M sulphuric acid. International Journal of Electrochemical Science. 2012;**7**:2983-2996

[11] Popoola API, Fayomi OSI. ZnO as corrosion inhibitor for dissolution of zinc electrodeposited mild steel in varying HCl concentration. International Journal of the Physical Sciences. 2011;**6**:2447-2454

[12] Sidharth AAP. Effect of pitting corrosion on ultimate strength and buckling strength of plates – A review. Digest Journal of Nanomaterials and Biostructures. 2009;**4**:783-788

[13] Volkan C. Handbook Engineering and Cathodic Protection Handbook. Wiley, Scrivener Publishing; 2017

[14] Roberge PR. Handbook of Corrosion Engineering, Corrosion of Oil and Gas Well Equipment, Book Two of Vocation Training Series. McGraw-Hill Education, LLC; 2000. pp. 1-30

[15] Roberge PR. Handbook of Corrosion Engineering. McGraw-Hill Professional; 1999

[16] Loto CA, Popoola API, Fayomi OS, Loto RT. Corrosion polarization behaviour of type 316 stainless steel in strong acids and acid chlorides. International Journal of Electrochemical Science. 2012;**7**:3787-3797

[17] Loto RT, Loto CA, Popoola API, Ranyaoa M. Corrosion resistance of austenitic stainless steel in sulphuric acid. International Journal of Physical Sciences. 2012;**7**:1677-1688

[18] Ashworth V. Industrial corrosion and corrosion control technology. In: Proceedings of the 2nd Arabian Corrosion Conference, Kuwait, October 12-15; 1996

[19] https://inspectioneering.com/news/2016-03-08/5202/nace-study-estimates-global-cost-of-corrosion-at-25-trillion-ann

[20] Xu WL, Yue TM, Man HC. Laser surface melting of aluminium alloy 6013 for improving pitting corrosion fatigue resistance. Journal of Materials Science. 2008;**43**:942-951

[21] Trydan U, Ocana JL, Grum J. Surface evaluation of laser shock processed aluminium alloy after pitting corrosion attack with optical 3-D metrology method. In: Proceedings of the 10th International Conference of the Slovenian Society for Non-destructive Testing Ljubljana, Slovenia, on 1-3 September; 2009. pp. 251-258

[22] Ali A, Kaveh E, Zenji H. Production of nanograined intermetallics using high-pressure torsion. Materials Research. 2013;**16**:672-678

[23] Zhao S, Wolfe DA, Huang TS, Frankel GS. Generalized model for IGC growth in aluminium alloys. Journal of Statistical Planning and Inference. 2007;**137**:2405-2412

[24] Popoola API, Abdulwahab M, Fayomi OSI. Corrosion inhibition of mild steel in *Sesamum indicum*-2M HCl/H_2SO_4 interface. International Journal of Electrochemical Science. 2012;**7**:5805-5816

[25] Singh A, Quraishi MA. Azwain (*Trachyspermum copticum*) seed extract as an efficient corrosion inhibitor for aluminium in NaOH solution. Research Journal of Recent Sciences. 2012;**1**:57-61

[26] Marsh J, Teh T. Conflicting views: CO_2 corrosion models, corrosion inhibitor availability philosophies, and the effect on subsea systems and pipeline design. In: SPE 109209 Presented at Offshore Europe, held in Aberdeen; 2007

[27] Ashry EH, El Nemr A, Essawy SA, Ragab S. Corrosion inhibitors part V: QSAR of benzimidazole and 2-substituted derivatives as corrosion inhibitors by using the quantum chemical parameters. Progress in Coatings. 2008;**61**:11-20

[28] Zhang DQ, Gao LX, Zhou GD. Inhibition of copper corrosion in aerated hydrochloric acid solution by heterocyclic compounds containing a mercapto group. Corrosion Science. 2004;**46**:3031-3040

[29] Zhang DQ, Gao LX, Zhou GD, Lee KY. Undecyl substitution in imidazole and its action on corrosion inhibition of copper in aerated acidic chloride media. Journal of Applied Electrochemistry. 2008;**38**:71-76

[30] Tebbji K, Aouniti A, Attayibat A, Hammouti B, Oudda H, Benkaddour M, Radi S, Nahle A. Inhibition efficiency of two bipyrazole derivatives on steel corrosion in hydrochloric acid media. Indian Journal of Chemical Technology. 2011;**18**:244

[31] Nahle A. Effect of triethanolamine on the electrochemical dissolution of solder in NaOH solution. Bulletin of Electrochemistry. 2002;**18**(3):105-110

[32] Talati JD, Desai MN, Shah NK. Meta-substituted aniline-N-salicylidenes as corrosion inhibitors of zinc in sulphuric acid. Materials Chemistry and Physics. 2005;**93**:54-64

[33] Qiu LG, Xie AJ, Shen YH. Understanding the effect of the spacer length on adsorption of gemini surfactants onto steel surface in acid medium. Applied Surface Science. 2005;**246**:1-5

[34] Nahle A, Abu-Abdoun I, Abdel-Rahman I. Electrochemical studies of the effect of trans-4-hydroxy- 4′-stilbazole on corrosion inhibition of mild steel in HCl solution. Anti-Corrosion Methods and Materials. 2007;**54**:244-248

[35] Nahlé A, Abu-Abdoun I, Abdel-Rahman I. Inhibition of carbon steel corrosion by 4-vinylbenzyl triphenyl phosphonium chloride in HCl solution. Anti-Corrosion Methods and Materials. 2008;**55**:217-224

[36] Amin MA, Abd El Rehim SS, El-Lithy AS. Pitting and pitting control of Al in gluconic acid solutions – Polarization, chronoamperometry and morphological studies. Corrosion Science. 2010;**52**:3099-3108

[37] Popoola API, Fayomi OSI, Abdulwahab M. Degradation behaviour of aluminium in 2M HCl/HNO_3 in the presence of *Arachis hypogeae* natural oil. International Journal of Electrochemical Science. 2012;**7**:5817-5827

[38] Umoren SA, Solomon MM. Effect of halide ions additives on the corrosion inhibition of aluminum in HCl by polyacrylamide. The Arabian Journal for Science and Engineering. 2010;**35**:115-129

[39] Obot IB, Obi-Egbedi NO. Fluconazole as an inhibitor for aluminium corrosion in 0.1M HCl. Colloids and Surfaces A: Physicochemical and Engineering Aspects. 2008;**330**:207-212

[40] Garcia SJ, Muster TH, Ozkanata O, Sherman N, Hughes AE, Terrn H, De Wita JHW, Mola JMC. The influence of pH on corrosion inhibitor selection for 2024-T3 aluminium alloy assessed by high-throughput multielectrode and potentiodynamic testing. Electrochimica Acta. 2010;**55**:2457-2465

[41] Avwiri GO, Igho FO. Inhibitive action of *Vernonia amygdalina* on the corrosion of aluminium alloys in acidic media. Materials Letters. 2003;**57**:3705-3711

[42] Umoren SA, Obot IB, Ebenso EE, Obi-Egbedi NO. The inhibition of aluminium corrosion in hydrochloric acid solution by exudate gum from *Raphia hookeri*. Desalination. 2009;**247**:561-572

[43] Halambek J, Jukic M, Berkovic K, Vorkapic-Furac J. Investigation of novel heterocyclic compounds as inhibitors of Al-3Mg alloy corrosion in hydrochloric acid solutions. International Journal of Electrochemical Science. 2012;**7**:1580-1601

[44] Savita NC, Singh VK, Quraishi MA. Corrosion inhibition performance of different bark extracts on aluminium in alkaline solution. Journal of the Association of Arab Universities for Basic and Applied Sciences. 2017;**22**:38-44

[45] Chaubey N, Yadav DK, Singh VK, Quraishi MA. A comparative study of leaves extracts for corrosion inhibition effect on aluminium alloy in alkaline medium. Ain Shams Engineering Journal. 2017;**8**:673-682

[46] Sanni O, Loto CA, Popoola API. Inhibitive behavior of zinc gluconate on aluminium alloy in 3.5% NaCl solution. Journal of Silicon. 2016;**8**:195-200. DOI: 10.1007/s12633.014-9180-8

[47] Sanni O. Evaluation of inorganic inhibitors for mild steel and aluminium alloy in acidic environments [M. Tech dissertation]. Department of Chemical and Metallurgical Engineering, Tshwane University of Technology, Pretoria, South Africa; 2013

[48] Sanni O, Popoola API, Loto CA. Inhibitive tendency of zinc gluconate for aluminium alloy in sulphuric acid solution. Polish Journal of Chemical Technology. 2013;**15**(4):60-64

[49] Popoola API, Sanni O, Loto CA, Popoola OM. Inhibitive action of ferrous gluconate on aluminum alloy in saline environment. Advances in Materials Science and Engineering. 2013;**2013**:1-8. http://dx.doi.org/10.1155/2013/639071

[50] Sanni O, Popoola API, Loto CA. Inhibitive effect of ferrous gluconate on the electrochemical corrosion of aluminium alloy in H_2SO_4 solution. In: Proc. ICCEM; 2012. pp. 1-6

[51] Sanni O, Popoola API, Loto CA. Inhibition effect of ferrous gluconate on the electrochemical corrosion behavior of aluminium alloy in H_2SO_4. In: Proceeding of International Conference on 'Clean Technology and Contemporary, Engineering Management, ICCEM 2012. Dept of Mechanical Eng, Covenant Univ., Ota, Nigeria; 2012

Chapter 10

Influence of Rapid Solidification on the Thermophysical and Fatigue Properties of Laser Additive Manufactured Ti-6Al-4V Alloy

Olawale Samuel Fatoba, Esther Titilayo Akinlabi and Mamookho Elizabeth Makhatha

Additional information is available at the end of the chapter

http://dx.doi.org/10.5772/intechopen.71697

Abstract

Modern industrial applications require materials with special surface properties such as high hardness, wear and corrosion resistance. The performance of material surface under wear and corrosion environments cannot be fulfilled by the conventional surface modifications and coatings. Therefore, different industrial sectors need an alternative technique for enhanced surface properties. The purpose of this is to change or enhance inherent properties of the materials to create new products or improve on existing ones. The most effective and economical engineering solution to prevent or minimize such surface region of a component is done by fiber lasers. Additive manufacturing (AM) is a breaking edge fabrication technique with the possibility of changing the perception of design and manufacturing as a whole. It is well suitable for the building and repairing applications in the aerospace industry which usually requires high level of accuracy and customization of parts which usually employ materials known to pose difficulties in fabrication such as titanium alloys. The current development focus of AM is to produce complex shaped functional metallic components, including metals, alloys and metal matrix composites (MMCs), to meet demanding requirements from aerospace, defense, and automotive industries.

Keywords: DLMD, solidification, fatigue, wear, Ti-6Al-4V alloy, microstructure, fiber laser, additive manufacturing

1. Introduction

Titanium is the ninth most abundant element on earth. It occurs as ilmenite, rutile and also present in titanates and iron ores. Titanium and its alloys are preferred choices for chemical, power generation, automobile, aerospace and airframe industries due to their excellent properties.

Titanium has high strength and low density (40% lower than steel). Titanium has many alloys with good properties. The most commonly used titanium alloy is the Ti-6Al-4V. It is also known as a workhorse. It is an alpha-beta alloy. Among other properties it possesses is that this alloy is heat treatable and has good hot forming qualities. However, its creep strength is not as good as the alpha alloys. Titanium and its alloys have low density, excellent combination of high specific ratio which is maintained at elevated temperatures, low modulus of elasticity and good corrosion resistance. Biocompatibility of the titanium alloy is excellent and makes it very good for biomedical applications due to its mechanical and corrosion properties. Titanium has a tenacious oxide which forms instantly upon exposure to air. This is the reason for its excellent corrosion resistance [1]. Titanium and it alloys are material of choice in a wide range of industrial applications owing to high strength-to-weight ratio, exceptional biocompatibility and corrosion resistance. However, the use of these materials is compromised due to high fabrication costs which are attributed to difficulty in the machining and loss of material during processing [2]. Estimate by research states that up to 50% of titanium cost are attributed to machining operations [3]. Fortunately, titanium is among the few privileged metallic materials that are fabricated by means of additive manufacturing which has the ability to fabricate high-quality parts with very little or no post-machining [4].

The application of laser surface modification to prolong the service life of engineering components exposed to aggressive environments has gained increasing acceptance in recent years and material processing by laser beams has been well established as an advanced manufacturing technology. The main mandate is the ability to precisely deposit a large amount of energy into a material over a short time range and in a spatially confined region near the surface. One of the major advantages of the laser as a tool for material processing is the ability to precisely control where in the material and at what rate energy is deposited. This control is exercised through the proper selection of laser processing parameters to achieve the desired material modification. Well known applications include the improvement of the wear resistance of diesel engine exhaust valves, the enhancement of the corrosion resistance of gas turbine blades and the repair of dies and inserts. The high-quality surface layers that can be produced by fiber laser only, make it a strategic technique. Surface engineering techniques especially by lasers are known to be cost-effective due to their precision and speed.

Material engineers continue to develop new and existing fabrication techniques in pursuit of improving or entirely replacing the conventional energy intensive, environmentally unfriendly and expensive methods [5]. Metal-based components are generally manufactured by subtractive and formative techniques; with the most recent additive fabrication which has been extensively explored since its development in the late 1980s [6]. The latter widens the perspective and boundaries of design as it provides new possibilities of fabricating complex geometries which are usually unattainable by traditional methods [7].

2. Direct laser metal deposition

The direct laser metal deposition (DLMD) is a recently developed technique for manufacturing solid parts, layer by layer, directly from powder. The process uses a high-power laser beam focused onto a metallic substrate to generate a molten pool to which a stream of powder is fed. This way, the material volume increases leading to the formation of a solid layer. The laser beam and powder nozzle are repeatedly scanned to accomplish a layered buildup of a solid part by additive manufacturing (AM). The key process parameters are laser power, scan speed, laser beam diameter, and powder feed rate. The parameters influence the thermal phenomena during the growth process of the solid part as well as the metallurgical and mechanical properties. This research will also describe the 2D & 3D multi-physics models to describe the physical mechanism of heat transfer, melting and solidification that take place during and post laser-powder interaction. The transient temperature, geometrical features of the generated structures and thermal cycles will be simulated. The transient temperature is critically important for determining the thermal stress distribution and residual stress state in additively manufactured parts. According to Patterson et al. [8], thermal history generated by a laser beam during additive manufacturing governs the microstructure, properties, residual stress and distortion in fabricated components. In an attempt to overcome or reduce the evolution of residual stresses and the resultant of their unwanted artifacts, modeling and simulation of the additive manufacturing processes is a way to go instead of the time intensive and economic demanding trial and error methods. Additive manufacturing is a process which employs laser radiation to join materials layer by layer to make objects from 3D model data, which is reverse of subtractive manufacturing method. Hanzl et al. [9] stated that different fabrication techniques even when producing components from similar materials result in different properties. Authors further stated that additive manufacturing produces finer grains compared to conventional processes; this is attributed to rapid solidification which is a result of rapid heat conduction from the localized molten zone in direction of the dispersed cool surroundings. Furthermore, additive fabrication provides various benefits including elimination of forming equipment, high material use efficiency and the flexibility of producing customized parts which require very little or no post process machining which makes a direct contribution to the reduction of production cost and delivery time [10] (**Figure 1**).

The past decade has seen a rapid development in the range of techniques which are available to modify the surfaces of engineering components. This in turn has led to the emergence of the new field of surface modification. It describes the interdisciplinary activities aimed at tailoring the surface properties of engineering materials. An engineering material usually fails when its surface cannot effectively withstand the external forces or environment to which it is subjected. Serviceable engineering components not only rely on their bulk material properties but also on the design and characteristics of their surface. The surface of these components may require treatment, to enhance the surface characteristics. Surface treatments that cause microstructural changes in the bulk material include heating and cooling/quenching through induction, flame, laser, and electron beam techniques, or mechanical treatments [12]. The behavior of a material is therefore greatly dependent on the surface of a material, surface contact area and the environment under which the material must operate. Surface engineering can cause physical and chemical effects on the bulk material, some beneficial and some detrimental. For example, stresses which may exist in the protective material can create problems; however careful monitoring and research may limit these effects, to hopefully produce quality and serviceable engineering components.

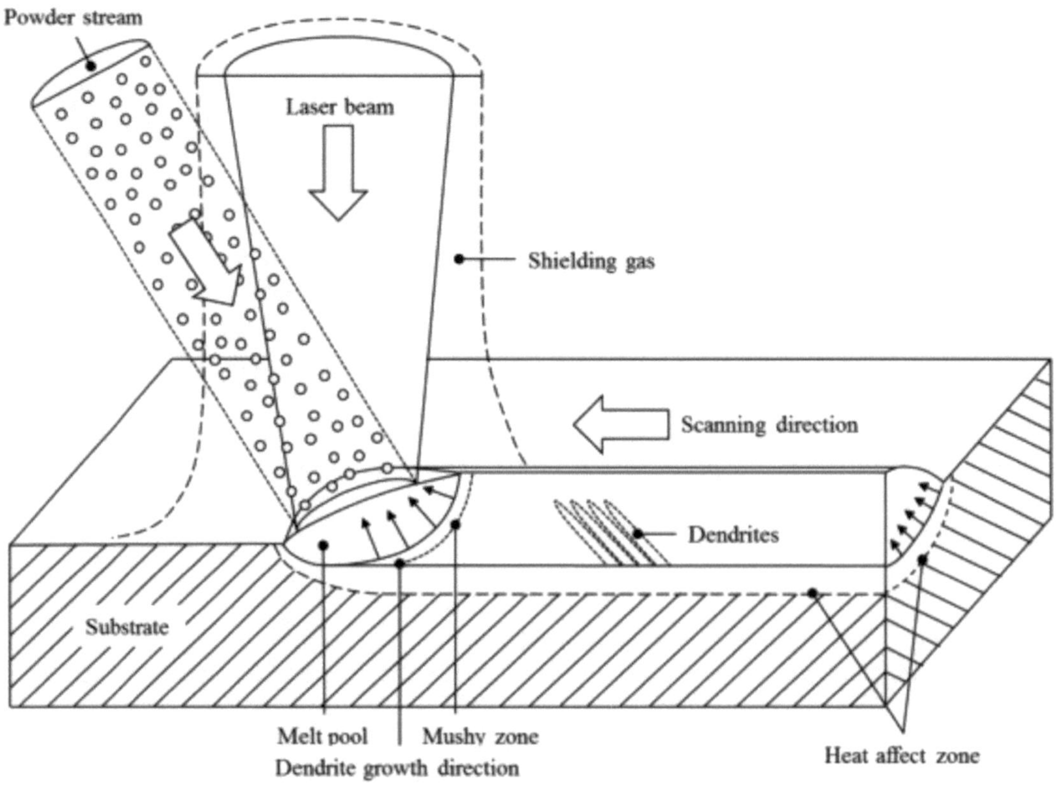

Figure 1. Schematic diagram illustrating the heat affected zone and melt zone created by laser surface melting [11].

The choice of a surface material with the appropriate thermal and sufficient resistance to wear and corrosion degradation is crucial to its functionality. The desired properties of surface-engineered components are improved corrosion resistance through barrier or sacrificial protection, improved wear resistance and mechanical properties such as fatigue and toughness. The surface modification of metals, using high intensity lasers, accepts various methods in making use of the heat from the laser beam in enhancing the morphology and microstructure in order to achieve numerous metallurgical outcomes [13]. Laser surface modification of materials entails an extensive composition of additive and subtractive methods. Metals such as magnesium, titanium and aluminum blend extremely desired properties with light weight without a significant alteration of the bulk material. This is the footing for high demand of laser surface modification techniques [14]. There are many laser surface treatment techniques responsible for fabrication of highly corrosion-resistant metallic surfaces, where the major ones used are laser surface melting, laser cladding and laser surface alloying [15, 16].

Laser surface alloying (LSA) modifies the surface morphology and near surface structure of components and its alloys with perfect adhesion to the interface of the bulk steel. The distinctive advantages of the LSA technique for surface modification include the refinement of the grain size because of rapid quench rates and the generation of meta-stable structures with novel properties that are not feasible by competing methods [17, 18]. With optimum laser processing parameters, a reliable coating that is free of cracks and pores can be produced on the matrix. LSA can rapidly provide a thick and crack-free layer in all instances with metallurgical bonds at the interface between the alloyed layer and the substrate [19]. In LSA, external alloying elements in form of powder (metal, alloy, ceramic, cermet or intermetallic) are introduced into the surface of a substrate, as pre-placed material or injected directly into the melt pool created on the substrate by a high-power laser beam. The melting of the substrate occurs rapidly only at the surface, while the bulk of the material remains cool, thus serving as an infinite heat sink.

The result of this is rapid self-quenching and resolidification of new alloy due to the large temperature gradients between the melted surface region and the underlying solid substrate. The evolution of a wide variety of microstructures as a result of the rapid cooling from the liquid phase is one of the consequences [20]. Hence, during laser alloying process, the synthesis of new alloy is possible by depositing a premixed ratio of elemental powders. Powders surfaced on new or worn working surfaces of components by LSA provides specific properties such as high abrasive wear resistance, erosion resistance, corrosion resistance, heat resistance and combinations of these properties. Consequently, improvements in machinery performance and safety in aerospace, automotive, can be realized by the method [21]. According to Poulon-Quintin et al. [22], laser beams, because of specific thermal characteristics induced by laser irradiation, can generate specific microstructures including metastable phases and nano-crystalline grains. Laser processing offers unique and significant quality and cost advantages over traditional techniques. These include high throughput speed, process compatibility, high process efficiency, low porosity and good surface uniformity. In addition, the rapid self-quenching in laser alloying results in a true metallurgical bond between the composite layer and the substrate, the formation of a non-equilibrium or amorphous phase as well as homogenization and refinement of the microstructure, all without affecting the bulk properties of the substrate [23–25].

3. Overview on application and improved properties of laser deposited and additive manufactured Ti-6Al-4V alloy

Xu et al. [26] investigated additive manufacturing ductile and strong Ti6Al4V by using selective laser melting through in situ martensite decomposition. Ti6Al4V powder was used to fabricate cubes and cylindrical bars by using a selective laser melting (SLM) facility with different process parameters. Different layer thicknesses were applied, 30, 60 and 90 μm and the process was carried out in an argon atmosphere. The samples were then subjected to isothermal treatment for 2 hours at temperatures ranging from 350 to 930°C. The tensile strength of the samples was evaluated and compared to Ti6Al4V manufactured by mill annealing and by solution treating and aging (STA). The microstructural analysis was completed by means of SEM and phase examination was carried out by X-ray diffraction. The results should that the finest columnar prior β grains were found in the 30 μm thick layer in comparison to the 60 and 90 μm thicknesses; this was attributed to fast cooling condition. It was found that a smaller focal offset distance (0–2 mm) resulted in an ultrafine lamellar ($\alpha + \beta$) microstructure, whereas a focal offset distance of 4 mm led to an acicular α' martensitic microstructure. This led to the authors concluding that the transformation of martensite to ultrafine lamellar structure is possible by manipulating the process parameters of selective laser melting. The high yield strength and elongation failure of the SLM fabricated Ti6Al4V with ultrafine lamellar is comparable with STA Ti6Al4V and was better than that of mill-annealed Ti6Al4V, with results of >1100 MPa yield strength and 11.4% failure elongation. The heat treatment of the samples revealed that decomposition of α' martensite into ultrafine lamellar took place from 400°C. It was therefore concluded by the authors that additive manufacturing of Ti6Al4V by SLM can improve the properties.

Balla et al. [27] studied the microstructure of laser surface melted Ti6Al4V alloy as well as its mechanical and wear properties. Ti6Al4V alloy is extensively used for implants due to their exceptional corrosion resistance and biocompatibility; however, its poor hardness and low wear resistance limit the extent of application. A continuous-wave Nd:YAG laser was used to melt the surface of Ti6Al4V alloy sheet, 250 W and 400 W were used for single and double passes. Light microscopy, SEM and XRD were used to analyze the microstructure and phase of the melted regions. Ball-on-disk wear testing and Vickers microhardness tester were employed to establish the wear and microhardness. All results of the laser treated alloy were compared to as-received Ti6Al4V. The results showed that the sample treated with 400 W and was doubled passed exhibited the highest hardness and the largest grains size as compared to the other samples. An average increase of 15–22% in hardness was obtained by the laser surface melted alloy, with the hardness of the substrate being 358 $HV_{0.3}$ and the hardness of the treated samples ranging from 413 $HV_{0.3}$ to 438 $HV_{0.3}$. Results from both microstructural and phase analysis show that the as-received Ti6Al4V alloy had equiaxed $\alpha + \beta$ phase, while the laser surface melted regions displayed needle-like acicular α (α' martensite) embedded in prior β matrix. The treated samples had a significantly increased relative concentration of α phase and a reduced β phase in comparison to the substrate. The lowest wear rate was obtained by the both 250 W samples, with a wear rate of 3.38×10^{-4} mm^3/Nm in comparison with 6.82×10^{-4} mm^3/Nm of the untreated substrate. The authors therefore concluded that the improvement of Ti6Al4V alloy for in vitro application was possible by laser surface melting.

Brandl et al. [28] investigated the microstructure, morphology and hardness of Ti6Al4V blocks manufactured by wire-feed additive layer manufacturing (ALM). When components are built up layer by layer, it is referred to as direct manufacturing, rapid manufacturing or additive layer manufacturing. When comparing ALM to conventional manufacturing techniques, it is more cost efficient and less time consuming. A wire-feed system with a Nd:YAG rod laser and a wire feeder was used to deposit Ti6Al4V powder on Ti6Al4V alloy substrate. The blocks produced were 7 layers high and 7 beads wide. The samples were subjected to heat treatment after laser processing. The heat treatment completed included stress-relieving at 600°C for 4 hours and then cooled in the furnace and heating to 1200°C for 2 hours and followed by furnace cooling. After processing, the samples were prepared for microstructural analysis by cold mounting, polishing and etching. Light microscopy was used to examine the microstructure, a Vickers hardness tester was used to determine the hardness of the surface and energy dispersive X-ray microanalysis was carried out. The morphology of the block exhibited columnar prior β-grains and the microstructure found within the β-grains was martensite and basket-weave α. The heat treatment of heating to 600°C for 4 hours and then furnace cooled did not change the morphology and microstructure but it did increase the hardness for 327

$HV_{0.5}$ to 342 $HV_{0.5}$. The samples subjected to heating to 1200°C for 2 hours and then cooled in a furnace had a lower hardness of 308 $HV_{0.5}$ due to change of the initial morphology to equiaxed prior β grains.

Makuch et al. [29] studied the laser surface alloying of commercially pure titanium using carbon and boron. The use of titanium alloys ranges over many industries but the application in friction and wear conditions is limited due to their high tendency to adhesive wear, poor hardness and low wear resistance. Thermo-chemical processes such as boriding, nitriding and carburizing have been applied to steels and Ti alloys in order to improve the wear performance, but the disadvantages are that the thermal diffusion process requires high temperatures to achieve an acceptable layer thickness and the duration of the process is long. The material to be laser surface alloyed with boron and carbon was ring-shaped commercially pure titanium Grade 2. Three different pastes were produced for alloying; boron, boron and carbon, and carbon were mixed with polyvinyl alcohol to produce the pastes. The external surface of the cylinders was coated by the three different pastes and then the surface was re-melted with a continuous-wave CO_2 laser. OPM and SEM were used to analyze the microstructure, XRD was used to examine the phases present, a Vickers hardness tester was used to determine the microhardness and a frictional pair was used to evaluate the abrasive wear of the layers. The greatest microhardness was achieved by the borided and borocarburized layers with 1250–1650 HV and 1200–1750 HV respectively, a hardness of 1000–1500 HV was obtained for the carburized layer. The increase in hardness was attributed to the formation of hard ceramic phases; TiB, TiB_2 and TiC. It was found that of the coatings, the borocarburized layer exhibited the worst wear and friction performance and the laser borided sample had the best friction behavior and the laser-carburized had the best wear performance. It was concluded by the authors that laser alloying with boron and carbon will improve the wear resistance of titanium alloys.

Yang et al. [30] investigated synthesizing, microhardness and microstructure distribution of Ti-Si-C-N/TiCN composite coating fabricated on Ti6Al4V by laser cladding. TiN and TiCN coatings have been broadly used to extend the lifetime of medical and cutting tools due to their high hardness, good wear performance, biocompatibility, excellent corrosion resistance and affordable price. Although the properties of TiCN coatings are desirable, their thermal stability is not adequate and can be improved by adding SiO_2 to the coating. A combination of titanium carbonitride and silica powders (20% SiO_2) was used to coat Ti6Al4V alloy by means of a YAG laser system. The laser beam scanning speed was varied at 1, 2, 3 and 4 mm/s in order to examine how the properties will be affected. XRD was used to analyze the phases, SEM-EDS was used to characterize the microstructure and a microhardness tester with a load of 200 g and a dwelling time of 10 s was used to measure the microhardness. The phases that were obtained were $TiC_{0.3}N_{0.7}$, Ti_3SiC_2 and Ti_2O. As the scan speed increased, the relative amount of $TiC_{0.3}N_{0.7}$ decreased, amounts Ti_3SiC_2 increased by more than two times at a speed of 2 mm/s from a speed of 1 mm/s and increased gradually after that, and the amount of Ti_2O was low at 1 mm/s, increased sharply at 2 mm/s and then decreased by almost half for the following speeds. The microhardness of the coatings increased from 1005 HV at 1 mm/s to 1400 HV at 3 mm/s was reached, for 4 mm/s, the hardness reduced. The authors therefore concluded that laser cladding Ti6Al4V at 3 mm/s will produced the most desired properties.

Chen et al. [31] studied coaxial laser cladding on Ti6Al4V alloy using Al_2O_3-13%TiO_2 powders. For applications that require wear, erosion and corrosion resistance, Al_2O_3-TiO_2 powders have been used in the plasma sprayed coatings, and the addition of TiO_2 improves the wear resistance and toughness of Al_2O_3. Laser cladding of these powders has seldom been stated in open literature. Al_2O_3-13%TiO_2 powders were laser cladded on Ti6Al4V alloy plate using Nd:YAG laser system. Optical microscopy and SEM were used to characterize the microstructure of the layer's cross-section and interface between the coating and substrate, EPMA was employed to analyze the element distribution of the cross-section and XRD was used to observe the phases present. After laser cladding, there was good metallurgical bonding between the coating and substrate and the coating exhibited no cracks. Three different regions of solidification microstructures were found from the bottom to the top surface of the melt pool. Region (I), which was closest to the substrate was found to be comprised of Ti columnar grains and uneven scattered ceramic particles. Region (II) consisted of fine equiaxed Al_2O_3 grains and region (III) showed Al_2O_3 grains that were coarse equiaxed and cellular with apparent isolation of TI and V elements. The ceramic phases that were present were α- Al_2O_3, some TiO_2 and Al_2TiO_5, and the cooling rate was too high to form sufficient γ-Al_2O_3. The authors therefore concluded that laser cladding of Al_2O_3-13%TiO_2 powders on Ti6Al4V alloy was completed successfully.

Weng et al. [32] investigated the microstructures and wear properties of Co-based composite coatings fabricated on Ti6Al4V by laser cladding. Titanium alloys are used in the automotive, aerospace, weapons and chemical industries due to their good corrosion resistance and high specific strength. Surface modification processes such as carburizing, nitriding, thermal spraying, physical vapor deposition and chemical vapor deposition are used to improve their surface properties in order to expand their application. The authors targeted improving the wear and friction properties of Ti6Al4V alloy by laser cladding. The substrate used was Ti6Al4V and the powders used were Co42, B_4C, SiC and Y_2O_3. Two mixtures of the powders were used for cladding; Co42-20B_4C-7SiC-1Y_2O_3 (specimen 1) and Co42-20B_4C-14SiC-1Y_2O_3 (specimen 2) in weight percentages. The powders were preplaced on the substrate and a cross flow CO_2 laser system was used for laser processing in an argon atmosphere. The microstructure was examined by SEM, the phase constituents was determined by XRD, a Vickers hardness tester was used to establish the microhardness and a disk wear tester was used to assess the wear resistance. The results revealed that the coatings mainly consisted of γ-Co/Ni solid solution and CoTi, $CoTi_2$, TiC, TiB_2, TiB, NiTi, Cr_7C_3 and Ti_5Si_3 phases. Specimen 1 was found to have a hardness of 1131 $HV_{0.2}$ while specimen 2 exhibited a hardness of 1314 $HV_{0.2}$ as compared to the hardness of the substrate, which was 350 $HV_{0.2}$. The wear tests showed that specimen had better wear performance. The authors concluded that the common wear mechanism was abrasive wear and the addition of 20% B_4C, 7% SiC, and 1% Y_2O_3 to the coating lead to excellent wear resistance. Too much SiC (14%) led to a coarser microstructure and the wear resistance is compromised as also reported by Fatoba et al. [33].

Carroll et al. [34] fabricated Ti6Al4V components by means of directed energy deposition additive deposition and examined the anisotropic behavior. Ti6Al4V has applications in the aerospace, medical device, sporting goods and petrochemical industries because of their high strength, low density and excellent corrosion resistance. Obtaining titanium is challenging hence the final products are costlier than their aluminum or steel counterparts. Powder-based additive manufacturing is a production technique in which a component is built on a work-piece by melting successive layers of metal feedstock. Additive manufacturing can produce a 3D component with complex parts and the same machine can be used to fabricate parts of different geometries. A Ti6Al4V plate was used as a substrate on which a Ti6Al4V cruciform was fabricated. A 2-kW laser was used to manufacture the laser build in an argon atmosphere. Tensile tests were conducted with a screw-actuated test frame, OPM was used to observe the microstructures and SEM was used to examine the fracture surface from the tensile tests and results were compared with those obtained with wrought Ti6Al4V. The yield strength of the additive manufactured was found to be 959 ± 22 MPa as compared to that of the wrought baseplate, which was established to be 973 ± 8 MPa and the ultimate tensile strength was determined to be 1064 ± 23 MPa for the additive manufactured versus 1050 ± 8 MPa of the wrought alloy. The authors concluded the mechanical properties of the additive manufactured part proportionate to those of the wrought material without any heat or pressure treatment after fabrication.

Gong et al. [35] examined how defects influenced the mechanical properties of Ti6Al4V parts manufactured by electron beam melting and selective laser melting. Additive manufacturing (AM) fabricates components from 3D CAD data, layer upon layer. The two most common AM processes based on powder-bed fusion are selective laser melting (SLM) and electron beam melting (EBM). The mechanical behavior of AM components can turn out greatly different from those that were manufactured conventionally due to their layered microstructure. The powers used were Ti6Al4V powders, the powder used for EBM was coarser (73 μm) and the one used for SLM (30 μm). Cylindrical bars were manufactured using both process and the parameters were varied for both processes. OPM was used to characterize the microstructure, the hardness was evaluated by a Rockwell C-Scale tester, tensile tests were completed using a tensile testing machine and fatigue tests were also conducted. The results showed that during SLM, defects were formed due to a low energy input and they caused the mechanical properties to be poor. Defects were also found in specimens that were fabricated with an excessive energy but they were not as detrimental as those produced by a lower energy. Large defects were found in EBM if the procedure used deviated from the optimum process parameters, which in turn led to poor mechanical properties. It was found that the microstructure formed depend mainly on the cooling rate rather than the defect generation as also reported by Makhatha et al. [36]. SLM and EBM-produced Ti6Al4V samples with comparable fatigue strength, tensile ductility and hardness but SLM produced slightly higher yield and tensile strengths.

Chikarakara et al. [37] worked on the high-speed laser surface modification of Ti6Al4V. The low density, high strength to weight ratio and biocompatibility affords Ti6Al4V application in biomedical engineering. However, wear and corrosion occurs when exposed to harsh conditions like the human due to some poor surface properties. When metallic ions are released and accumulate in the human body, discoloration of the nearby tissue will occur, or inflammatory reactions that lead to pain and osteolysis. The mechanical and tribological properties of titanium and its alloys are known to be improved by laser surface modification. Laser surface modification can serve multiple roles; its makes the surface more resistant to corrosion and wear, hardens the surface alloy, makes the surface more bioactive and stimulates bone growth because of an enhancement in wettability. A CO_2 continuous wave was used to irradiate Ti6Al4V alloy. The laser parameters were varied; the laser powers used were 100, 130 and 170 W, irradiances applied were 15.72, 20.44, 26.72 kW/m^2, scanning speeds of 2500, 3750 and 5000 mm/in and residence time of 2.16, 1.44 and 1.08 s. OPM and SEM were used to observe the microstructures, surface roughness was evaluated by a stylus profilometer, XRD was used to identify phase constituents and microhardness was determined with the aid of a Vickers hardness tester. The results showed that the surface roughness of the samples decreased when the irradiance and residence time increased. XRD showed that after laser processing, the α-Ti phase changed into an acicular structure cased within the aged β matrix. The highest microhardness was obtained with the combination of the highest irradiance and lowest residence time, which was 760 HV as compared to 460 ± 13 HV of the as-received alloy.

Yang et al. [38] investigated the corrosion behavior of additive manufactured Ti6Al4V alloy in NaCl solution. Because of the combination of low density, excellent strength, high biocompatibility and high fracture toughness, Ti6Al4V alloy is used in the marine, aerospace, energy and medical implant fields. Ti6Al4V alloy specimens produced by selective laser melting (SLM), selective laser treatment followed by heat treatment (SLM-HT), and wire and arc additive manufacturing (WAAM) had their microstructures observed and their electrochemical behaviors were tested. The heat-treated samples were heated to two different temperatures; namely 750°C (SLM-HT1) and 1020°C (SLM-HT2). A standard three-electrode electrochemical cell was used to evaluate the corrosion performance of the samples, OPM and SEM were utilized for microstructural examination and XRD was used to identify the phases present in the sample after they were laser processed. Microstructural analysis revealed that the SLM sample had comprised of α' martensites within columnar β grains. In SLM-HT1, the martensites decomposed into a fine lamellar combination of α and β phases inside the columnar prior β grains. However, SLM-HT2 exhibited coarse α and β phases and no columnar prior β grains. WAAM consisted of fine α lamellae in $\alpha + \beta$ matrix. The electrochemical tests showed that the best corrosion resistance was obtained by SLM-HT1, followed by WAAM, SLM-HT2 and lastly SLM. The leading form of corrosion failure for the samples was pitting corrosion. It was therefore concluded by the authors that the heat treatment of SLM samples can improve the corrosion behavior as also reported by Popoola et al. [39].

Farayibi et al. [40] studied laser deposition of Ti6Al4V wire with WC powder for functionally graded components. Titanium alloys are used in the chemical, oil and gas, petrochemical, and ground transport industries because of their corrosion and fatigue resistance, high strength to weight ratio, ductility and toughness. The tribological properties are poor due to low hardness and high friction coefficients and therefore limit their applications abrasive and sliding wear conditions. Ti6Al4V wire and WC powder were deposited on Ti6Al4V alloy substrate by means of a 2-kW ytterbium-doped continuous-wave fiber laser. Other process parameters were kept constant while the powder feed rate was varied. The parameters were set to a power of 1800 W, a transverse speed of 172 mm/min, wire feed rate of 800 mm/min and the powder rates used were 10, 15, 20, 25, 30 and 40 g/min. Vickers hardness tests were done on the surfaces of the samples, SEM was used to characterize the microstructure and XRD was employed to identify the phases present. The increase in the powder feed rate led to a decrease in the clad width and an increase in the height. It is important that this effect be understood for proper control of volumetric addition. The phases found in all the cladded samples were WC, W, TiC and β-TiC. The hardness increased with the increase in powder feed rate. An average range of 600–1030 $HV_{0.2}$ was found for rates with arrange of 10–40 g/min, which is a significant improvement from the hardness of the substrate (350 HV). The increase in hardness was attributed to the presence of W and TiC precipitates.

Mahamood et al. [41] examined the effect of laser power on the microhardness and microstructure during laser metal deposition of Ti6Al4V. The most commonly produced titanium alloy, Ti6Al4V, is commonly used in the aerospace industry. Titanium and its alloys are generally difficult to machine and the high temperatures and galling caused by the interaction of the tool and alloy leads to a shortened lifespan for the tool. Additive manufacturing is a technique of processing a material by adding the material in a layer by layer manner. Laser metal deposition is an additive manufacturing method achieved by feeding a powder into a melt pool formed by a focused laser beam on a substrate. A 4.4 kW fiber Nd-YAG laser was used to deposit Ti6Al4V powder on a Ti6Al4V substrate. The scanning speed used was 0.005 m/s, the powder feed rate was kept constant at 1.44 g/min and the laser powers used were 0.8, 1.2, 1.6, 2.0, 2.4, 2.8, 3.0 kW.

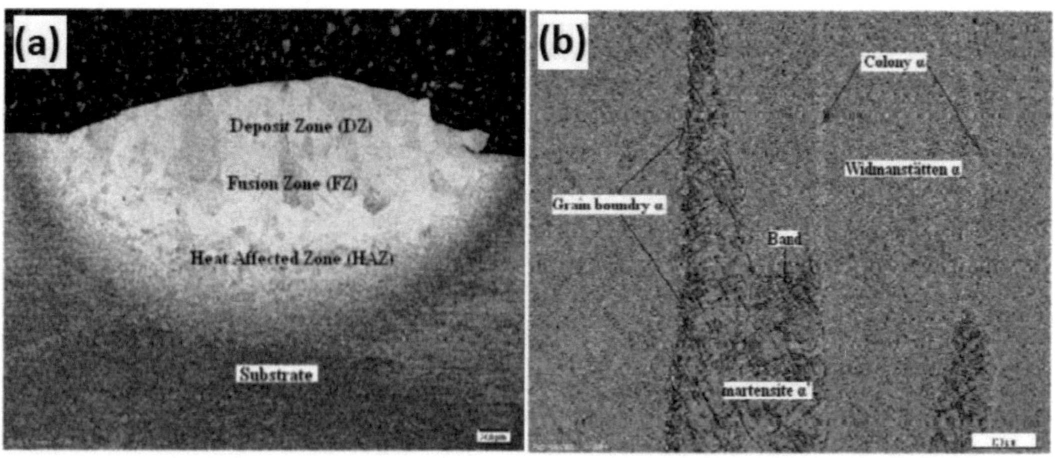

Figure 2. (a) Morphology of Sample A showing the different zones and (b) microstructure of the fusion zone of Sample G showing different α grains [41].

Sample designation	Laser power (kW)	Scanning speed (m/s)	Powder flow rate (g/min)	Gas flow rate (l/min)
A	0.8	0.005	1.44	4
B	1.2	0.005	1.44	4
C	1.6	0.005	1.44	4
D	2.0	0.005	1.44	4
E	2.4	0.005	1.44	4
F	2.8	0.005	1.44	4
G	3.0	0.005	1.44	4

Table 1. Processing parameters [41].

OPM was used to study the microstructure of the samples and the microhardness was evaluated by a Vickers hardness tester with a load of 500 g and a dwelling time of 5 s. the tests displayed that as the laser power increases, the microstructure of the heat affected zone became finer and the globular primary alpha phase became coarser. When the power increased, the microstructure change from fine martensite to thick martensite. The microhardness of the samples was found to increase with the increase in laser power. The authors therefore concluded that increasing the laser power of Ti6Al4V laser deposition will improve the properties (**Figure 2**, **Table 1**).

4. Overview on the rapid solidification, surface and fatigue properties of additive manufactured Ti-6Al-4V alloy

Rafi et al. [42] compared the mechanical properties and microstructures of Ti6Al4V components manufactured by electron beam melting and selective laser melting. Electron beam (EBM) and selective laser melting (SLM) are powder-bed fusion method of additive manufacturing used to produce metallic parts. SLM and EBM are advantageous over other conventional methods due to the freedom to manufacture complex shapes, elimination of expensive tooling and optimum material usage. SLM uses a laser heat source and its main process parameters are scan velocity, laser power, layer thickness and hatch spacing. EBM utilizes an electron beam to melt the powder layers. Ti6Al4V parts were fabricated using an SLM machine and an EBM machine by using Ti6Al4V powder. After manufacturing, the samples were taken for microstructural characterization by means of OPM and SEM. Phase composition was evaluated by XRD, hardness was determined by using a Rockwell hardness tester, tensile test were done at room temperature by means of a tensile testing machine and a fatigue testing machine was used to complete the fatigue tests. The surface finish of the parts showed that the EB samples had a rougher surface. The microstructure observed for the SL-produced parts consisted in martensitic α' microstructures while EBM-produced samples resulted in an α phase while the β phase separating the α lamellae. The highest tensile strength was obtained by the SLM-produced parts and the EBM-parts had higher ductility. The high strength was caused by the presence of martensitic α' microstructure in the SLM samples, while the lamellar α phase led to the high ductility. The lamellar phase in the EBM-produced parts resulted in a fatigue limit of 340 MPa and the SLM sample reached 550 MPa. The authors concluded that the selection of the additive manufacturing process will depend on the application but both methods exhibited excellent mechanical properties (**Figure 3**, **Table 2**).

Vrancken et al. [43] examined the microstructure and mechanical properties of heat-treated Ti6Al4V alloy produced by selective laser melting. Selective laser melting (SLM) is one of additive manufacturing techniques which has advantages over conventional methods, such as a high level of flexibility, near net shape production, reduced production steps and high material use efficiency. Ti6Al4V powder was used as a base powder for selective melting processing and it was hot forged and mill annealed. SLM was done with an SLM machine equipped with SMYb:YAG fiber laser, with a laser power of 250 W and a scanning speed of 1600 mm/s. The samples were heat treated in a vertical tube furnace and then three different cooling methods were employed. Furnace cooling was done by turning the furnace off, air

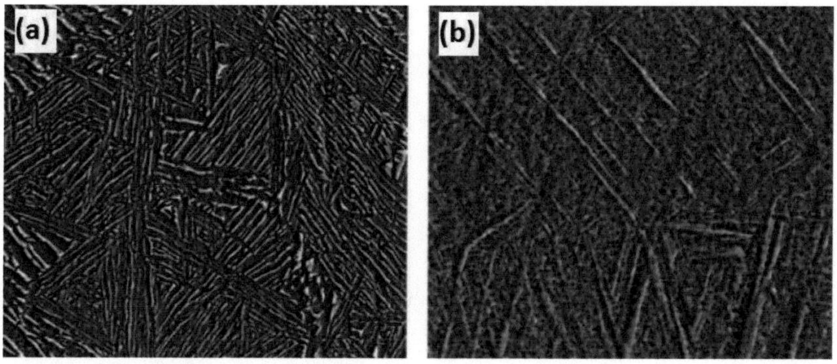

Figure 3. SEM images of (a) showing Widmanstatten structure in EBM-produced Ti-6Al-4V sample and (b) SLM-produced Ti-6Al-4V alloy sample [42].

	Stress at yield (offset 0.2%), MPa	Ultimate tensile stress, MPa	Strain at break, %
EBM (vertically built and machined)	869 (SD: 7.2)	928 (SD: 9.8)	9.9 (SD: 1.7)
SLM (vertically built and machined)	1143 (SD: 30)	1219 (SD: 20)	4.89 (SD: 0.6)
% Increase	31	31	−50
EBM (horizontally built and machined)	899 (SD: 4.7)	978 (SD: 3.2)	9.5 (SD: 1.2)
SLM (horizontally built and machined)	1195 (SD: 19)	1269 (SD: 9)	5 (SD: 0.5)
% Increase	33	30	−47
ASM handbook (Ref. [15]) (cast and annealed)	885	930	

SD: standard deviation.

Table 2. Tensile results for SLM-produced and EBM-produced Ti-6Al-4V alloy samples [42].

cooling at room temperature and water quenching. The heat treatment was performed on Ti6Al4V with α' microstructure, when heated, α phase was found to precipitate at the boundaries of α'. At the maximum temperatures below β transus, grain growth was prevented by the mixture of α and β phases formed. Microstructural revealed that for furnace cooling, the β grains altered to lamellar α + β phases, air cooling resulted in the β grains forming α-Widmanstätten clusters and water quenching caused the β grains to change into α' martensite. SLM samples had higher ultimate tensile strength and yield strength than the substrate. The best mechanical properties were produced by furnace cooling. It was therefore concluded that the mechanical properties depend on the heat treatment performed (**Figure 4**).

Hu et al. [44] studied how laser texturing of Ti6Al4V affected its tribological behavior. Titanium and its alloys have industrial and biomedical applications because of their excellent mechanical and chemical properties. Their applications are limited by their poor tribological performance; this is caused by their severe adhesive wear, high friction coefficients and sensitivity to fretting wear. Surface texturing has been introduced into a sliding contact surface and

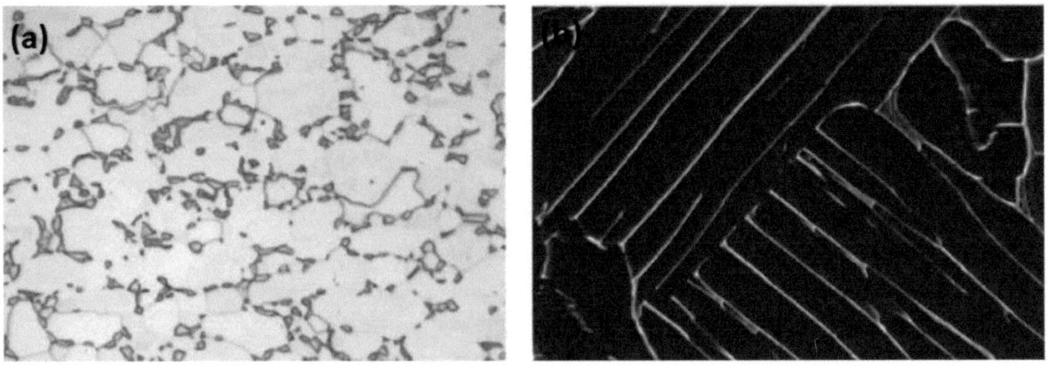

Figure 4. Microstructures of (a) the reference material (b) Ti-6Al-4V alloy produced by SLM after heat treatment [43].

was reported to remarkably affect the tribological properties. A Nd:YAG laser system was used to texture Ti6Al4V blocks and three different dimple diameters were fabricated; 45, 160 and 300 μm and the depth was 25 μm for all diameters. The surfaces of the textured and untextured samples were observed using a scanning electron microscope and an optical surface profiler, the microhardness was determined by means of a Vickers hardness tester and the friction tests were done with a pin-on-disk tribometer with the presence of a low viscosity oil lubricant. The hardness tests showed that of the textured and untextured samples was found to be the same, at a hardness of $295 \pm 15\ HV_{100}$. The friction test results showed that at a lower load (2 N), the friction coefficient of the textured samples was lower than that of the untextured Ti6Al4V, and the lowest coefficient was obtained by the 160 μm dimples. However, at a higher load (10 N), the friction coefficients of the samples were generally the same. The authors concluded that dimple texture results in a lower friction coefficient and the best dimple diameter for friction reduction was 160 μm.

According to Gong et al. [45] additive manufactured parts are subject to inclusion of defects which negatively affects the materials mechanical properties, development of these artifacts is dependent on the choice of process parameter. Authors reported the defects generation mechanism of Ti-6Al-4V parts fabricated by selective laser melting (SLM) and electron beam melting (EBM) at varied process parameters. The results revealed a significant effect of energy input on defect generation on both the above-mentioned processes. However, different defect generation mechanism was observed when excessively high or low power density was used. In SLM, the use of high energy density leads to over-melting of powders and the vaporization of the melt pool result in defects generation. Whereas for EBM processes no bubbles or pores were observed, this was attributed to the ability of autonomously avoiding over-melting and evaporation. In conclusion, it was pointed out that the choice of process parameters governs the generation of defect. Fatoba et al. [46] and Aigbodion et al. [47] also agreed that optimizing process parameters are very necessary in avoiding cracks and pores.

Hrabe et al. [48] investigated the effects of residual stresses and internal defects (pores and voids) on fatigue performance of Ti-6Al-4V fabricated by electron beam melting. The fatigue properties of the as-fabricated, stress-relieved and the hot isostatic pressed (HIPed) samples were comparatively studied. The results obtained displayed insignificant residual stress in all samples which was attributed to high process temperatures of over 600°C. The chemical compositions of all samples were unchanged, with only fatigue strength of HIPed being the different than in other conditions. However, the minimal microstructural coarsening, reduced porosity/void density and an increase in fatigue strength were observed in HIPed samples. It was concluded that voids were the initiation sites for the fatigue crack, which meant reducing their frequency and density plays a role in improving fatigue strength (**Figures 5** and **6**).

Cunningham et al. [49] investigated the effect of processing parameters on porosity of electron beam melted Ti-6Al-4V. The fatigue properties of a material are affected by porosity; these are points of weakness where cracks are initiated. Synchrotron-based X-ray microtomography characterization was carried out to determine the pore size, shape and spatial distribution of samples fabricated at various process conditions. The average diameter of all present pores was found to be less than 10 microns. However, some pores of about 50 microns in size were present at the surface of the samples of large melt pools. The volume of spherical pores which are associated with trapped gases displayed decrease with decrease in speed function. The authors further pointed out that the decrease in melt pool area significantly increases the lack of fusion porosity.

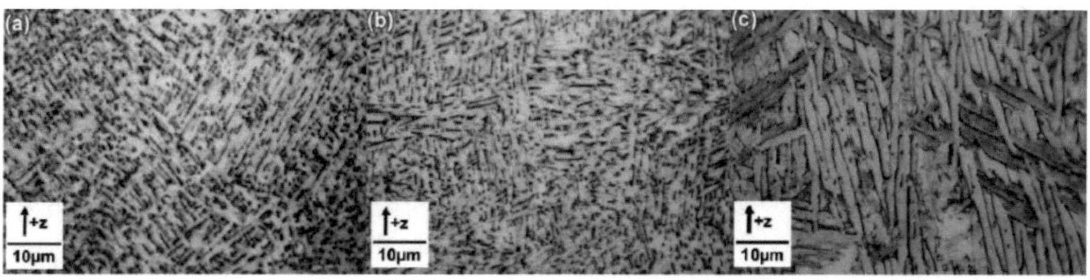

Figure 5. Optical microscope images for the (a) as-built (b) stress-relieved and (c) HIPed conditions showing expected equilibrium acicular or Widmanstatten microstructure [48].

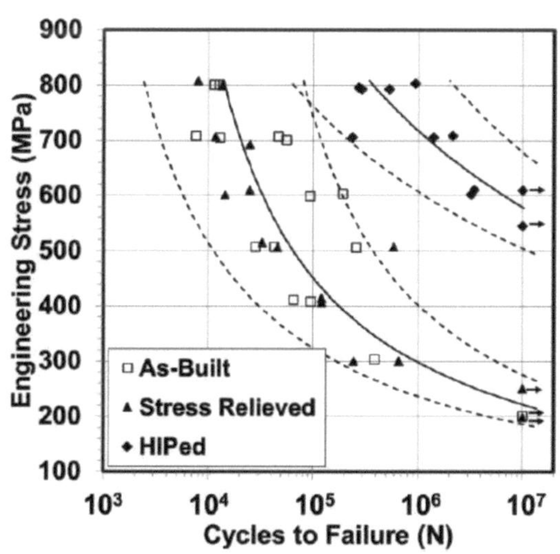

Figure 6. S-N curve fatigue results for all three conditions [48].

Wang et al. [50] successfully fabricated Ti-6Al-4V by laser additive manufacturing with the aim of investigating the build direction on microhardness and Tribology behavior. The phase analyses of the fabricated samples obtained by XRD revealed alpha, alpha martensite and beta, which is uncharacteristic of the conventional phases found in typical Ti-6Al-4V which are alpha + beta phases co-existing. The peak microhardness value was obtained on the vertically built sample and the same sample displayed best wear resistance improvement at higher applied loads. Authors also pointed out that slow cooling rates at the middle of the built samples attributed to phases responsible for high hardness in this region. This corrobates the works of Adebiyi et al. [51] and Fatoba & Makhatha [52].

Zhao et al. [53] studied the evolution of plastic deformation and the effect it has on mechanical properties of extra low industrial Ti-6Al-4V repaired by laser additive technique. The microstructure of the laser additive manufactured samples presented a superfine microstructure within the columnar grains while the wrought substrate that the fabrication was done on displayed a Widmanstatten microstructure with coarse equiaxial grains as also reported by Adesina et al. [54]. The micro-hardness profile displays an increase of the microhardness property from the substrate toward the direction of build which is attributed to the rapid cooling rate of the molten powders. The tensile strength and ductility of the laser additive manufactured sample were both greater than that of the wrought titanium substrate; however, the opposite can be said with elongation.

Denlinger et al. [55] studied the effect of the inter-layer dwell time times (period whereby the laser is at rest) on distortion and residual stresses induced on titanium and nickel alloys during additive manufacturing. The dwell of 0, 20 and 40 s were used for each material while keeping all other processing parameters constant. The obtained results displayed that the accumulated distortion increases with increased dwell time for Ti-6Al-4V, whereas it remained relatively constant for the nickel alloy. Furthermore, authors pointed out that majority of distortion occurred in the first two layers of the Ti-6Al-4V build, which drastically reduces with each layer and becomes more dwell time dependant. In conclusion, less distortion was attained in the Ti-6Al-4V fabricated with no dwell time and all the nickel alloys having greater distortion when compared to the Ti-6Al-4V counterparts.

5. Conclusion

The success of laser AM will rely on the detailed understanding of the relationship between the process parameters and the resulting material properties. This is because of the exceptionally high solidification rates caused by the unusually high laser heat input, which results in highly non-equilibrium microstructures. While these microstructures are mostly beneficial to the mechanical properties of the component, they cannot be predicted from conventional phase diagrams. It means they must be studied experimentally and numerically. Moreover, an understanding of the relationship between process parameters and material defects such as cracks, voids, inclusions and lack of fusion must be developed experimentally. Solving these problems requires new approaches to the understanding of the laser-matter interaction and the manufacturing processes of the near net shaped components. Hence, tailored parts and components can be manufactured, repaired, and modified by AM. Near-net-shape AM techniques offer new opportunities about a sustainable protection of (material and energy) resources due to a very high material efficiency and very low energy consumption.

Author details

Olawale Samuel Fatoba[1]*, Esther Titilayo Akinlabi[1] and Mamookho Elizabeth Makhatha[2]

*Address all correspondence to: drfatobasameni@gmail.com

1 Department of Mechanical Engineering Science, Faculty of Engineering and the Built Environment, University of Johannesburg, South Africa

2 Department of Metallurgy, Faculty of Engineering and the Built Environment, University of Johannesburg, South Africa

References

[1] Adesina OS, Popoola API, Fatoba OS. Laser surface modification — A focus on the wear degradation of titanium alloy. In: Paul M, editor. Fiber Laser. InTech; 2016. DOI: 10.5772/61737. Available from: http://www.intechopen.com/books/fiber-laser/

[2] Attar H, Prashanth KG, Chaubey AK, Calin M, Zhang LC, Scudino S, Eckert J. Comparison of wear properties of commercially pure titanium prepared by selective laser melting and casting processes. Materials Letters. 2015;**142**:38-41

[3] Froes FH, Gungor MN, Imam MA. The Minerals, Metals and Materials Society JOM. 2007;**59**:28-31

[4] Vrancken B, Thijs L, Kruth J-P, Van Humbeeck J. Heat treatment of Ti6Al4V produced by Selective Laser Melting: Microstructure and mechanical properties. Journal of Alloys and Compounds. 2012;**541**:177-185

[5] Baumer M, Dickens P, Tuck C, Hague R. The cost of additive manufacturing: Machine productivity, economies of scales and technology-push. Technological Forecasting and Social Change. 2016;**102**:193-201

[6] Ferdinand J-P, Petschow U, Dickel S. The Decentralized and Network Future of Value Creation: 3D Printing and its Implications for Society, Industry and Sustainable Development. City (Berlin): Springer International Publishing AG Switzerland; 2016. pp. 76-77

[7] Pinkerton AJ. [INVITED] Lasers in additive manufacturing. Optics & Laser Technology. 2016;**78**:25-32

[8] Patterson AE, Messimer SL, Farrington PA. Overhanging features and the SLM/DMLS residual stresses problem: Review and future research need. Technologies. 2017;**5**(2):15

[9] Hanzl P et al. The influence of processing parameters on the mechanical properties of SLM parts. Procedia Engineering. 2015;**100**:1405-1413

[10] Francois MM, Sun A, King WE, Henson NJ, Tourret D, Bronkhorst CA, Carlson NN, Newman CK, Haut T, Bakosi J, Gibbs JW, Livescu V, Vander Wiel SA, Clarke AJ, Schraad MW, Blacker T, Lim H, Rodgers T, Owen S, Abdeljawad F, Madison J, Anderson AT, Fattebert J-L, Ferencz RM, Hodge NE, Khairallah SA, Walton O. Modeling of additive manufacturing processes for metals: Challenges and opportunities. Current Opinion in Solid State and Material Science. 2017

[11] Vora D, Srinisvasan G, Rajamure NDR. Laser Alloyed AL-W Coatings on Aluminium for Enhanced Corrosion. Elsevier; December 2014. p. 10

[12] Bhushan B, Gupta BK. Handbook of Tribology: Materials, Coatings, and Surface Treatments; 1991

[13] de Oliveira U, Ocelik V, De Hosson JTM. Analysis of coaxial laser cladding processing conditions. Surface and Coatings Technology. 2000;**197**:127-136

[14] Betts JT. Practical methods for optimal control and estimation using nonlinear programming. Society for Industrial and Applied Mathematics. SIAM Review. 2010;**53**(1):183–185

[15] Watkins KG, McMahon MA, Steen WM. Microstructure and corrosion properties of laser surface processed aluminium alloys: A review. Journal of Material Science and Engineering. 1997;**231**:55-61

[16] Fatoba OS, Popoola API, Farotade GA, Pityana SL. Computational dynamics of laser alloyed metallic materials for improved corrosion performance: Computational dynamics of laser alloyed metallic materials; 2016. DOI: 10.4018/978-1-5225-03293.ch008. http://www.igiglobal.com/chapter/computational-dynamics-of-laser-alloyed-metallic-materials-for-improved-corrosion performance/149842

[17] Kwok CT, Cheng FT, Man HC. Cavitation erosion and corrosion behaviour of laser-aluminized mild steel. Surface and Coatings Technology. 2006;**200**:3544-3552

[18] Fatoba OS, Popoola API, Fedotova T, Pityana SL. Electrochemical studies on the corrosion behaviour of laser alloyed Zn-Sn coatings on UNS G10150 steel in 1M HCl solution. Silicon. 2015;**7**(4):357-369

[19] Fagagnolo JB, Pallone E, Martin DR, Kiminami CS, Bolfarini C, Botta W. Processing of Al matrix composites reinforces with Al-Ni compounds and Al_2O_3 by reactive milling and reactive sintering. Journal of Alloys and Compounds. 2009;**471**:448-452

[20] Adebiyi DI, Popoola API, Pityana SL. Microstructural evolution at the overlap zones of 12Cr martensitic stainless steel laser alloyed with TiC. Optics and Laser Technology. 2014;**61**:15-23

[21] Yakovlev A, Bertrand PH, Smurov I. Laser cladding of wear resistant metal matrix composite coatings. Thin Solid Films. 2004;**453**:133-138

[22] Poulon-Quintin A, Watanabe I, Watanabe E, Bertrand C. Mechanical and microstructural properties of surface treated cast titanium with Nd: YAG laser. Dental Materials. 2012;**24**: 2769-2783

[23] Zhou R, Sun GF, Chen KK, Tong YQ. Effect of tempering on microstructure mechanical properties of cast iron rolls laser alloyed with C-B-W-Cr. Materials Science and Engineering. 2014;**62**:1-7

[24] Fatoba OS, Popoola API, Pityana SL, Adesina OS. Computational dynamics of anti-corrosion performance of laser alloyed metallic materials. In: Paul M, editor. Fiber Laser. InTech; 2016. DOI: 10.5772/62334. Available from: http://www.intechopen.com/books/fiber-laser/computational-dynamics-of-anti-corrosion-performance-of-laser-alloyed-metallic-materials

[25] Popoola API, Fatoba OS, Aigbodion VS, Popoola OM. Tribological evaluation of mild steel with ternary alloy of Zn-Al-Sn by laser deposition. International Journal of Advanced Manufacturing Technology. 2016:1-7. DOI: 10.1007/s00170-016-9170-7

[26] Xu W, Sun S, Elambasseril J, Liu Q, Brandt M, Qian M. Ti-6Al-4V additively manufactured by selective laser melting with superior mechanical properties. The Minerals, Metals & Materials Society. 2015;**67**:668-673

[27] Balla VK, Sonderlind J, Bose S, Bandyopadhyay A. Microstructure, mechanical and wear properties of laser surface melted Ti6Al4V alloy. Journal of the Mechanical Behaviour of Biomedical Materials. 2014;**32**:335-344

[28] Brandl E, Schoberth A, Leyens C. Morphology, microstructure, and hardness of titanium (Ti-6Al-4V) blocks deposited by wire-feed additive layer manufacturing (ALM). Materials Science and Engineering A. 2012;**532**:295-305

[29] Makuch N, Kulla M, Dziarski P, Przestacki D. Laser surface alloying of commercially pure titanium with boron and carbon. Optics and Lasers in Engineering. 2014;**57**:64-81

[30] Yang Y, Guo N, Li J. Synthesizing, microstructure and microhardness distribution of Ti–Si–C–N/TiCN composite coating on Ti–6Al–4V by laser cladding. Surface & Coatings Technology. 2013;**219**:1-7

[31] Chen Y, Wu D, Ma G, Lu W, Guo D. Coaxial laser cladding Al_2O_3-13%TiO_2 powders on Ti6Al4V alloy. Surface & Coatings Technology. 2013;**228**:S452-S455

[32] Weng F, Yu H, Chen C, Dai J. Microstructures and wear properties of laser cladding Co-based composite coatings on Ti–6Al–4V. Materials and Design. 2015;**80**:174-181

[33] Fatoba OS, Popoola O, Popoola API. The effects of silicon carbide reinforcement on the properties of Cu/SiCp composites. Silicon. 2015;**7**:351-356

[34] Carroll BE, Palmer TA, Beese AM. Anisotropic tensile behavior of Ti–6Al–4V components fabricated with directed energy deposition additive manufacturing. Acta Materialia. 2015;**87**:309-320

[35] Gong H, Rafi K, Gu H, Ram GDJ, Starr T, Stucker B. Influence of defects on mechanical properties of Ti–6Al–4 V components produced by selective laser melting and electron beam melting. Materials and Design. 2015;**86**:545-554

[36] Makhatha ME, Fatoba OS, Akinlabi ET. Effects of rapid solidification on the microstructure and surface analyses of laser-deposited Al-Sn coatings on AISI 1015 steel. The International Journal of Advanced Manufacturing Technology. 2017:1-15. https://doi.org/10.1007/s00170-017-0876-y

[37] Chikarakara E, Naher S, Brabazon D. High speed laser surface modification of Ti–6Al–4V. Surface & Coatings Technology. 2012;**206**:3223-3229

[38] Yang J, Yang H, Yu H, Wang Z, Zeng X. Corrosion behavior of additive manufactured Ti-6Al-4V alloy in NaCl solution. Metallurgical and Materials Transactions A. 2017. DOI: 10.1007/s11661-017-4087-9

[39] Popoola API, Fatoba OS, Popoola OM, Pityana SL. The influence of heat treatment and process parameter optimization on hardness and corrosion properties of laser alloyed X12CrNiMo steel. Silicon. 2016;**8**(4):579-589

[40] Farayibi PK, Folkes JA, Clare T. Laser deposition of Ti-6Al-4V wire with WC powder for functionally graded components. Materials and Manufacturing Processes. 2013;**28**: 514-518

[41] Mahamood RM, Akinlabi ET, Shukla M, Pityana S. Laser metal deposition of Ti6Al4V: A study on the effect of laser power on microstructure and microhardness. In: Proceedings of the International MultiConference of Engineers and Computer Scientists. Vol. II; 2013

[42] Rafi HK, Karthik NV, Gong H, Starr TL, Stucker BE. Microstructures and mechanical properties of Ti6Al4V parts fabricated by selective laser melting and electron beam melting. Journal of Materials Engineering and Performance. 2013;**22**:3872-3883

[43] Vrancken B, Thijs L, Kruth J-P, Van Humbeeck J. Microstructure and mechanical properties of a novel beta titanium metallic composite by selective laser melting. Acta Materialia. 2014; **68**:150-158

[44] Hu T, Hu L, Ding Q. Effective solution for the tribological problems of Ti-6Al-4V: Combination of laser surface texturing and solid lubricant film. Surface & Coating Technology. 2012;**206**:5060-5066

[45] Gong H, Rafi K, Gu H, Starr T, Stucker B. Analysis of defect generation in Ti–6Al–4V parts made using powder bed fusion additive manufacturing processes. Additive Manufacturing. 2014;**1-4**:87-98

[46] Fatoba OS, Popoola API, Aigbodion VS. Experimental study of hardness values and corrosion behaviour of laser alloyed Zn-Sn-Ti coatings of UNS G10150 mild steel. Journal of Alloys and Compounds. 2016;**658**:248-254

[47] Aigbodion VS, Popoola API, Fatoba OS. Evaluation of hardness values and corrosion behaviour of laser alloyed 20Al-20Sn-60Ti coatings of UNS G10150 mild steel. International

Journal of Advanced Manufacturing Technology. 2015;**81**:1-11. DOI: 10.1007/s00170-015-8111-1

[48] Hrabe N, Gnäupel-Herold T, Quinn T. Effects of internal defects and residual stress on fatigue properties of a titanium alloy (Ti-6Al-4V) fabricated via electron beam melting (EBM). In: World Materials Research Institutes Forum International Workshop for Young Scientists; 2016

[49] Cunningham R, Nicolas A, Madsen J, Fodran E, Anagnostou E, Sangid MD, Rollett AD. Analyzing the effects of powder and post-processing on porosity and properties of electron beam melted Ti-6Al-4V. Materials Research Letters. 2017. DOI: 10.1080/21663831.2017.1340911

[50] Wang P et al. Microstructure and mechanical properties of a newly developed low Young's modulus Ti–15Zr–5Cr–2Al biomedical alloy. Materials Science and Engineering: C. 2017;**72**:536-542

[51] Adebiyi DI, Fatoba OS, Pityana SL, Popoola API. Parameters optimization, microstructure and microhardness of silicon carbide laser deposited on titanium alloy. In: Proceedings of International Conference on Surface Modification Technologies; 29th June-1st July, 2016; Milan, Italy; 2016. pp. 1-6

[52] Fatoba OS, Makhatha ME. Improved corrosion, mechanical and tribological properties of Ti-6Al-4V alloy by direct laser deposition. In: Paul M, editor. Nova Science Publishers Book Chapter; 2017. https://www.novapublishers.com/catalog/product_info.php?products_id=62603

[53] Zhao Z, Chen J, Tan H, Lin X, Huang W. Evolution of plastic deformation and its effect on mechanical properties of laser additive repaired Ti64ELI titanium alloy. Optics & Laser Technology. 2017;**92**:36-43

[54] Adesina OS, Popoola API, Fatoba OS. Laser surface modification—A focus on the wear degradation of titanium alloy. In: Paul M, editor. Fiber Laser. InTech; 2016. DOI: 10.5772/61737. Available from: http://www.intechopen.com/books/fiber-laser/

[55] Denlinger ER, Heigelb JC, Michalerisb P, Palmerc TA. Effect of inter-layer dwell time on distortion and residual stress in additive manufacturing of titanium and nickel alloys. Journal of Materials Processing Technology. 2015;**215**:123-131

Chapter 11

Aluminum Sheet Metal Damage Mechanisms Application to Trimming and Hemming

Helmut Klöcker and Christopher Yukna

Additional information is available at the end of the chapter

http://dx.doi.org/10.5772/intechopen.70687

Abstract

Aluminum alloys are increasingly used in automotive structural applications thanks to their combination of low density, high strength, and good formability. Appropriate strength levels for dent resistant automotive aluminum panels may be achieved by a combination of alloy design and thermo-mechanical treatments. Unfortunately, increased strength (hardness) generally limits homogeneous plastic deformation. Thus, when shaping complex components large deformation and eventually strain localization, necking and final failure is more likely to occur. In this chapter, we endeavor to explore the best compromises between increased hardening and satisfying formability. Sheet metal shaping generates large strains and eventually strain localization. The accurate measure of metal behavior is discussed first. The following section is dedicated to the kinetics of damage development during shaping. Then metal ductility at low stress triaxiality and consequent strain localization are discussed. Finally, trimming and hemming, the most complex shaping operations to perform efficiently are analyzed providing a powerful tool for alloy design.

Keywords: plane strain compression, extended Bridgman model, damage criterion, trimming, hemming, aluminum automotive alloy design

1. Introduction

Aluminum alloys are increasingly used in automotive structural applications thanks to their combination of low density, high strength, good formability and the bonus of better corrosion resistance. Appropriate hardness levels for dent resistant automotive aluminum panels (doors, hoods) may be achieved by a combination of alloy design and thermo-mechanical treatment. To insure adequate hardness values, most aluminum automotive alloys contain large amounts of hard intermetallic particles. In ductile metals, voids are first nucleated by decohesion or cracking of these second phase particles and then grow until they eventually coalescence to form a macroscopic crack. Hence, increased strength (hardness) generally limits homogeneous plastic deformation. In the early 70s a new branch of material science ("the local approach of ductile fracture"), was initiated by the pioneering contributions of Gurland and Plateau [1], McClintock [2], Rice and Tracey [3], Koplik and Needleman [4], Gurson [5], Le Roy et al. [6], Hutchinson and Tvergaard [7]. One outcome of all these publications is the major role played by stress triaxiality. Furthermore, Marciniak and Kuczynski [8] highlighted the tremendous effect of strain localization when sheet metal is submitted to large strains. Unfortunately, shaping complex components imposes large deformations and eventually strain localization.

In Section 2, we outline the accurate characterization of the sheet material behavior at large strains by adequate compression tests. Next, the measurement of the macroscopic strain to failure by standard tensile tests is offered. In the same section, the relation between this macroscopically determined characteristic and the local strain and stress triaxiality field is analyzed. Then, the kinetics of strain localization causing damage is presented. As aforementioned, we end with two particularly discriminating forming operations, trimming and hemming, which are examined in detail.

2. Characterization of sheet metal behavior

During deep drawing or cutting operations of sheet metal local strains up to 200% are reached. Hence, large deformation compression and tensile tests on rolled sheets reveal the necessity for finite element simulations of machining and forming operations. In Section 2.1, the most commonly used compression tests providing sheet metal characterization are shown. These determine material behavior under large strains without significant damage. As sheet shaping and forming operations are limited by necking and consequent failure, Section 2.2 will demonstrate how to determine the failure strain by classic tensile tests.

2.1. Measuring stress-strain curves to large strains on sheet metal

Due to necking, only comparatively small strains can be attained in standard tensile tests. Standard uniaxial compression tests on cold rolled sheets (typically 1 mm thickness) are also impossible due to friction. Finally, torsion tests on cold rolled 1-mm-thick sheets would lead to extremely small sample sections. In this section the experimental techniques particularly suited for high strain characterization will be presented.

Half a century ago, Watts and Ford [9] introduced the **P**lane **S**train **C**ompression punching technique to determine the room temperature behavior of strip metal by compressing by means of a punching tool which is longer than the test-piece. However using this technique, the sheets can only be compressed along the normal axis. The maximum strain is much less than unity and the deformed sample volume is very small. Recently, Vegter et al. [10] determined the initial yield stress of sheet metal by several tests (tension, plane strain tension, compression of layered samples and shear tests). The tension and compression tests are limited to strains smaller than 0.2 and the shear test is restricted to strains smaller than 0.4.

The ***channel-die test*** is an alternative plane strain compression test without any lateral spreading, but which was then relatively unknown and has so far been relatively little employed. Maurice and Driver [11, 12] have developed original hot channel-die equipment which has the advantage of imposing true plane strain compression without major friction as shown by Bacha et al. [13]. **Figure 1** shows a schematic of the channel-die apparatus first described by Maurice and Driver [11]. The channel is composed of 3 separate steel plates of which the center plate has the same thickness (7 mm) as the punch width. The equipment of **Figure 1** was installed in an a Schenck 100 kN servo-hydraulic machine whose electronic and hydraulic control systems enable constant plastic strain rate tests up to about 20 s^{-1} (according to sample dimensions). Loads are measured by standard load cells and displacements by means of a linear variable differential transformer (LVDT transducer) situated between the crosshead and the base of the channel-die. More details concerning the channel-die equipment can be found in the original papers [11, 12]. Channel-die tests have also been used for room temperature testing on mono-block samples [14].

To our knowledge, Bacha et al. [15] were the first to apply PSC to layered sheet specimens (**Figure 2**). The sheets were machined to obtain seven strips 1 mm thick 9.01 mm high and 10.01 mm long and bonded with standard cyanoacrylate glue. The sample preparation is crucial to insure plane mutually orthogonal surfaces. The specifics can be found in Bacha et al. [15]. The area reduction due to the glue layer is thus smaller than 1%. Since four major surfaces of the sample are in contact with the tools, it is wrapped in a PTFE (Teflon) ribbon (50 μm thickness).

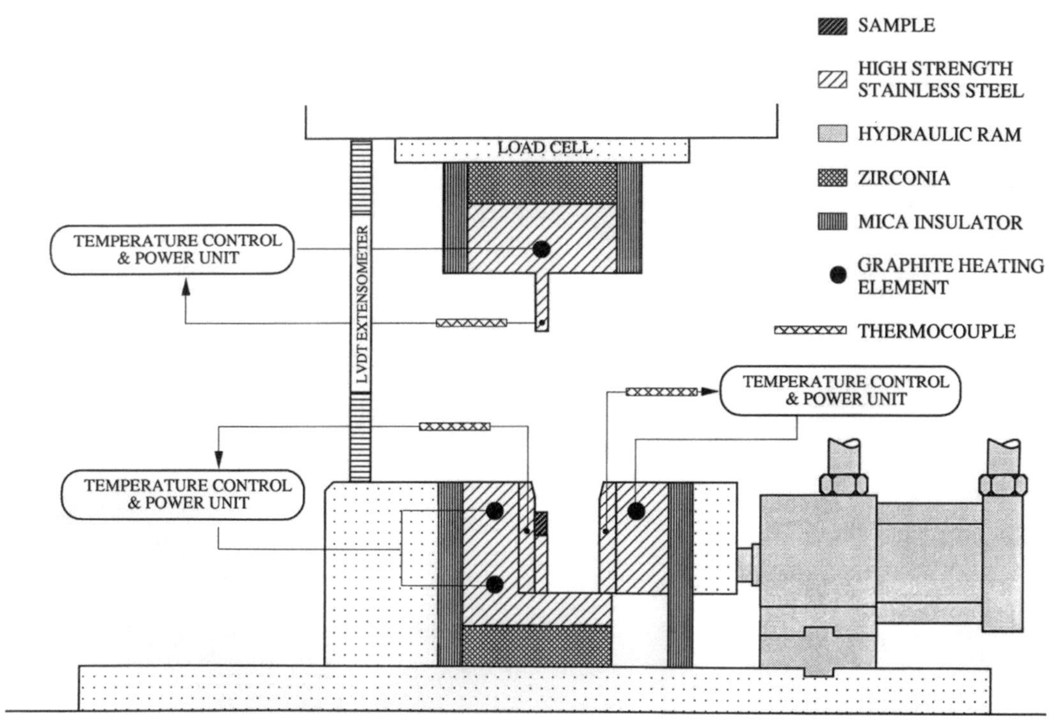

Figure 1. Schematic of the channel-die equipment [11].

Bacha et al. [15] compressed two aluminum alloys, AA5182 and AA6016, widely used in sheet forming particularly in the automotive industry, at a constant strain rate of 10^{-1} s^{-1}, by channel-die compression. For the AA6016 alloy two different grades (***a, b***) were considered. **Table 1** gives the chemical composition of the alloys.

Brunet and Morestin [16] determined the Langford coefficient of the AA6016 sheets by tensile tests and showed that the anisotropy is negligible for the two grades of AA6016. The 5182 alloy was compressed along rolling and transverse directions and revealed in plane isotropy.

To determine reliable stress-strain curves, particular attention should be paid to the influence of the PTFE-layer and the compliance of the test rig. The displacement was corrected to take into account the deformation of the test rig and the PTFE-layer [17]:

$$d_{corrected} = \left(d_{measured} - \frac{Q_{measured}}{k}\right)\left(1 - \frac{h_{PTFE}}{h_0}\right) \quad (1)$$

h_0 and h_{PTFE} designate the initial sample and PTFE-layer thicknesses. $d_{measured}$ is the displacement measured by the LVDT and $d_{corrected}$ takes into account the deformation of the PTFE layer. $Q_{measured}$ is the measured load. k is the stiffness of the test rig and was determined by measuring the punch displacement without a sample present in the channel (k = 91 kN/mm). The measured load has been corrected to take into account friction effects [17] as follows:

$$Q_{corrected} = \frac{Q_{measured}}{1 + \frac{\dot{W}_F}{\dot{W}_P}} \quad (2)$$

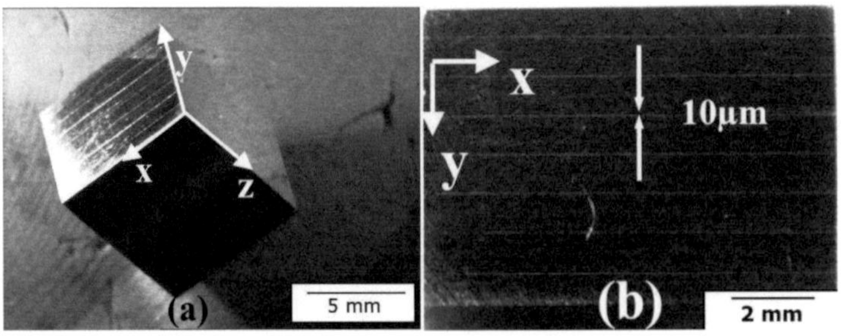

Figure 2. Layered sheet sample (a) 3D view, (b) view along the direction of the compression z.

Grade	Mg	Si	Cu	Fe	Mn
5182	4.25	0.09	0.06	0.29	0.35
6016 (a)	0.33	1.03	0.12	0.3	0.11
6016 (b)	0.43	0.98	0.15	0.3	0.11

Table 1. Chemical composition of the different grades.

where $Q_{corrected}$ is the corrected value of the applied load $Q_{measured}$. $\dot{W}_F$ and $\dot{W}_P$ are the work rates associated respectively to friction and to plastic deformation. The ratio between these work rates was estimated by [17] as

$$\frac{\dot{W}_F}{\dot{W}_P} = \frac{\overline{m}}{12 l_0 h_0 e_0} \left[\frac{l^3}{4} \ln\left(\frac{Z+2h}{Z-2h}\right) + 2h^3 \ln\left(\frac{Z+l}{Z-l}\right) + 2lhZ + l^2 e_0 \right] \tag{3a}$$

$$Z = \sqrt{l^2 + 4h^2} \tag{3b}$$

where $\overline{m}$ is the Tresca friction coefficient .h_0, l_0, and h, l are respectively the initial and current sample height and length. e_0 corresponds to the sample and channel width (7 mm).

To determine the equivalent stress $\overline{\sigma}$ and the equivalent strain $\overline{\varepsilon}$, the overall load was corrected by Eqs. (2), (3a) and (3b) with a value of $\overline{m} = 0.02$. However the shear stresses and strains due to friction were neglected, i.e., the stress and the strain rate tensors were assumed diagonal. Thus, the equivalent von Mises stress and strain may be written:

$$\overline{\sigma} = \frac{\sqrt{3}}{2} \sqrt{\frac{Q_{corrected}}{e_0 l_0} \left(1 - \frac{d_{corrected}}{h_0}\right)}, \quad \overline{\varepsilon} = \frac{2}{\sqrt{3}} \ln\left(\frac{h_0}{h_0 - d_{corrected}}\right) \tag{4}$$

Figure 2 shows equivalent stress *vs.* equivalent strain curves for all the grades. To represent the material flow stress, σ_Y, at large deformations an extended Voce model [18, 19] has been chosen.

$$\begin{cases} \sigma_Y = \sigma_{Yo} + (\sigma_\infty - \sigma_{Yo})(1 - \exp(-\delta \overline{\varepsilon}_p)) + \alpha \overline{\varepsilon}_p \\ \overline{\varepsilon}_p = \overline{\varepsilon} - \dfrac{\sigma_Y}{E} \end{cases} \tag{5}$$

E is the Young's modulus. $\overline{\varepsilon}$ and $\overline{\varepsilon}_p$ are respectively the total and the plastic equivalent strain. σ_{Yo} is the initial flow stress. σ_∞ corresponds to the stress at very large plastic strains in the Voce model. $\alpha \overline{\varepsilon}_p$ is an additional term accounting for small positive strain hardening at very large plastic deformation. The initial flow stress σ_{Yo} was determined previously in simple tension. **Table 2** summarizes the coefficients of the different Voce models. The most significant results are the different values of the α parameter for 5182 and 6016 alloys (**Figure 3**).

To assess the validity of the experimental method described above, Bacha et al. used the flow rule (5) and reproduced the experimental load displacement curves by finite element model [15]. Moreover, the finite element calculations showed that, using a Teflon film, the friction

Grade	E (GPa)	σ_o (MPa)	σ_∞ (MPa)	δ	α (MPa)
AA5182	70	130	264	14.6	101
AA6016a	70	130	310	16	15
AA6016b	70	100	248	12	14

Table 2. Voce model parameters for the AA5182 and AA6016 grades.

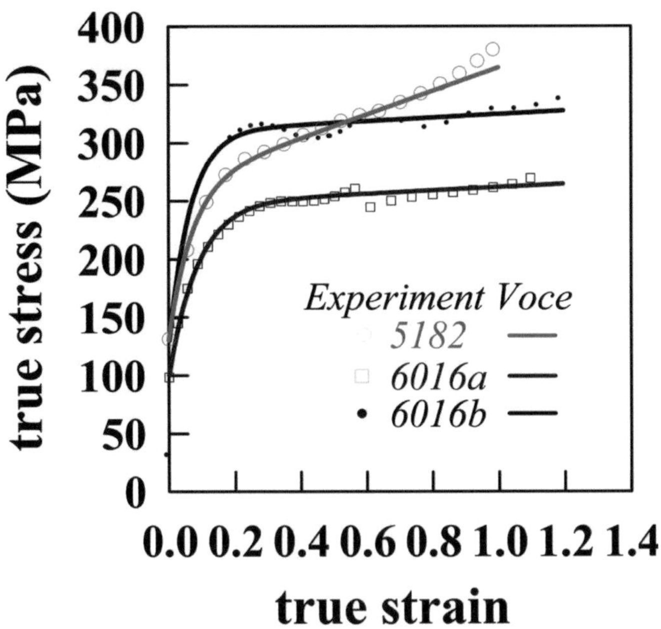

Figure 3. Measured true stress-strain curves compared to extended Voce model.

coefficient between the test sample and the tool is less than 0.02 and the friction coefficient between different layers of the sample only play a very minor role. The sample deforms homogeneously over more than 80% of its volume irrespective of interlaminate (glue) behavior.

Barlat et al. [20] measured the initial yield surface by biaxial compression tests on cubic specimens made from laminated binary Al-Mg alloy sheet samples. They used an expensive biaxial compression testing machine. Because of the physical constraints of the loading dies, their deformation step was limited to a strain of about 0.1. This technique was more costly than the simple one presented here.

To our knowledge all previous methods (tension, shear, and compression) measure the initial yield strength but are not very useful for analyzing the strain hardening of the material. The PSC method is therefore complementary to these techniques, and hence particularly useful for obtaining large-strain material behavior. However, very thin sheets (of thickness less than 1 mm) are also widely used in industry. The stress-strain curves of these sheets are even more difficult to obtain than the stress-strain curves of the larger 1 mm ones. Determining the minimum layer thickness that avoids instabilities and gives reliable results in plane strain compression would be interesting for ultra-thin sheets.

For ductile materials, damage consists of void nucleation around intermetallic particles. These cavities grow and coalesce up to the development of a crack and sample failure. Historically, hole growth and coalescence have been analyzed by unit cell calculations consisting of finite element simulations on a representative material element containing a single void [4, 21, 22]. During deep drawing or cutting operations local tensile and shearing strains up to 200% are reached. As a consequence, to satisfactorily simulate and thoroughly analyze shaping operations, the following material description is needed:

1. The σ(ε) curves for large deformations up to 200%;
2. The local strain at failure in tension;
3. A definition of a damage variable and the determination of its critical value for complex stress-strain paths accurately corresponding to shaping operations.

The present section outlined some experimental procedures for determining large-strain σ(ε) curves. The following will offer an approach for measuring the strain to failure by simple tensile test on sheet material specimens [23–25].

2.2. The strain to failure in sheet metal

Historically, ductility is defined by the area reduction at failure ($\ln(A_r/A_0)$) in a tensile test. A_r and A_0 designate respectively the initial and final cross-sectional area. Siebel and Pomp [23] and Bridgman [26] developed expressions for the true stress-strain curves in the neck of axisymmetric tensile specimens. Subsequently Needleman [27] and Le Roy et al. [6] studied the necking of axisymmetric specimens by finite element analysis. In uniaxial tensile tests on ductile rate-independent sheet material, plastic instability and flow localization occurs just after the maximum load when diffuse necking starts. But, after the onset of a diffuse neck, deformation continues under a falling load until the development of a localized neck then ductile fracture [28]. Secondary strain localization depends on the shape factor (width/thickness) of the tensile test-piece and the material strain hardening [29]. Large width/thickness ratios and small strain hardening parameters lead to localization in the width of the specimen. Whereas, small width/thickness ratios and large-strain hardening parameters cause through-thickness localization.

As mentioned previously, most aluminum automotive alloys contain large amounts of intermetallic phases. **Figure 4** shows typical micrographs of the two AA6016 grades [30]. These contain Fe_mAl_n or Mg_2Si particles. Void nucleation and growth around intermetallic particles may occur [31] and eventually failure happen.

Ductile failure is largely controlled by stress triaxiality at the core of the sheet even though this is often considered in two dimensions. The stress triaxiality T corresponds to the ratio between the mean stress σ_m and the equivalent von Mises stress $\overline{\sigma}$

$$T = \frac{\sigma_m}{\overline{\sigma}} \tag{6}$$

Void growth during overall straining was shown to increase exponentially with the stress triaxiality. Early models describe the growth of a single cavity in an infinite elastic plastic solid [2, 3] or an infinite viscoplastic solid [32, 33]. Finite void volume fractions were addressed first by the plane strain analysis of Needleman [27] representing a square array of cylindrical voids. Gurson [5] determined an analytical expression for the stress potential of a spherical volume of elastoplastic material containing a concentric spherical void. Tvergaard [34] used an

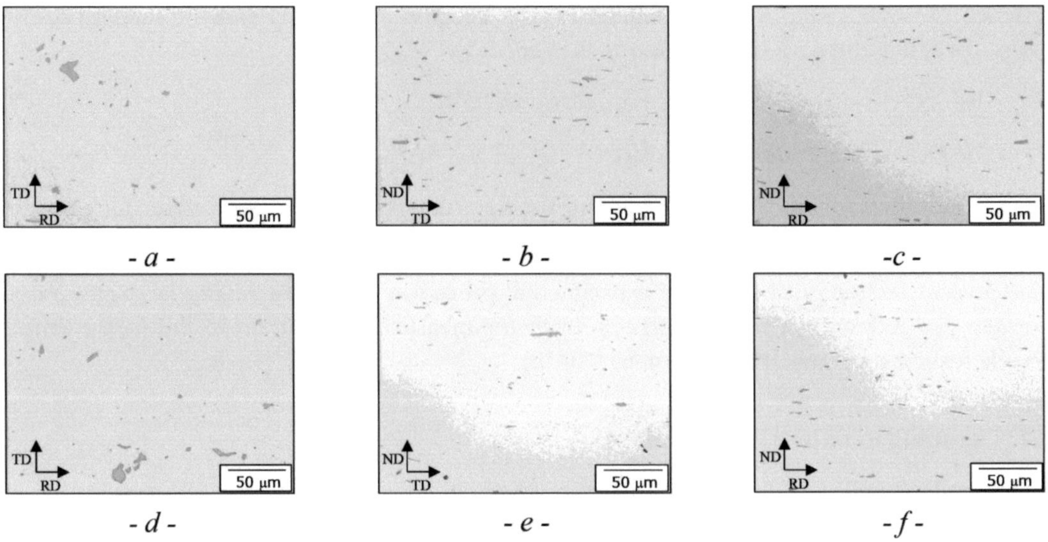

Figure 4. FEG-SEM pictures showing the typical microstructure of the two AA6016 grades: ***a*** (a,b,c) and ***b*** (d,e,f).

axisymmetric cell model to represent a periodic array of spherical holes. All these models can be summarized by expressing the void volume growth as follows [35]:

$$\frac{V(\varepsilon_p)}{V_0} = A\exp\left(\alpha\exp\left(T - 1/3\right)\varepsilon_p\right) \tag{7}$$

ε_p is the equivalent plastic strain of the material surrounding the void, V_0 the initial and V the current void volume. A is constant depending on the model chosen. The coefficient α may take rather complex expressions in the different models. Nevertheless, the cavities grow exponentially with the plastic strain when stress triaxiality T is larger than 1/3. The fundamental role of the stress triaxiality parameter will be discussed hereafter.

Bacha et al. [36] ran tensile tests on the previously mentioned AA6016 alloys. **Figure 5** shows the initial sample geometry and the shape at failure. Tensile tests on sheet metal (rectangular cross-section) obviously cause heterogeneous deformation in the cross-section. Due to strain localization in the minimum section of test samples with a rectangular cross-section, the macroscopic measurement does not provide satisfactory results. This section presents a simple method for measuring true strain to failure by simple tensile tests on sheet metal.

The method to determine the macroscopic strain at failure is outlined in Bacha et al. [36]. The macroscopic strains to failure for both alloys were found equal to 0.67 for grade ***a*** & 0.87 for grade ***b*** respectively. In the same work, the local strain field was determined by FE simulation based on the material behavior characterized by PSC. **Figure 6** shows the FE-predicted geometry of grade ***a*** sample at failure and equivalent plastic strain (ε_p) in the minimum cross-section. The center of the specimen exhibits a maximum plastic strain of 0.91, much higher than the macroscopic value. The second illustration in **Figure 5** corresponds to the stress triaxiality distribution.

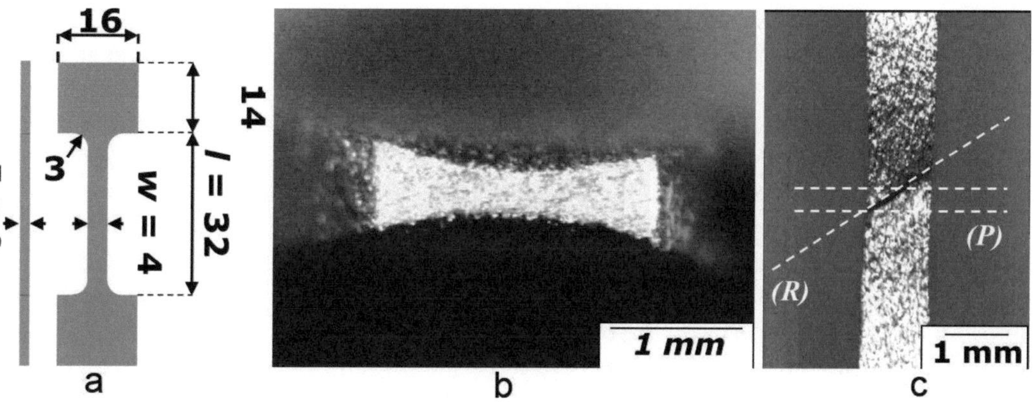

Figure 5. Tension specimen (width w = 1 mm), the thickness (t = 1 mm) corresponds to the thickness of the rolled sheet and height (l = 32 mm). (a) Initial shape, (b) optical view of cross-section after failure and (c) specimen close to failure.

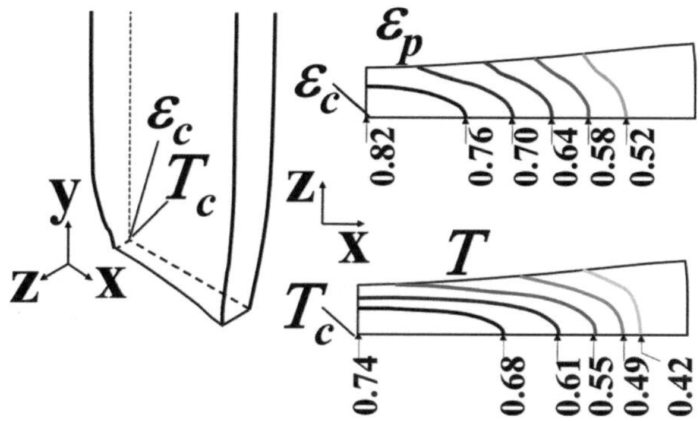

Figure 6. AA6016 grade *a*. Finite element simulation results: overall specimen shape at failure. Equivalent strain and stress triaxiality maps in the minimum section.

Seeing the primary importance of this parameter on damage severity, the variation in the cross-section highlights the damage development in the center of the specimen.

As ductile void growth rate increases exponentially with stress triaxiality, small errors in the estimation of triaxiality lead to large differences in the cavity growth rate and poorly predict material ductility. Even for materials with constant strain hardening, Bridgeman's analysis of the stress triaxiality in axisymmetric tensile specimens may not be extended to tensile specimens with rectangular cross-sections.

According to the present analysis, the strain and stress gradient in rectangular cross-section tensile specimens of is clearly larger than in axisymmetric samples. For material ductility analysis, the local stress-strain in the central area has to be known precisely. The classic Bridgman model was developed for axisymmetric specimens and therefore underestimates the stress triaxiality at the center.

The center stress and strain depend on the pre and post-necking material behavior. The influence of hardening and the post-necking strain on test-piece center stress and strain may be analyzed separately only for constant strain hardening. Only for these academic material laws, the area reduction may be used to predict the center stress and strain. For actual materials with variable strain hardening, the influence of the post-necking strain and material behavior on the stress and strain field in the specimen center may not be separated. Hence, the macroscopic strain given by the area reduction in the minimum section may not be used to characterize the strain and stress in the center of the piece, nor to define material ductility. The extrapolation of $\sigma(\varepsilon)$ determined by tensile tests to large-strain behavior predicts erroneously ductility. To determine correctly the true ductility of the material in question, the $\sigma(\varepsilon)$ curves have to be measured previously by compression tests (Section 2.1) and the tensile test has to be analyzed by FE-calculations.

Obviously, the values predicted for the strain at coalescence and the experimental values of the failure strain are very close. Including damage during the tensile test prior to cavity coalescence by a Gurson-like model is possible in principle. But damage prior to void coalescence only slightly affects the projected values of strain at coalescence. In addition, the parameters required as input for a modified Gurson model are not yet well established. Moreover, in present alloys, secondary void nucleation and growth around smaller particles may be observed and have more influence. Precisely taking into account this secondary cavity coalescence requires a robust model for void nucleation. With a minimum of reasonable and verified assumptions, the local strain in the specimen center at failure should be predicted very accurately. Sheet metal large-strain behavior was obtained by PSC (Section 2.1), and the strain to failure by a simple tensile test (Section 2.2). Most shaping or forming operations correspond to straining under small stress triaxiality. The ductility (strain to failure) under such loading conditions will be analyzed next.

3. Damage development in sheet metal trimming and hemming

3.1. Metal ductility at low stress triaxiality

Large stress triaxiality values ($T > 0.4$) decrease the fracture strain (ductility). Thus, during the last 20 years, most research efforts dealt essentially with medium to high stress triaxiality loadings. But, in many forming operations, the material is submitted to large compressive or small tensile loads corresponding to negative or small stress triaxiality loadings. In this context, Bao and Wierzbicki [37] deformed an AA2024 alloy under various stress-strain paths covering small and large values of the stress triaxiality; their results are summarized on **Figure 7** [30]. The high stress triaxiality results are well described by the damage models outlined in Section 2. However, when stress triaxiality is small, these approaches significantly over-estimate the strain to failure and hence cannot forecast forming limit diagrams.

Bacha et al. [30, 38, 40] analyzed shearing of the two AA6016 sheets characterized previously in PSC. During trimming or hemming of heterogeneous materials, the second phase particles undergo large displacements. **Figure 8a** and **b** illustrate the microstructural changes observed

http://dx.doi.org/10.5772/intechopen.70687

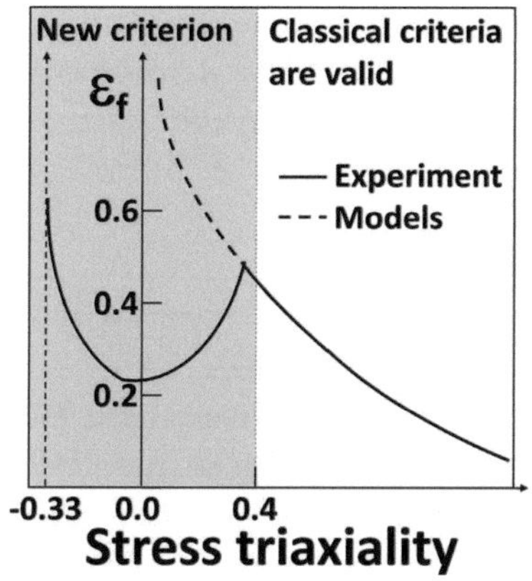

Figure 7. Strain to failure as a function of the stress triaxiality as measured by Bao and Wierzbicki [37] and predicted by Bacha et al. [30].

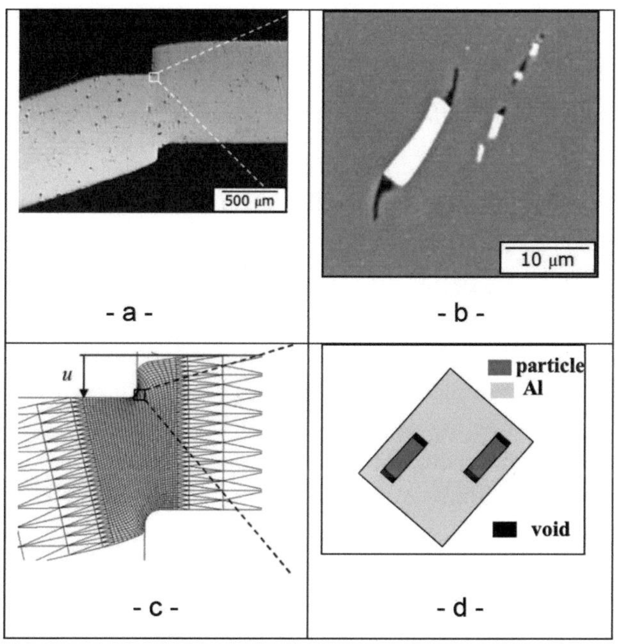

Figure 8. Shearing of an AA6016 grade **a** sheet. (a) Overall shape, (b) intermetallic particle orientation, (c) simulated overall shape, and (d) simulated intermetallic particle orientation.

and subsequently simulated during cropping of a AA6016 grade ***a*** sheet. The intermetallic particles (Mg_2Si) are initially aligned in the rolling or transverse directions (**Figure 4**). The shearing (**Figure 8a**) generates significant rotations and smaller particle distances (**Figure 8b**). Nevertheless, the broken particle matrix interface lost its load carrying capacity. Hence, subsequent failure occurs due to the reduction of inter-particle spacing and subsequent strain localization. Based on Thomason's void coalescence model [41], Bacha and co-workers [30] developed a method to describe strain localization due to reduced inter-particle spacing after large material rotations. **Figure 8** highlights the comparison between experimental and FE-predicted particle orientation. The corresponding damage criterion, valid at small stress triaxiality and large material rotations, is applied to trimming in the third section.

3.2. Trimming of aluminum automotive sheets

After stamping, the final shape of car body panels is achieved by cutting or trimming. Metal cutting is a complex three-dimensional process [42]. However, the physics of this process can be understood by a 2D model. Early work on blanking used macroscopic analysis. Deformation localizes in a small band below the blade. Its width is controlled by the material. At the beginning of the cutting process, the stress in the sheet corresponds to simple shear. During the cutting process, tension stress increases [43, 44]. However, to properly analyze the stress-strain concentrations close to the blades observed in the actual process, a finite element model is essential. The three principal parts of a trimming machine are the blade, sheet and die (**Figure 9**). Other elements may also be added like a blank holder, and sometimes a scrap holder to prevent sheet bending. The following parameters describe the geometry of the plane cutting process:

- Sheet thickness T,
- Cutting angle α, i.e., the angle between the sheet normal and the direction of blade displacement
- Clearance C between the blade and the die, i.e., the distance between the cutting section and the die
- Blade (R_{blade}) and die radii (R_{die}),
- Blade displacement U

Cutting a sheet generates a work piece and scrap. Here, the work piece is maintained under the blank holder after cutting. Trimming or cutting an infinitely ductile material with no clearance, would generate a straight cutting path. Actually, the presence of the clearance and finite material ductility produce more complex cutting paths with various characteristic zones (**Figure 10**):

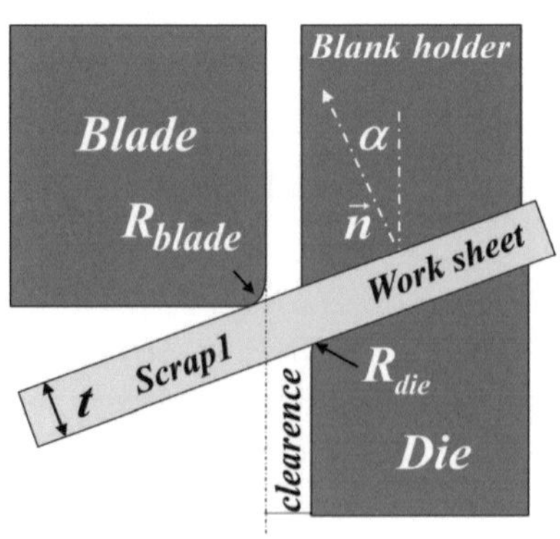

Figure 9. Schematic of the planar sheet trimming process.

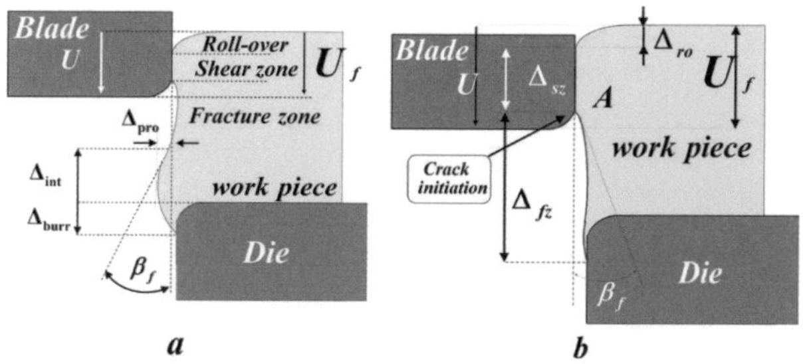

Figure 10. Schematic representation of plane trimming process. Cut generating slivers (a), cut profile without slivers (b) [40].

- Roll-over (height Δ_{ro}) corresponds to elastoplastic shearing of the sheet material,
- Sheared zone (height Δ_{sz}) corresponds to indentation of the sheet material,
- Fractured zone (height Δ_{fz}),
- Burr (height Δ_{burr}) depending on the crack path at the final stage of the process.

Two other crucial geometrical parameters associated with the cutting path are:

- Blade displacement at crack initiation U_f,
- Angle between the fractured zone and the blade displacement direction β_f

During the trimming process, small filaments of hardened material called slivers may be produced (**Figure 10b**). Li and co-workers observed sliver generation when cutting AA6111T4 and AA6022T4 alloys [45–47]. **Figure 11** shows slivers observed when shearing the AA6016 (grade ***a***) sheet [38].

Bacha et al. [38–40] characterized the trimming of two steel alloys (1 and 2), plus a 5000 series and three 6000 series aluminum alloys. In order to analyze the influence of the material behavior on the crack path, the tool geometry was aligned to a reference configuration. The values of the parameters of the following industrial tool set-up are frequently documented. The clearance C, the blade (R_{blade}) and die (R_{die}) radii were respectively 5, 5 and 10% of the sheet thickness T. ({c, R_{blade}, R_{die}} = {0.05, 0.05, 0.10}). The cutting angle α was zero, and no scrap holder was used. **Figure 12** shows the corresponding fracture surfaces. The results prove that sliver generation is related to small values of the sheared zone Δ_{sz}.

Cutting AA6016 (grade ***a*** and ***c***) alloys, in the reference tool configuration, induces slivers. Thus, the parameters controlling the tool geometry were changed. The results show that enlarging the blade radius (R_{blade} = 0.3) as well as an increase of the cutting angle (α = 20°) or the use of a scrap holder completely prevent convex cracking (**Figure 13**). Increasing the clearance does decrease the blade-bulge interaction but does not completely prevent slivers.

The clearance and the scrap holder influence the specimen sheet geometry. All other parameters influence directly the height of the sheared zone Δ_{sz}. The height of the sheared zone Δsz depends on the blade displacement at crack initiation U_f:

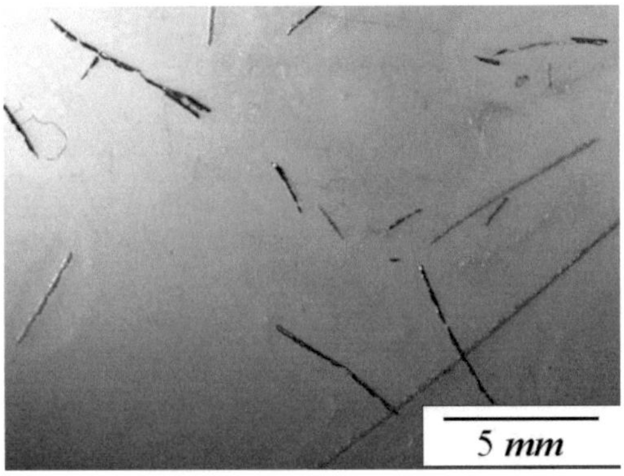

Figure 11. Example of slivers generated on a AA6016 (grade a) sheet.

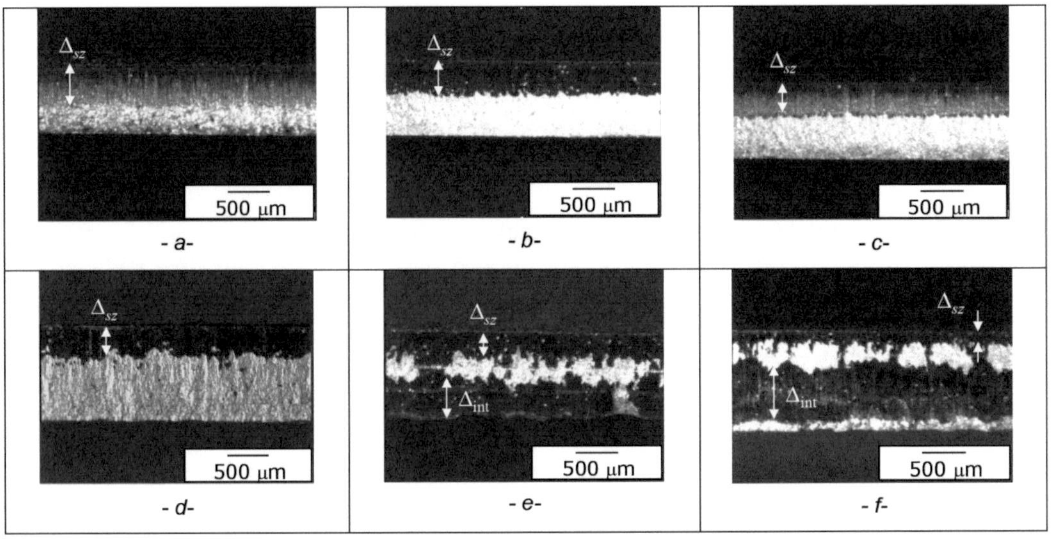

Figure 12. Fracture surfaces of trimmed alloys. The shear zone height decreases from a to f: (a) steel 1, (b) AA5182, (c) steel 2, (d) AA6016 (grade **b**), (e) AA6016 (grade **c**), (f) AA6016 (grade **a**).

$$u_f = \frac{U_f}{t} = \frac{\Delta_{sz} + \Delta_{ro} + R_{blade}}{t} \tag{8}$$

The experimental results reveal that small values of the blade displacement at crack initiation ($u_f < 0.3$) lead to convex cracking and slivers proliferation. Conversely, if u_f is larger than the threshold 0.3, no convex cracking is observed. Bacha et al. [38–40] confirmed this by the finite element model outlined in Section 3.1.

During trimming, the material undergoes locally severe rotations leading to a rearrangement of the intermetallic particles and consequent strain localization. This localization controls the blade displacement at crack initiation and hence both the crack path and sliver production.

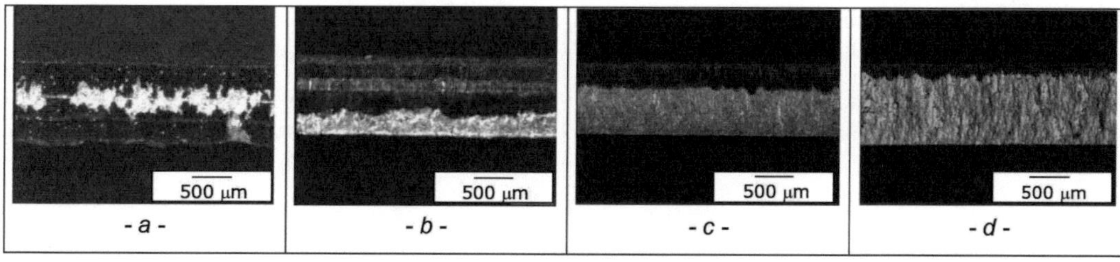

Figure 13. Influence of the tool geometry on the fracture surface of AA6xxx-**b**: (a) reference geometry, (b) increase of the blade radius, (c) increase of the cutting angle, (d) use of a scrap holder.

In the next section, we address hemming, another process essential to sheet forming that leads to significant strain localization.

3.3. Hemming of aluminum automotive sheets

Bending is important in the shaping of complex parts and also critical in the attachment of outer skins to inner sheet panels by hemming. Appropriate strength levels for automotive aluminum panels are achieved by a combination of alloy conception and thermo-mechanical treatments [48, 49]. Several experimental studies underlined the influence of alloy composition on bend performance of Al 6xxx automotive sheet [50, 51]. The bendability has been shown to decrease with increasing copper, iron and Si content.

Most theories attribute bending damage to the development of surface roughness and intense strain localization. Several authors conclude that shear bands initiate at points of strain concentration induced by initial thickness heterogeneity [7, 52, 53]. Other authors have analyzed the effects of thermal softening on shear banding. Crystal plasticity simulations have also been used, essentially to analyze the effect of particular textures on the formability and bendability of sheet metal [28, 52, 54–56]. Probably the chief consensus of these studies is the close relationship between surface roughening and strain localization. But strain localization is difficult to characterize experimentally. Therefore, most experimental studies have compared different alloys based on macroscopic measurements (bend angle), whereas theoretical damage analyses focus on local behavior (shear band pattern).

Several tests to characterize bend performance have been described in the literature [57, 58]. Mattei and co-workers [59, 60] characterized bendability of 6016T4 sheet metal by the standardized Daimler-Chrysler (DIN 50 111). The test set-up is shown in **Figure 14**. During the bend test the applied force F and the punch displacement U were recorded continuously.

Figure 15a shows a typical load displacement curve for the 6016T4 sheet with a punch displacement at failure U_f of 12.8 mm and a minimum bending angle β_f of 28°. The apparent rigidity (dF/dU) plotted in **Figure 15b** exhibits a clear maximum at 0.92 U_f. A crack develops "suddenly" at a punch displacement of 0.92 U_f corresponding to the maximum of the overall rigidity [60]. Crack formation during a standard bend test occurs when the overall rigidity attains a maximum.

Mattei et al. [60] compared the bendability of samples with an industrial surface finish and mirror polished samples. Initial surface roughness (undulations) favors the strain localization. But eliminating the surface roughness at an intermediate bending angle does not enhance the bendability compared to an initial mirror surface finish. **Figure 16** shows an optical

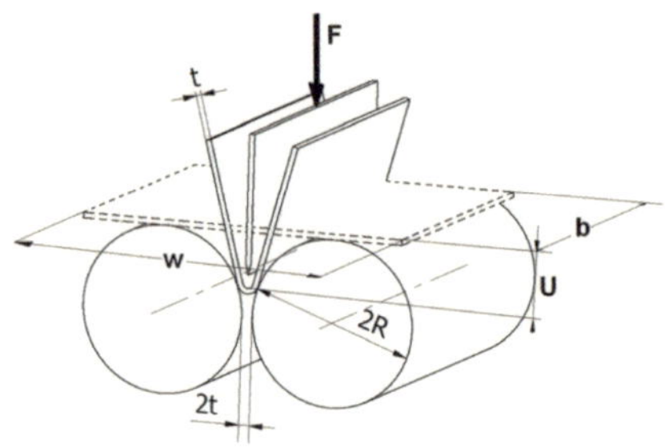

Figure 14. Bending test overall view according to DIN 50 111.

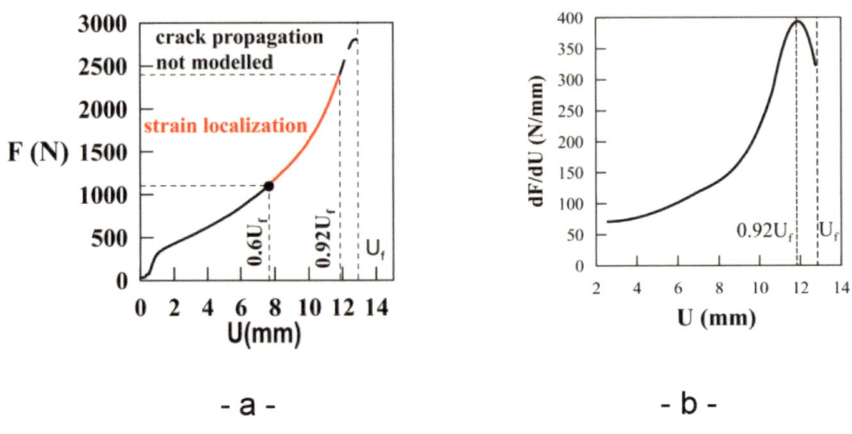

Figure 15. Standard bending test on AA6016 1 mm sheet. (a) Typical load displacement and (b) rigidity displacement curve.

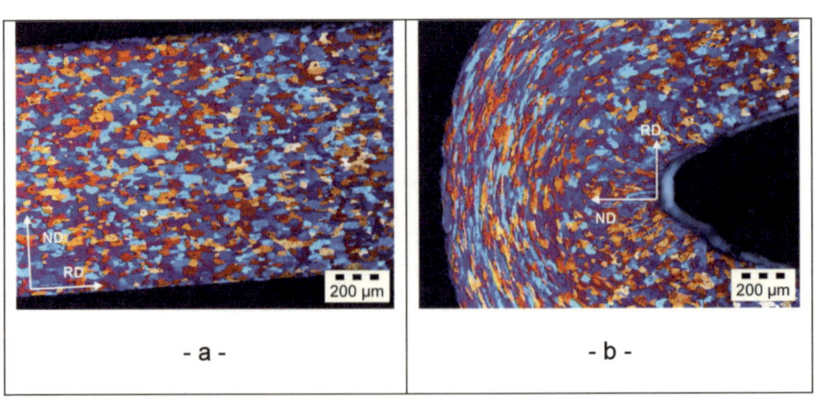

Figure 16. Optical micrograph of the initial microstructure (a) and after bending to failure (b).

micrograph of a sample bent to the minimum bend angle. Strain localization leads to significant changes in grain shape and orientation. Grain rotation and deformation at the outer surface lead to surface undulations. This experimental work clearly shows that through-thickness strain localization controls damage development during AA6016 sheet bending.

Mattei et al. [61] developed a finite element microstructure based model of the standard bending test to predict strain localization. The sheet metal is modeled as a grain aggregate, each grain with its own flow stress. This simple model facilitates industrial AA6xxx sheet alloys conception by correctly describing the respective influences of sheet thickness, grain size and shape, and work hardening. In particular, this model brings out the primary importance of large-strain hardening and the flow stress spatial distribution. It can be used to furnish simple guidelines for designing highly bendable sheet metal.

4. Conclusion

First, we showed PSC, a relatively easy and reliable method for determining large-strain (σ,ε) curves on sheet metal. In other words, layered sheet metal samples can be deformed in channel-die (plane strain) compression up to strains of 120%. Then, the strain to failure was discussed, based on an extension of the well-known Bridgman's model, to specimens drawn from sheet metal, exhibiting a rectangular cross-section. This thorough, but non exhaustive discussion of the material behavior, was followed by an analysis of strain localization during sheet forming. The main factors controlling strain localization and consequent damage were offered.

A new damage criterion by Bacha et al., particularly suited for small stress triaxiality and large material rotations was introduced. This predicts perfectly hemming behavior. Finally, the chapter finished with a novel analysis of hemming. Clearly, strain localization defines hemming limits. This hemming model (based on sheet geometry, grain size, and PSC measured large-strain behavior) is a powerful tool for alloy design.

Acknowledgements

H. Klöcker and Ch. Yukna would like to express their highest regards for Professor's J. H. Driver work. We would like to thank Professor Driver for his 30 year-long counsel founded in our deepest respect.

Author details

Helmut Klöcker[1]* and Christopher Yukna[2]

*Address all correspondence to: klocker@emse.fr

1 Materials Science and Mechanical Engineering, École des Mines de Saint-Étienne, France

2 Department of Languages, École des Mines de Saint-Étienne, France

References

[1] Gurland J, Plateau J. The mechanism of ductile rupture of metals containing inclusions. Transactions of American Society for Metals. 1963;**56**:442-454

[2] McClintock FA. A criterion for ductile fracture by the growth of holes. Journal of Applied Mechanics. 1968;**35**:363-371

[3] Rice JB, Tracey DM. On the ductile enlargement of voids in triaxial stress fields. Journal of the Mechanics and Physics of Solids. 1969;**17**:201-217

[4] Koplik J, Needleman A. Void growth and coalescence in porous plastic solids. International Journal of Solids and Structures. 1988;**24**:835-853

[5] Gurson AL. Continuum theory of ductile rupture by void nucleation and growth: Part I – 10 Yield criteria and flow rules for porous ductile media. Journal of Engineering Materials and Technology. 1977;**99**:2-15

[6] Le Roy G, Embury JD, Edwards G, Ashby MF. Model of ductile fracture based on the nucleation and growth of voids. Acta Metallurgica. 1981;**29**:1509-1522

[7] Hutchinson JW, Tvergaard V. Shear band formation in plane strain. International Journal of Solids Structures. 1981;**17**:451-470

[8] Marciniak Z, Kuczynski K. Limit strains in the processes of stretch-forming sheet metal. International Journal of Mechanical Sciences. 1967;**9**:609-620

[9] Watts AB, Ford H. On the basic yield stress curve for a metal. Proceedings of the Institution of Mechanical Engineers. 1955;**169**(14):1141-1156

[10] Vegter H, ten Hor CHLJ, An Y, Atzema EH, Pijlman HH, van den Boogaard TH, Huetink H. Characterisation and modelling of the plastic material behaviour and its application in sheet metal forming simulation. In: Onate E, Owen DR, editors. VII International Conference on Computational Plasticity, Complas VII; Barcelona; 2003. pp. 1-20

[11] Maurice CI, Driver JH. High temperature plane strain compression of cube-oriented aluminium crystals. Acta Metallurgica et Materialia. 1993;**41**:1653-1664

[12] Maurice CI, Piot D, Klöcker H, Driver JH. Hot plane strain compression testing of aluminium alloys by channel-die compression. Metallurgical and Materials Transactions A. 2005;**36A**:1039-1047

[13] Bacha A, Maurice Cl, Klöcker H, Driver JH. The large strain flow stress behaviour of aluminium alloys as measured by channel-die compression (20-500°C). Materials Science Forum. 2006;**519-521**:783-788

[14] Orlans-Joliet B, Driver JH, Montheillet F. Plane strain compression of silicon-iron single crystals. Acta Metallurgica et Materialia. 1990;**38**:581-594

[15] Bacha A, Daniel D, Klöcker H. On the determination of true stress triaxiality in sheet metal. Journal of Materials Processing Technology. 2007;**184**:272-287

[16] Brunet M, Morestin F. Experimental and analytical necking studies of anisotropic sheet metals. Journal of Materials Processing Technology. 2001;**112**:214-226

[17] Driver JH, Skalli A. L'essai de compression plane de monocristaux encastrés: méthode d'étude du comportement d'un cristal soumis à une déformation plastique imposée. Revue de Physique Appliquee. 1983;**17**:447-451

[18] Voce E. The relationship between stress and strain for homogeneous deformation. Journal of Institute of Metals. 1948;**74**:537-562

[19] Voce E. True stress-strain curves and their application to cold-working processes. Metal Treatment. 1948;**15**:53-60

[20] Barlat F, Maeda Y, Chung K, Yangawa M, Brem JC, Hayashida Y, Lege DJ, Matsui K, Murtha SJ, Hattori S, Becker RC, Makosey S. Yield function development for aluminum alloy sheets. Journal of the Mechanics and Physics of Solids. 1997;**45**(11/12):1727-1763

[21] Klöcker H, Tvergaard V. Void growth and coalescence in metals deformed at elevated temperature. International Journal of Fracture. 2000;**106**:259-276

[22] Pardoen T, Hutchinson JW. An extended model for void growth and coalescence. Journal of the Mechanics and Physics of Solids. 1999;**48**(12):2467-2512

[23] Siebel E, Pomp A. Mitteilungen Kaiser-Wilhelm Inst. für Eisenforschung. Vol. 9; 1927. p. 157-165

[24] Leblond JB, Perrin G, Suquet P. Exact results and approximate models for porous viscoplastic solids. International Journal of Plasticity. 1994;**10**:213-235

[25] Tvergaard V. Influence of void nucleation on ductile shear fracture at a free surface. Journal of the Mechanics and Physics of Solids. 1984;**32**:373-393

[26] Bridgman W. The stress distribution at the neck of a tension specimen. Transactions of American Society for Metals. 1944;**32**:553-574

[27] Needleman A. A numerical study of necking in circular cylindrical bars. Journal of the Mechanics and Physics of Solids. 1972;**20**:111-127

[28] Tvergaard V. Necking in tensile bars with rectangular cross-section. Computer Methods in Applied Mechanics and Engineering. 1993;**103**:273-290

[29] Zhang ZL, Hauge M, Ødegård J, Thaulow C. Determining material true stress-strain curves from tensile specimens with rectangular cross-section. International Journal of Solids and Structures. 1999;**36**:3497-3516

[30] Bacha A, Daniel D, Klocker H, editors. Jürgen Hirsch, Birgit Skrotzki, Günter Gottstein, trimming of aluminium sheets: experimental and numerical investigations. ICAA11; 2008. pp. 1829-1835

[31] Benzarga AA, Besson J, Batisse R, Pineau A. Synergistic effects of plastic anisotropy and void coalescence on fracture mode in plane strain. Modelling and Simulation in Materials Science and Engineering. 2002;**10**:73-102

[32] Budiansky B, Hutchinson JW, Slutsky S. Void growth and collapse in viscous solids. In: Hopkins HG, Sewell MJ, editors. Mechanics of Solids, The R. Hill 60th Anniversary Volume; 1982. pp. 13-44

[33] Huang Y. Accurate dilatation rates for spherical voids in triaxial stress fields. Transactions of the ASME, Journal of Applied Mechanics. 1991;**58**:1084-1085

[34] Tvergaard V. On localisation in ductile materials containing spherical voids. International Journal of Fracture. 1982;**18**:237-252

[35] Tvergaard V, Needleman A. Analysis of the cup-cone fracture in a round tensile bar. Acta Metallurgica. 1984;**32**:157-169

[36] Bacha A, Feuerstein M, Desrayaud Ch, Klöcker H. Measuring stress strain curves to large strain on sheet metal. Journal of Testing and Evaluation. 2007;**35**(2):157-166

[37] Bao Y, Wierzbicki T. On fracture locus in the equivalent strain and stress triaxiality space. International Journal of Mechanical Sciences. 2004;**46**:81-98

[38] Bacha A, Daniel D, Klöcker H. Metal ductility at low stress triaxiality application to sheet trimming. Journal of Materials Processing Technology. 2008;**203**:480-497

[39] Bacha A, Daniel D, Klöcker H., ed. Jürgen Hirsch, Birgit Skrotzki, Günter Gottstein, Crack deviation during trimming of aluminium automotive sheets. ICAA11. 2008

[40] Bacha A, Daniel D, Klöcker H. Crack deviation during trimming of aluminium automotive sheets. Journal of Materials Processing Technology. 2010;**210**(14):1885-1897

[41] Thomason PF. Ductile Fracture of Metals. Oxford: Pergamon Press; 1990

[42] Gosh S, Li M, Khadke A. 3D modelling of shear slitting process for aluminium alloys. Journal of Material Processing Technology. 2005;**167**:91-102

[43] Atkins AG. On cropping and related processes. International Journal of Mechanical Sciences. 1980;**22**:215-225

[44] Zhou Q, Wierzbicki T. A tension zone model of blanking and tearing of ductile metal plates. International Journal of Mechanical Sciences. 1996;**38**:303-324

[45] Li M, Fata G. Sliver reduction in trimming aluminium autobody sheet. SAE Technical Paper Series, ed. European Aluminium Association; 1999-01-0661; 1999

[46] Li M. Method and apparatus for trimming aluminium sheet, United States Patent Application Publication, US2002/0017173A1; 2000

[47] Li M. An experimental investigation on cut surface and burr in trimming aluminium autobody sheet. International Journal of Mechanical Sciences. 2002;**42**:889-906

[48] Lloyd DJ, Gupta AK. In: Chandra T, Sakai T, editors. On the precipitation-hardening behavior of the Al-Mg-Si-Cu alloy AA6111. THERMEC'97, TMS; 1997. pp. 99-107

[49] Miller WS, Zhuang I, Bottema J, Wittebrood AJ, De Smet P, Haszler A, Vieregge A. Recent development in aluminium alloys for the automotive industry. Materials Science and Engineering: A. 2000;**A280**:37-49

[50] Minoda T, Asano M, Yoshida H. Influence of iron content on the mechanical properties of AA6016 alloy sheet, Materials Science Forum. 2006;**519-521**:859-864

[51] Sarkar J, Kutty TRG, Wilkinson DS, Embury JD, Lloyd DJ. Tensile properties and bendability of T4 treated AA6111 aluminium alloys. Materials Science and Engineering: A. 2004;**A369**:258-266

[52] Kuroda M, Tvergaard V. A phenomenological plasticity model with non-normality effects representing observations in crystal plasticity. Journal of the Mechanics and Physics of Solids. 2001;**49**:1239-1263

[53] Triantafyllidis N, Needleman A. On the development of shear bands in pure bending. International Journal of Solids and Structures. 1982;**18**(2):121-138

[54] Kuroda M, Tvergaard V. Effects of texture on shear band formation in plane strain tension/compression and bending. International Journal of Plasticity. 2007;**23**(2):244-272

[55] Becker R. Effects of strain localization on surface roughening during sheet forming. Acta Metallurgica. 1998;**46**(4):1385-1401

[56] Dao M, Li M. A micromechanics study on strain localization induced fracture initiation in bending using crystal plasticity models. Philosophical Magazine A. 2001;**81**(8):1997-2020

[57] Borland JC. Fundamentals of solidification cracking in welds.-II. Welding and Metal Fabrication. 1979;**33**:505-508

[58] Selcuk A, Rawlings RD. A Cantilever-type bend test technique for formability analysis of strip/plate metals. Journal of Testing and Evaluation. 1991;**19**:349-358

[59] Mattei L, Daniel D, Guiglionda G, Klöcker H. Hemming of aluminium automotive sheet. In: Hirsch J, Skrotzki B, Gottstein G, editors. ICAA 11, Aluminium Alloys, Their Physical and Mechanical Properties; 2008. pp. 1829-1835

[60] Mattei L, Daniel D, Guiglionda G, Klöcker H, Driver JH. Strain localization and damage mechanisms during bending of AA6016 sheet. Materials Science and Engineering: Sheet. Materials Science and Engineering A. 2012;**559**:812-821

[61] Mattei L, Daniel D, Guiglionda G, Moulin N, Klöcker H, Driver JH. Grain scale modelling of the bendability of AA6xxx Al alloy sheet. Materials Science and Engineering: A. 2013;**583**:96-104

Chapter 12

Microbiologically Influenced Corrosion in Aluminium Alloys 7075 and 2024

Vejar V. Nelson, Orrego T. Maria,
Sancy V. Mamiè and Paez C. Maritza

Additional information is available at the end of the chapter

http://dx.doi.org/10.5772/intechopen.70735

Abstract

Aluminum and its alloys are central materials for the aircraft industry. Aluminum alloys (AA) 7075 and 2024 are widely used both in the structures and in brittle sections of the airplanes. The presence of the alloying elements in these metals makes them susceptible to localized corrosion at the same time vulnerable to bacterial attachment. A great number of reports on aircraft deterioration are related to microbial growth by contamination inside fuel storage tanks and aircraft wing tanks; this phenomenon is known as microbiologically influenced corrosion (MIC). As expected, corrosion and biocorrosion increase maintenance costs and time of the aircraft in the hangar. Therefore, the growing interest is to shed light on these issues and develop future inhibition methods. In this chapter, we will give an overview of microbiologically influenced corrosion associated with AA 2024 and 7075 by consortia and bacteria. Three mechanisms of biocorrosion in aluminium alloys have been described. In addition, some alternatives methods to battle the effect of biocorrosion will be shown, these methods are based on green compound which blocking of attached of bacteria and promote the detachment of biofilm, being these a tendency of the last innovation way to inhibit this kind of phenomenon.

Keywords: microbiologically influenced corrosion, aluminum alloy, mechanism, 2024, 7075, inhibition of biocorrosion

1. Introduction

Aluminum and its alloys have a good corrosion resistance owing to aluminum oxide passive films. Unfortunately, the susceptibility of these materials to localized corrosion makes them particularly vulnerable to microbiologically influence corrosion (MIC). The phenomenon of MIC involves the acceleration and/or alteration of corrosion processes resulting from the presence and activities of microorganisms with the generating a biofilm at the metal surface [1]. This phenomenon has been documented for metals exposed to different media, for instance, groundwater [2] and industrial waters [3]. The MIC is not a new kind of corrosion, although the deterioration of metal substrate is caused by an increase in the kinetic of the corrosion process by the presence and the metabolic activity of the microorganisms. These effects can be critical and therefore are important to consider the conditions and roles of microorganisms in the corrosion [4]. The consequences of microbiologically influenced corrosion are pitting corrosion, crevice corrosion, and de-alloying corrosion and correspond to localized corrosion type.

The kinetic of corrosion is determined by the characteristic of the physical chemistry environment at the interface. This environment can be clout by the parameters of concentration of oxygen, salts, pH, redox potential, and conductivity. All these factors are, at the same time, influenced by growing microorganisms [4]. The organisms can attach to surfaces, embed themselves in slime and so-called extracellular polymeric substances (EPSs) [5]. The EPS, usually called "biofilms," plays an important function of cell adhesion [6], biofilm formation [7], and protection of microorganism from adverse environment [8]. Biofilm formation consists of a sequence of steps and begins with adsorption of macromolecules (proteins, polysaccharides, and humic acids) and smaller molecules (fatty acids and lipids) at surfaces [9]. Both adsorbed molecules and biofilms can change the physical chemistry characteristics of the interface, including surface hydrophobicity and electrical charge. The amount of adsorbed organic material is a function of ionic strength and can be enhanced on metal surfaces by polarization. The biofilms formed on a metallic surface can be very thin (monolayers) but can reach the thickness of centimeters [4]. Thus, biofilms, characterized by a strong heterogeneity, are very complex system and require analysis in order to characterize and clarify which component could be available to influence some steps of the redox reaction or how could be the change of characteristic of the oxide passive films.

The characteristics of the surface of metallic substratum play an essential role in developing the biofilm during the early stages of EPS accumulation and may influence the rate of cell accumulation and distribution [10]. There is a vast amount of literature investigating the influence of metallic surface properties on the interaction with microorganisms, of which are, chemical composition, roughness, wettability, and polarization.

The microorganism requires metal ions for it is development. The colonization of a metal surface is likely affected by the presence and free availability of metal ion or salts [11]. Gerchakov et al. evaluated the chemical composition of metals substratum which impacted the microfouling formation rate and the cell distribution in samples exposure to seawater environment. This influence was markedly detected during the first hours of exposition [12]. Moreover, the chemistry of the metallic substrate alters the bacteria viability on the metal surface in favor or opposing, because the chemistry can improve the attachment of cell by the interaction of biofilm [13], and on the other hand, preventing the colonization as a result of the toxic element, for instead, chrome [14].

Other parameters that may influence the rating process of bacterial accumulation on metal are the surface free energy, which is as reflected by hydrophobicity or hydrophilicity and surface roughness. Mueller et al. determined the rate coefficient of adsorption, desorption, and growth for early bacterial colonization on different metal surfaces (copper, silicon, 316 stainless steel, and glass) with different roughness. They found that this parameter correlated positively with biological adsorption process [15]. The topography of surface on microbial attachment has been studied by Nickels et al. in silicate grain shape on the distribution of a microbial community. The roughness is related to the pattern and texture of a surface, and it is inherent to both obtaining, by incorporation of alloying elements and aging process of aluminum alloys. The higher surface roughness could increase the quantity of bacterial accumulation due to the better adhesion that took place at surface irregularities; nevertheless, other authors found reduced adhesion to rougher surfaces [16].

It is well known that the metabolic activity of biofilm clusters can change the pH value for more than three units locally. This means that directly at the interface, where the corrosion process is actually taking place, the pH value can differ significantly from that in the water phase [4]. Thus, water sample pH values do not reflect such effects [17].

However, these phenomenon not are the total of observation, for instead, in some cases, the interaction of bacteria with metal surfaces causes increased corrosion rates, but it has been observed too that many bacteria can reduce corrosion rates of different metals and alloys in many corrosive environments [18], but it is dependent on exposure time because the positive or negative effect is closely related to metabolic activity and mature biofilm.

This chapter is intended for readers interested in the impact of microorganisms on the corrosion processes. A review of the main mechanisms associated to microbiologically influenced aluminum alloy corrosion will be given. Therefore, it is expected that this text can be used for those requesting an introduction to the subject of MIC related to aluminum alloys. On the other hand, some attractive mitigation methods of biocorrosion in aluminum alloy will be presented, thinking of a new methodology with a "green" approach.

2. MIC related to aluminum alloy 2024 and 7075

Most reports of MIC found on aluminum alloys (AA) 2024 and 7075 are used in aircraft or in underground fuel storage tanks. The microorganisms are introduced in to fuel tanks from airborne contaminants, and through water, that enters the tank [19]. The accumulated water inside fuel tanks (building with AA 2024-T3) induces the growth of microorganisms and may lead to microbiologically induced corrosion (MIC) of these tank structures. In the past, jet fuel contaminants have been reported to include a diverse group of bacteria and fungi, with the most common contaminant being the fungus *Hormoconis resinae* [20]. However, Rauch et al. [21] investigated microbial contamination in the United States Air Force (USAF) aviation fuel tanks. They found 12 genera, including four *Bacillus* species and two *Staphylococcus* species.

The most interesting part of this work is related to the kind of bacteria found in comparison to another study carried out in 1950 [22]. Changes in the microbial community constitution may demonstrate microbial adaptation to variations in fuel composition, particularly additives and biocides, enabling the material to be dominated by bacteria. Microbial proliferation in aircraft fuel storage tanks leads to deterioration of fuel quality and corrosion of aluminum alloys [23].

In contrast, the corrosion of AA 7075 has been evaluated in the presence of fungi *Cladosporium resinae* [24] and bacteria. Hagenauer et al. [25] in a joint project of different European aircraft manufacturers studied the involvement of microorganisms in aircraft corrosion damage. This study found that a total of 208 microorganisms among them 158 bacteria, 36 yeasts, and 14 fungi are collected from corroded sites of 7 different airplanes. The results show that the corrosive effect of the isolated microorganisms of the genera *Micrococcus, Enterococcus, Staphylococcus, Aerococcus, Bacillus, Aspergillus,* and *Penicillium* induced strong corrosion toward AA 7075.

Material degradation and/or metallic corrosion involve the transfer of electrons that release metal ions into the surrounding medium but, how far microorganisms promote corrosion in airplanes in practice is hard to prove, and more investigations concerning the microflora of corroded sites as well as the underlying corrosion promoting mechanisms have to be conducted. Today's scientific challenge rests in clarifying the interactions and roles of the metallic surface with the microbial activity that affects the kinetics of electron transfer reactions such as cathodic and/or anodic reactions, resulting in material breakdown.

Different mechanisms have been proposed to explain the microbial corrosion process of aircraft aluminum alloys: (i) oxygen concentration cell, (ii) organic acids, and (iii) extracellular enzyme activity.

2.1. Oxygen concentration cells

A mature biofilm restricts the diffusion of oxygen toward cathodic sites. The oxygen concentration cells formed between the zones covered by the microorganisms and the uncovered zones as a consequence of the fungus respiration have been reported by Miller et al. [26]. Iverson et al. [27] reported the presence of tubercles on the bottom of jet fuels tanks containing sulfate reducing-bacteria which corrode materials through the formation of electrochemical oxygen concentration cells, and cathodic depolarization due to the bacterial sulfate reducers. **Figure 1** shows the adsorbed cells and biofilm formation on the metallic surface. The non-uniform bacterial colonization results in local anode and cathode zones with the formation of differential aeration cell. Under aerobic environment, areas under respiring colonies become anodic reaction (metal dissolution), and surrounding areas become cathodic (oxygen reduction reaction) [28].

Therefore, the formation of differential aeration cell is intrinsic to the development of the film and is one of the first mechanisms to be postulated because it does not depend on the metabolic wastes or activity of the microorganisms, since the only presence of the biofilms on the metal surface has the potential to produce this kind of mechanism by itself.

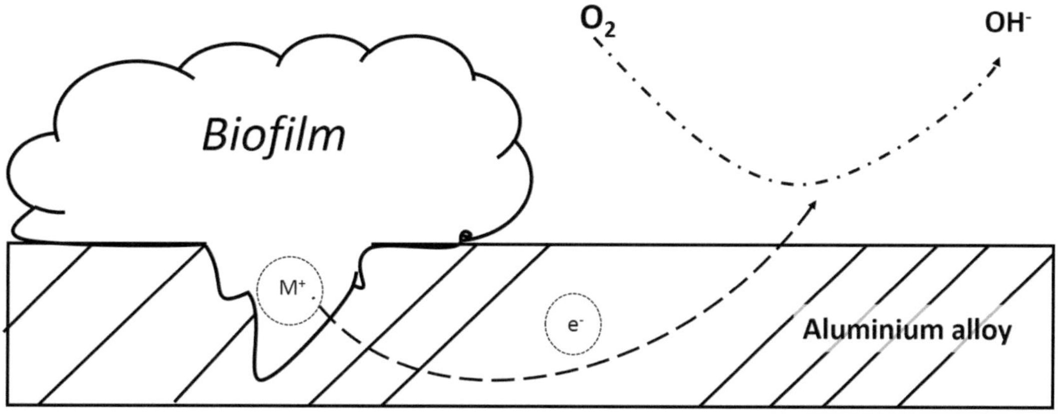

Figure 1. The physical presence of microbial cells on a metal surface as well as their metabolic activities, modify electrochemical processes. The adsorbed cells grow and reproduce, forming colonies that constitute physical anomalies on a metal surface, resulting in the formation of local cathodes or anodes.

2.2. Organic acids

Another mechanism related to microbial corrosion of aluminum alloys is the production of organic acids, which are metabolized by the microorganism. Most heterotrophic bacteria secrete organic acids during fermentation of organic substrates. The kind and amount of acids depend on the type of microorganisms and the available substrate molecules. Organic acids may force a shift in the tendency for corrosion to occur. The impact of acidic metabolites is intensified when they are trapped at the biofilm/metal interface. Furthermore, fungi are ubiquitous in atmospheric and aquatic environments where they assimilate organic material and produce organic acids including oxalic, lactic, formic, acetic, and citric acids. McKenzie et al. [29] found the production of citric, cis-aconitic, isocitric, and α-Ketoglutaric acid with *C. resinae* growing in the jet fuel. These acids are able to facilitate an alloy pitting corrosion through a decrease in the pitting potential of aluminum alloys [30].

The organic anions produced by metabolic activity are able to activate the passive aluminum. Schiapparelli and Meybaum [31] proposed a chemical reaction between the metal surface and organic acid, which could be controlling the corrosion process, this reaction is expressed in the next Eq. (1);

$$n\,X^{-} + Al \;\rightarrow\; Al\,X_{n}^{-(n-3)} \xrightarrow{H_2O} Al\,(OH)_3 + 3\,H^{+} + n\,X^{-} \tag{1}$$

Where X is the aggressive anion of a metabolic product from bacteria. The compound formed by the reaction between the anion and the aluminum would be of the non-stoichiometric type and produced pitting and is independent of the electrolyte used. Eq. (2) shows the kinetics of the pit growth that approaches the Engell-Stolica relation [32].

$$i = At^{b} \tag{2}$$

From Eq. (2), "i" is the current density measured at the whole electrode, and "A" and "b" are constants. Therefore, the presence of anions does not change the kinetics of pit growth, showing an exponential coefficient approximately equal to 1. It is generally accepted that the rate of pit propagation follows Eq. (2) and, the exponential coefficient b is a constant dependent on pit geometry.

Citrate and oxalate anions are able to form soluble complexes with the aluminum ions. The experimental results using citrate, oxalate, and acetate anions show that these anions act as pitting inhibitors increasing the pitting potential of 2024 alloys in neutral chloride solutions toward more positive values [30]. Similar results were reported by Rudd and Scully for aluminum in chloride solutions containing citrate anions. The inhibitory effect of acetate and oxalate anions in neutral media can be explained by the formation of a precipitation type compound [33]. On the contrary, acetic and oxalic acid increase the corrosion rate of aluminum which depends on the pH value. In the acid solutions, pitting occurs at more negative values than in neutral chloride solutions. The acidity can prevent the re-passivation process facilitating pitting. Besides, in the acid solutions, pitting can be accomplished by the formation of soluble complexes between the organic acid anions and the aluminum cations, thus increasing the dissolution rate [34]. Moreover, another investigation also reported an increase of the pitting potential of aluminum in presence of chloride solutions as the acetate concentration increases [35].

2.3. Enzymatic activity

Two mechanisms for microbiologically influenced corrosion of aluminum alloys has been largely documented, which are (a) production of water-soluble organic acids and (b) formation of differential aeration cells. Nonetheless, a recent study has inquired a mechanism related to enzymatic activity as metabolic product of bacteria.

Enzymatic activity is evidenced in the formation of corrosion products and has been considered one of the most important processes and a key to explain the microbial corrosion of aluminum alloys. Hedrick et al. [36] found that the transformation or dissolution of certain metallic atoms can be linked to the presence of extracellular enzyme by comparing the degree of attack to the magnesium content of different in aluminum alloys, given the importance of magnesium ions in microorganism growth.

In a recent work carried out by our group, we isolated and identified an aggressive bacteria to AA 7075-T6. The bacteria were found in corrosion product of a drain valve from military aircraft, and it is closely related to *Bacillus megaterium* species [37]. *B megaterium* bacteria have been declared before, by other authors, as fuel tank microbial contaminant of the United States Air Force aviation [21]. The presence of this type of bacteria in fuel tanks and drain valves [23] could be caused by changes in refinery practices and chemical composition of fuels or possibly by the increased use of fuel additives [20]. In the laboratory study using *B. megaterium* and AA 7075-T6, the results showed, after 14 days, an increase of the corrosion current nearly to three orders of magnitude in comparison with sterile control. By electrochemical impedance spectroscopy (EIS), we found that this microorganism can change the pH in the interface. The enzymatic activity could be related to this change.

Bacteria have evolved specific mechanisms to prevent production and adverse effects of reactive oxygen species (ROS). Catalase is an integral component of the bacterial cell response to oxidative stress, and together with superoxide dismutase and alkyl hydroperoxidases, it is thought to limit the accumulation of reactive oxygen species, such as hydrogen peroxide (H_2O_2) by catalyzing its decomposition to water and oxygen [38]. An adapted schematic representation of the proposed catalase mechanism by Busalmen et al. is shown in **Figure 2** [39]. It is commonly accepted that oxygen reduction can proceed through a parallel mechanism involving the direct reduction via four electrons and a pathway where intermediate peroxide is formed. In this last case, oxygen electro-reduction proceeds by the transference of two electrons producing H_2O_2 as an intermediate compound and then to hydroxyl ions by the transference of two more electrons and ultimately increasing the local pH to alkaline conditions [40].

A schematic representation of the proposed catalase mechanism in a metallic surface is shown in **Figure 2**. It is commonly accepted that oxygen reduction can proceed through a parallel

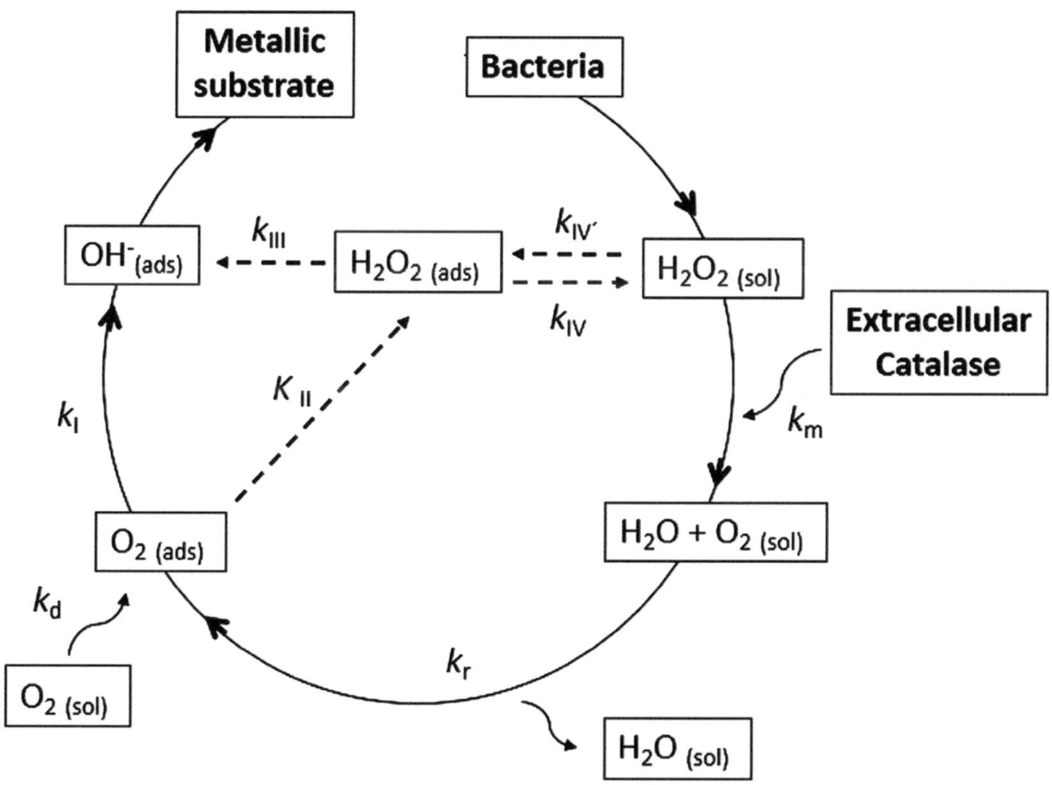

Figure 2. Proposal of microbial corrosion mechanism for catalase. The electrochemical reduction of oxygen to OH^- is coupled with the enzymatic decomposition of hydrogen peroxide by an autocatalytic cycle. *Kn* is the rate constant at their respective step (n). Species in solution (sol) or adsorbed to the metallic surface (ads) are indicated.

mechanism involving the direct reduction via four electrons and a pathway where intermediate peroxide is formed. In this last case, oxygen electro-reduction proceeds by the transference of two electrons, with the production of H_2O_2 as an intermediate compound (k_{II}), and then to hydroxyl ions by the transference of two more electrons. In the figure, the rate constant k_{III} represents the second two electrons transferred, yielding low levels of H_2O_2 in solution. The electro-reduction of peroxide can be inhibited depending on the composition of the surface oxides, favoring its desorption to the solution (k_{IV}). The catalase mechanism is based on the enzymatic conversion of electrochemically produced H_2O_2 to water and oxygen in the proximity of the surface (Km).

3. Future in mitigation methods

Among the strategies to mitigate or control the effects of MIC in material using aircraft industry, biocides are the most widely applied. This approach aims to reduce microorganisms in both, number and type within a given system. In general, biocides may be inorganic compounds such as ozone, chlorine, and bromine, and organic compounds such as isothiazolone, quaternary ammonium, and aldehydes (glutaraldehyde and acrolein), which have the ability to disrupt cellular physiology by disrupting the membrane of the cell, interfering with cellular metabolic processes, or by interrupting the ability of the cell to generate energy in the form of ATP [41]. One of the problems associated with the use of biocides is the inability to effectively penetrate biofilms. Furthermore, due to the high toxicity of antibacterial applied in the industry and in view of complex environmental, ecological, and economic constraints placed by international organizations, great efforts have been focused on the development of benign biocides to prevent bacterial colonization and subsequently the formation of biofilms [42]. In recent years, a number of biocides that could be used to mitigate biocorrosion in the aeronautical industry have been proposed.

3.1. Enzymatic activity inhibition

One of the most interesting strategies of biocides is their action as enzymatic activity inhibitors. Although these compounds do not eliminate the bacteria, they do however reduce their metabolic activities rendering them harmless. A compound with potential development and application in the aeronautical industry is capsaicin compound. The capsaicin (**Figure 3**) is known as the main component of red pepper species (Capsicum species). It is composed of three major functional moieties, namely, a vanilloid, an amide, and a hydrophobic side chain. Capsaicin and its synthetic derivatives have been extensively investigated in the fields of pharmacy [43], neuroscience [44], and antimicrobial drugs [45]. At present, following the evaluation of inhibition of zebra mussel attachment, capsaicin showed the most promising activity, being highly antifouling (EC50 13.7 mM) but weakly toxic toward mussels and water fleas [46] and consequently may be more promising as an environmentally friendly antifoulant [47].

Studies carried out using microalgae *P. tricornutum* showed that the reactive oxygen species (ROS) levels were significantly increased in nonivamide (Capsaicin)-treated algae [48]. Algal

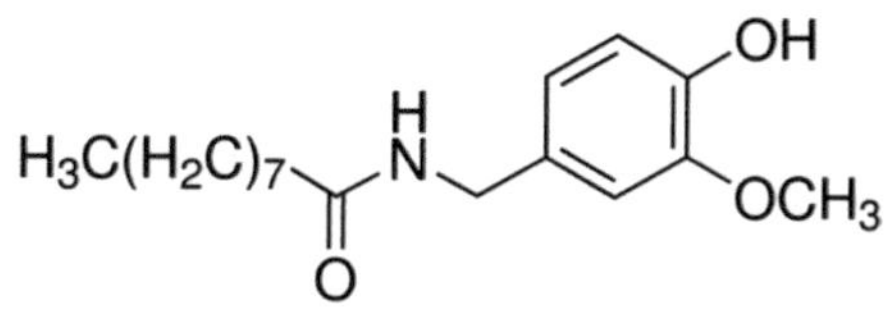

Figure 3. Capsaicin structure.

antioxidants, including catalases (CAT), peroxidases (POD), superoxide dismutase (SOD), and glutathione (GSH), were all stimulated by the ROS burst. The overproduction of ROS substances can consume large amounts of metabolic energy and also lead to the loss of photosynthetic pigments and lipid membrane damages, thereby inhibiting algal growth.

3.2. Colonization and biofilms formation inhibition

The composition of the biofilms is due to the combination of the following: active secretion, shedding of cell surface material, cell lysis, and/or adsorption of substances from the environment like organic and inorganic compounds. EPSs usually include organic acid with functional groups, such as carboxylic and amino acids that readily bind metal ions. EPSs are also related to increased biofilm resistance to biocides and other antimicrobial compounds [49]. Therefore, the adherence of bacteria on the metallic surface could be the key to blocking many metabolic activities of the bacteria that ultimately lead to aluminum alloy corrosion. There are currently two methods by which it possible to inhibit the action of bacteria to colonize the surface, (a) the modification of the metallic surface toward a super-hydrophobic surface or (b) use D-amino acids to prevent the biofilms formation. The first case, the super-hydrophobic modified metallic surface could prevent the attachment of microorganism on the aluminum alloy. The second case, the use of D-amino acids could lead a breakdown of a biofilm formed on a metallic substrate, and/or the presence of specific quantities could avoid the development of biofilm.

3.3. Super-hydrophobic surface

The sorption of microorganisms to surfaces is an important factor in microbial adhesion on the metallic surface. This task is accomplished by an initial reversible phase of bacterial sorption to the solid surfaces which are time-dependent, following a cell growth, and finally an irreversible phase of sorption of some of the bacteria [50]. After that, the development and growth of biofilms take place.

The considerable importance of characteristic of the surfaces in corrosion resistance, the super-hydrophobic modified surface, inspired by nature, is one the most interesting strategies to avoid corrosion [51]. Superhydrophobic surfaces, with water contact angles (CA) greater than 150°, are recently interest due to their self-cleaning character. In order to generate such a modified surface, including micro- and/or nanostructures, or low surface energy coating should be considered. Many attentions are now focused on the design of microstructure surface with the advancement of nanotechniques, such as chemical vapor deposition, chemical etching, template-based extrusion, and electro-deposition [40].

Surface wetting behavior can generally be broken down into different regimes, based on the value of water contact angle (WCA). Two traditional regimes are the hydrophilic and hydrophobic regimes, defined as WCAs in the range of $10° < \theta < 90°$ and $90° < \theta < 150°$, respectively. Hydrophobic coatings are intensively used in plenty of engineering applications; moreover, they are widely used in paint and varnish industries. Although hydrophobic and hydrophilic regimes comprise a vast amount applications, superhydrophobic and superhydropholic regimes, which describe the extremes surface wetting behavior, are much more interesting. Superhydrophilicity, which is characterized by WCAs in the range of $\theta < 10°$, within 1 s of the initial wetting, describes nearly perfect wetting. In contrast, superhydrophobicity, described by WCAs of $\theta > 150°$, describes a state of nearly perfect nonwetting (**Figure 4**) [52].

Corrosion resistance can be improved by introducing the anticorrosive coatings on the surface of aluminum and its alloys, e.g., preparing the chromate-based coatings or surface anodizing [53]. Chromate-based coatings can provide highly effective corrosion protection, but environmental regulations are increasingly restricting their use, while the anodization will increase the weight of the aluminum and its alloys. Given the strong water repulsive property of the superhydrophobic surfaces, they make a promising technology for slowing the breakdown of the native aluminum oxide layer and thereby retarding corrosion of the aluminum layer underneath. As a result, the anticorrosion performance of aluminum and its alloys may be improved due to superhydrophobic surfaces which inhibit the contact of a surface with water, environmental humidity, and corrosive medium. However, in these cases, either special equipment/complex process control is required, or caustic or costly reagents, for instance, HNO_3, HF, H_2O_2, or fluoride are needed. Consequently, these methods/procedures may result problematic as either pollution or expensive methods. Therefore, a simple, inexpensive, and environment friendly fabrication method is quite needed, since it is very advantageous for industrial large scale production. Herein, work with a novel method for the preparation of super-hydrophobic surfaces on the aluminum alloys is very important.

Some work has been reported on the fabrication of super-hydrophobic aluminum surfaces using chemical etching techniques. The aforesaid chemical etching technique is an inexpensive and simple method to produce artificial super-hydrophobic coatings. Recently,

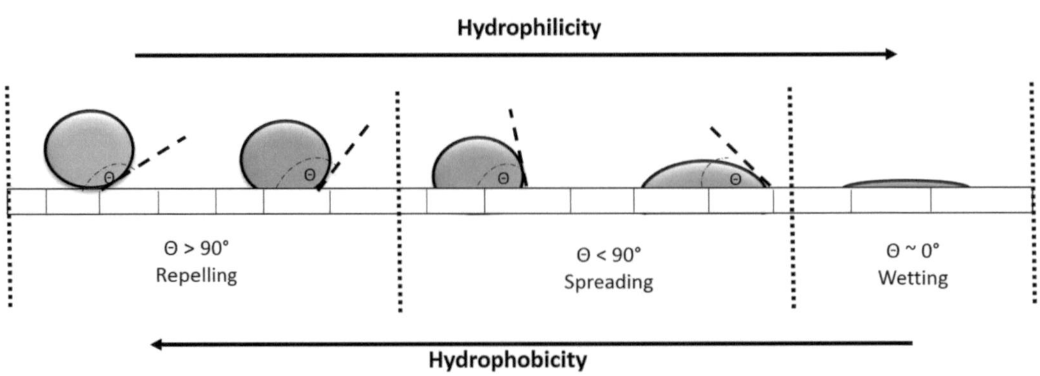

Figure 4. Schematic of contact angle (CA) for a water drop placed on surfaces of different hydrophobicity.

Xie et al. [54] developed super-hydrophobic aluminum surface by a chemical etching process using NaOH solution as an etchant and lauric acid as a low surface energy material. Wang et al. [55] worked on aluminum alloy using stearic acid. Their results showed long time stability and excellent resistance to corrosive liquids including acidic, basic, and salt solutions. The procedures are fairly facile to carry out, and no special technique or equipment is required. Moreover, only the ordinary reagents, such as the boiling water, ethanol, stearic acid, etc., are used in the procedure (**Figure 5**). The methods are also environmental friendly, while the as-fabricated super-hydrophobic aluminum alloy surface could confer excellent corrosion resistance. Therefore, the methods using super-hydrophobic modified surface are a promising alternative due to involving the main step of attachment of bacteria on a metallic surfaces of aluminum alloys.

3.4. D-Amino acids

In all life kingdoms, organisms predominantly use and synthesize L-type enantiomers of amino acids. On the other hand, D-type amino acids are accumulated in millimolar quantities in the supernatant of bacterial cultures in stationary phase, and whose function is to modulate the synthesis of peptidoglycan (PG). PG is a strong and elastic polymer and main component structural supports of bacterial cell wall, controlling the turgor and stability of cell wall in front to change of the environment.

About of the application of D-aminoacids, as both, corrosion and biocorrosion, not much evidence that show to these amino acids as corrosion promoters [10]; on the contrary, authors suggest that D-amino acids have inhibitors effect [17]. Zhuo et al. studied peptides of 12 amino acids in length which are bind to aluminum inhibiting the corrosion [56].

Lam et al. found that bacteria *Vibrio cholerae* produce D-methionine and D-leucine, while *Bacillus subtilis* produce D-tyrosine and D-phenylalanine [57]. These amino acids produced by bacteria are a common strategy to adapt in changing environmental conditions. Moreover, a mix of D-amino acids can control the biofilms formation or breakdown. Kolodkin-Gal et al. found that *Bacillus subtilis*, *Staphylococcus aureus*, and *Pseudomonas aeruginosa* produce a mix of D-amino acids between leucine, methionine, tyrosine, and tryptophan, which prevented the biofilms formation and could break down existing biofilms [58]. Cava et al. evaluated how extracytoplasmic processes regulate the biofilm formation; the authors also showed that D-amino acids are secreted outside of bacteria and perform to regulate the biofilms [59]. **Figure 6** shows a representative mechanism of the D-amino acids proposed by Cava et al.

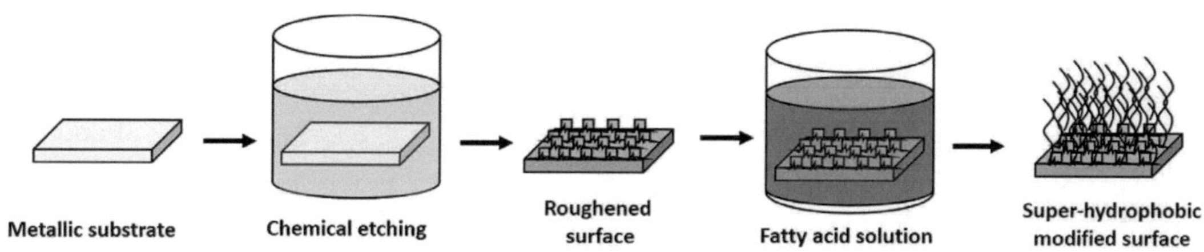

Figure 5. Schematic diagram of simple method to obtain super-hydrophobic modified surface.

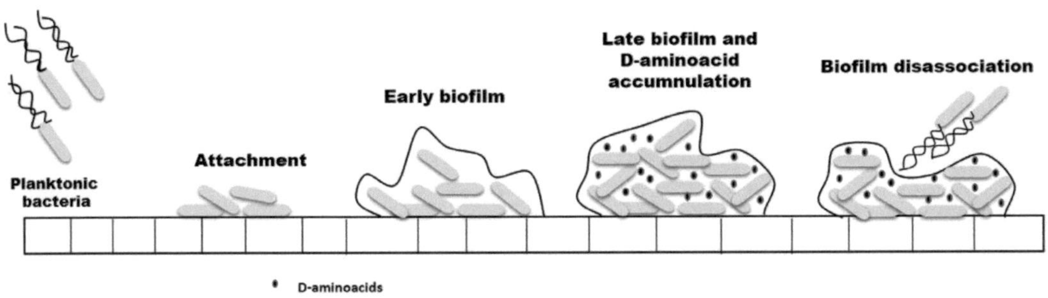

Figure 6. Schematic representation of the mechanism of D-amino acids in the breakdown of biofilms.

The detachment of the biofilm from the substrate has been studied. Xu and Liu evaluated the D-tyrosine in the detachment behavior of biofilms of *Vibrio harveyi* [60]. These studies showed that D-tyrosine has two effects, inhibited microbial attachment and promoted biofilm detachment, and in addition, the production of EPS was significantly suppressed.

Considering the potential of D-amino acids as inhibitors in biocorrosion is due to their low toxicity. Studies have used mixtures of D-amino acids together with corrosion inhibitors. Xu et al. tested a mixture of D-methionine and tetrakis hydroxymethyl phosphonium sulfate (THPS) for biocorrosion inhibition of carbon steel exposure to *Desulfovibrio vulgaris* [61]. The author concludes that the binary mixture of D-amino acid and THPS prevented biofilm formation and removed pre-established biofilms from surfaces, and it also reduced MIC pitting of the carbon steel coupon. Further investigation by Xu et al. showed that applying a mixture of D-amino acids and THPS is more effective in developing biofilm inhibition than using either compound on its own. The amino acid cocktail in question is being constituted by D-tyrosine, D-methionine, D-tryptophan, and D-leucine [62]. These results offer significant candidates in the search for efficient biocides, since one of the problems that arise with the use of these treatments is the deactivation of biocides by the biofilms.

4. Conclusions

Aluminum and its alloys are susceptible to microorganism attached, as both fungus and bacteria. The susceptibilities have been related to the chemical composition of the alloy and the morphology and topography of the surface associated with roughness parameters. Report of attaching of microorganism on aluminum alloy is fuel line system, mainly due to the fuel is a source food but is not enough to microorganism growth and development of biofilms, also some salts are necessary, for instead, magnesium salts. The magnesium could be obtained from the metallic substrate. On the other hands, the role in the bacterial activity of another metal present in the surface of an alloy is not clear. In fact, the knowledge related to antibacterial properties of the intermetallic (θ phases, rich in copper) in AA 2024-T3 is poor, and at the same time, the role of the intermetallic containing zinc metal in AA 7075-T6.

The interactions of the alloys and bacteria, in most of the cases studied, cause a detrimental by the influence of the microorganisms in corrosion phenomenon. The biocorrosion has been widely studied, and the research shows a correlation of microorganism activity and corrosion acceleration. The microbiologically influenced corrosion in aluminum and its alloy are related to three mechanisms: (a) Differential oxygen cell, (b) weakening of the aluminum oxide films by the presence of organic acids, and (c) extracellular enzyme activity which including the catalase enzyme. However, some articles indicate that sulfate reducing bacteria as responsible for corrosion in aluminum and copper containing alloys such as Monel [63]. The anaerobic bacteria activity in biocorrosion of aluminum alloys is not clear, and it is very interesting to evaluate the anaerobic bacteria in consequences of degradation of organic materials as fuel and corrosion of aluminum alloys.

The primary effect of the microorganisms is the destabilization of the aluminum protective oxide film either mechanically or chemically. Once biofilm is formed on the metallic substrate, a local corrosion is developed. The local corrosion of aluminum alloy is a catastrophic problem in the aircraft industry.

The interdisciplinary studies are essential to understand the biocorrosion phenomenon and the main proposed mitigation strategies and methodology including different scientific area as engineering, chemical, and biological. Effective methodologies to inhibit the adverse effect of the microbiologically influenced-corrosion have not been developed. The creation of effective methodologies is a great challenge, due to the biocorrosion phenomenon is a complex with many variables, and moreover, the plan of action should include green methods (reagent and procedures).

Acknowledgements

The authors thank FONDECYT (Grant 1,160,604), PIA-CONICYT (Grant ACT-1412), and AFOSR (Grant FA 9550-16-1-0063) for financial support.

Author details

Vejar V. Nelson[1]*, Orrego T. Maria[3], Sancy V. Mamiè[2] and Paez C. Maritza[3]

*Address all correspondence to: nelsonvejar.v@gmail.com

1 Aerospace Sciences Research and Development Centre (CIDCA), Chilean Air Force, San Bernardo, Chile

2 Civil Construction School, Faculty of Engineering, Pontifical Catholic University of Chile, Chile

3 Department of Materials Science, Faculty of Chemistry and Biology, University of Santiago of Chile, Estación Central, Chile

References

[1] Rajasekar A, Ting Y. Microbial corrosion of aluminum 2024 aeronautical alloy by hydrocarbon degrading bacteria Bacillus cereus ACE4 and Serratia Marcescens ACE2. Industrial and Engineering Chemistry Research. 2010;**49**:6054-6061. DOI: 10.1021/ie100078u

[2] Beech IB. Corrosion of technical materials in the presence of biofilms: Current understanding and state-of-the art methods of study. International Biodeterioration & Biodegradation. 2004;**3**(53):177-183. DOI: 10.1016/S0964-8305(03)00092-1

[3] Beech IB, Sunner J. Biocorrosion: Towards understanding interactions between biofilms and metals. Current Opinion in Biotechnology. 2004;**15**:181-186. DOI: 10.1016/j.copbio.2004.05.001

[4] Beech I, Bergel A, Mollica A, Flemming H, Scotto V, Sand W. Simple methods for the investigation of the role of biofilms in corrosion. Brite-Euram III Themat. Netw. N° ERB BRRT-CT98-5084. 2000;5084

[5] Costerton AJW, Stewart PS, Greenberg EP. Bacterial biofilms : A common cause of persistent infections. American Association for the Advancement of Science Linked references are available on JSTOR. 1999;**284**:1318-1322. DOI: 10.1126/science.284.5418.1318

[6] Frank BP, Belfort G. Polysaccharides and sticky membrane surfaces: Critical ionic effects. Journal of Membrane Science. 2003;**212**:205-212. DOI: 10.1016/S0376-7388(02)00502-1

[7] Cescutti P, Toffanin R, Pollesello P, Sutherland IW. Structural determination of the acidic exopolysaccharide produced by a Pseudomonas sp. strain. Carbohydrate Research. 1999;**315**:159-168. DOI: 10.1016/S0008-6215(98)00318-8

[8] Looijesteijn PJ, Trapet L, De Vries E, Abee T, Hugenholtz J. Physiological function of exopolysaccharides produced by *Lactococcus lactis*. International Journal of Food Microbiology. 2001;**64**:71-80. DOI: 10.1016/S0168-1605(00)00437-2

[9] Little BJ, Lee JS. Microbiologically Influenced Corrosion. New Jersey: John Wiley & Sons; 2007. pp. 280 p. DOI: 10.1002/047011245X

[10] Little B, Wagner P, Mansfeld F. Microbiologically influenced corrosion of metals and alloys. International Materials Review. 1991;**36**:253-272. DOI: 10.1179/imr.1991.36.1.253

[11] Beech IB, Campbell SA, Walsh FC. The role of surface chemistry in SRB influenced corrosion of steel. International Journal of Marine Biology and Oceanography. 1993;**13**:243-240

[12] Gerchakov S, Roth F, Sallman B, Udey L, Marszalek D. Observations on microfouling applicable to OTEC systems. In: Biofouling Corros. Symp; Seattle, WA. 1977. pp. 63-75.

[13] Beech I. Biofilm formation on metal surfaces [thesis]. City London Polytech. CNAA; 1990

[14] Percival SL, Beech IB, Edyvean RGJ, Knapp JS, Wales DS. Biofilm development on 304 and 316 stainless steels in a potable water system. Journal of the Chartered Institution of Water and Environmental Management. 1997;**11**:289-294

[15] Mueller RF, Characklis WG, Jones WL, Sears JT. Characterization of initial events in bacterial surface colonization by two Psuedomonas species using image analysis. Biotechnology and Bioengineering. 1992;**39**:1161-1170

[16] Nickels JS, Bobbie RJ, Martz RF, Smith GA, White DC, Richards NL. Effect of silicate grain shape, structure, and location on the biomass and community structure of colonizing marine microbiota. Applied and Environmental Microbiology. 1981;**41**:1262-1268

[17] Alasvand Zarasvand K, Rai VR. Microorganisms: Induction and inhibition of corrosion in metals. International Biodeterioration and Biodegradation. 2014;**87**:66-74. DOI: 10.1016/j.ibiod.2013.10.023

[18] Mansfeld F. The interaction of bacteria and metal surfaces. Electrochimica Acta. 2007;**52**:7670-7680. DOI: 10.1016/j.electacta.2007.05.006

[19] Smith RN. Developments in fuel microbiology. Biodeterioration and Biodegradation. 1991;**8**:112-124

[20] McNamara CJ, Perry TD, Leard R, Bearce K, Dante J, Mitchell R. Corrosion of aluminum alloy 2024 by microorganisms isolated from aircraft fuel tanks. Biofouling. 2005;**21**:257-265. DOI: 10.1080/0892701050038992.

[21] Rauch ME, Graef HW, Rozenzhak SM, Jones SE, Bleckmann CA, Kruger RL, Naik RR, Stone MO. Characterization of microbial contamination in United States Air Force aviation fuel tanks. Journal of Industrial Microbiology & Biotechnology. 2006;**33**:29-36. DOI: 10.1007/s10295-005-0023-x

[22] Bakanauskas S. Bacterial Activity in JP-4 Fuel. Wright-Patterson Air Force Base, OH. Wright Air Development Center; 1958 March. Report nr WADC Technical Report. 32-58

[23] M.B. Stowe, Microbially-induced corrosion and fuel-tank protection. Aircr eng. Aerosp. Tec. 1995;**67**:5-9

[24] Ayllon ES, Rosales BM. Corrosion of Aa 7075 aluminum alloy in media contaminated with Cladosporium resinae. Corrosion. 1988;**44**:638-643

[25] Hagenauer A, Hilpert R, Hack T. Microbiological investigations of corrosion damages in aircraft. Materials and Corrosion. 1994;**45**:355-360. DOI: 10.1002/maco.19940450606

[26] Miller R, Herron WC, Krwrens AG, Cameron L, Terry BM. Microorganisms cause corrosion in air craft fuel tanks. Materials Protection. 1964;**3**:60-69

[27] Iverson WP. A possible role for sulfate reducers in the corrosion of aluminum alloys. Electrochemical Technology. 1967;**5**:77-79

[28] Wagner P, Little B. Impact of Alloying on Microbiologically Influenced Corrosion. A Review. No. NRL/JA/7333--93-0007. Stennis Space Center, MS: Naval Research Lab; 1993

[29] McKenzie P, Akbar AS, Miller JD. Fungal corrosion of aircraft fuel tank alloys, Inst. Pet. (Technical Pap). 1977;**37**.

[30] Salvarezza R, Mele M, Videla H. Mechanisms of the microbial corrosion of aluminum alloys. Journal of Industrial Pollution Control. 2007;**23**:197-208. DOI: 10.5006/1.3580810.

[31] Schiapparelli E, Meybaum B. V Congr. Int. Sobre Corros. Mar. E Inscrustaciones. Barcelona: Spaña; 1980. p. 1

[32] Albrimi YA, Eddib A, Douch J, Berghoute Y, Hamdani M, and Souto R. Electrochemical behaviour of AISI 316 austenitic stainless steel in acidic media containing chloride ions. *Int. J. Electrochem. Sci*, 2011;**6**:4614-4627

[33] Ruddy WJ, Scully JC. The function of the repassivation process in the inhibition of pitting corrosion on aluminium * pitting corrosion is one of the main types of corrosion that is commonly found in aluminium. Its occurrence can be minimized in a number of ways. Corrosion Science. 1980;**20**:611-631

[34] Samuels BW, Sotoudeh K, Foley RT. Inhibition and acceleration of aluminum corrosion. Corrosion. 1981;**37**:92-97. DOI: 10.5006/1.3593852

[35] Böhni H and Uhlig HJ. Environmental factors affecting the critical pitting potential of aluminum Electrochem. Soc.1959;**116**:906

[36] Hedrick H, Reynolds R, Crum M. Microbiological corrosion of aircraft metal alloys. Developments in Industrial Microbiology. 1969;**10**:228-233

[37] Vejar ND. Enhanced corrosion of 7075 alloy by the presence of bacillus megaterium. International Journal of Electrochemical Science. 2016;**11**:9723-9733. DOI: 10.20964/2016.11.33

[38] Lai ME, Bergel A. Electrochemical reduction of oxygen on glassy carbon: Catalysis by catalase. Journal of Electroanalytical Chemistry. 2000;**494**:30-40. DOI: 10.1016/S0022-0728(00)00307-7

[39] Busalmen JP, Vázquez M, De Sánchez SR. New evidences on the catalase mechanism of microbial corrosion. Electrochimica Acta. 2002;**47**:1857-1865. DOI: 10.1016/S0013-4686(01)00899-4

[40] Vazquez MV, de Sanchez SR, Calvo EJ, Schiffrin DJ. The electrochemical reduction of oxygen on polycrystalline copper in borax buffer. Journal of Electroanalytical Chemistry. 1994;**374**:189-197. DOI: 10.1016/0022-0728(94)03343-9

[41] Lee MHP, Caffrey SM, Voordouw JK, Voordouw G. Effects of biocides on gene expression in the sulfate-reducing bacterium Desulfovibrio Vulgaris Hildenborough. Applied Microbiology and Biotechnology. 2010;**87**:1109-1118. DOI: 10.1007/s00253-010-2596-1

[42] Kähkönen E, Nordström K. Toward a nontoxic poison: Current trends in (European Union) biocides regulation. Integrated Environmental Assessment and Management. 2008;**4**:471-477. DOI: 10.1897/IEAM_2008-021.1

[43] Degim IT, Uslu A, Hadgraft J, Atay T, Akay C, Cevheroglu S. The effects of Azone and capsaicin on the permeation of naproxen through human skin. International Journal of Pharmaceutics. 1999;**179**:21-25. DOI: 10.1016/S0378-5173(98)00353-6

[44] Terashima SI, Ogawa K. Degeneration of infrared receptor terminals of snakes caused by capsaicin. Brain Research. 2002;**958**:468-471. DOI: 10.1016/S0006-8993(02)03746-0

[45] Cichewicz RH, Thorpe PA. The antimicrobial properties of chile peppers (capsicum species) and their uses in Mayan medicine. Journal of Ethnopharmacology. 1996;**52**:61-70. DOI: 10.1016/0378-8741(96)01384-0

[46] Angarano M-B, McMahon RF, Hawkins DL, a Schetz J. Exploration of structure-antifouling relationships of capsaicin-like compounds that inhibit zebra mussel (*Dreissena polymorpha*) macrofouling. Biofouling. 2007;**23**:295-305. DOI: 10.1080/08927010701371439

[47] Qian P-Y, Xu Y, Fusetani N. Natural products as antifouling compounds: Recent progress and future perspectives. Biofouling. 2010;**26**:223-234. DOI: 10.1080/08927010903470815

[48] Zhou J, Yang C, Wang J, Sun P, Fan P, Tian K, Liu S, Xia C. Toxic effects of environment-friendly antifoulant nonivamide on *Phaeodactylum tricornutum*. Environmental Toxicology and Chemistry. 2013;**32**:802-809. DOI: 10.1002/etc.2132

[49] Lawrence JR, Korber DR, Hoyle BD, Costerton JW, Caldwell DE. Optical sectioning of microbial biofilms. Journal of Bacteriology. 1991;**173**:6558-6567. DOI: 10.1128/jb.173.20.6558-6567.1991

[50] Marshall KC, Cruickshank RH. Cell surface hydrophobicity and the orientation of certain bacteria at interfaces. Archiv für Mikrobiologie. 1973;**91**:29-40. DOI: 10.1007/BF00409536

[51] Zhang D, Wang L, Qian H, Li X. Superhydrophobic surfaces for corrosion protection: A review of recent progresses and future directions. Journal of Coatings Technology and Research. 2016;**13**:11-29. DOI: 10.1007/s11998-015-9744-6

[52] Zhang X, Jin M, Liu Z, Nishimoto S, Saito H, Murakami T, Fujishima A. Preparation and photocatalytic wettability conversion of TiO2-based superhydrophobic surfaces. Langmuir. 2006;**22**:9477-9479. DOI: 10.1021/la0618869

[53] Krasowska M, Zawala J, Malysa K. Air at hydrophobic surfaces and kinetics of three phase contact formation. Advances in Colloid and Interface Science. 2009;**148**:155-169. DOI: 10.1016/j.cis.2008.10.003

[54] Xie D, Li W. A novel simple approach to preparation of superhydrophobic surfaces of aluminum alloys. Applied Surface Science. 2011;**258**:1004-1007. DOI: 10.1016/j.apsusc.2011.07.104

[55] Wang Q, Zhang B, Qu M, Zhang J, He D. Fabrication of superhydrophobic surfaces on engineering material surfaces with stearic acid. Applied Surface Science. 2008;**254**:2009-2012. DOI: 10.1016/j.apsusc.2007.08.039

[56] Zuo R, Örnek D, Wood TK. Aluminum- and mild steel-binding peptides from phage display. Applied Microbiology and Biotechnology. 2005;**68**:505-509. DOI: 10.1007/s00253-005-1922-5

[57] Lam H, Oh D-C, Cava F, Takacs C, Clardy J, de Pedro M, Waldor MK. D-amino acids govern stationary phase. Science. 2009;**325**:1552-1555

[58] Bramley PM, Gilbert LE, Raven PH. References and Notes. 2010;**1**:627-629.

[59] Cava F, Lam H, De Pedro MA, Waldor MK. Emerging knowledge of regulatory roles of d-amino acids in bacteria. Cellular and Molecular Life Sciences. 2011;**68**:817-831. DOI: 10.1007/s00018-010-0571-8

[60] Xu H, Liu Y. D-amino acid mitigated membrane biofouling and promoted biofilm detachment. Journal of Membrane Science. 2011;**376**:266-274. DOI: 10.1016/j.memsci.2011.04.030

[61] Xu D, Li Y, Gu T. D methionine as a biofilm dispersal signaling molecule enhanced tetrakis hydroxymethyl phosphonium sulfate mitigation of Desulfovibrio vulgaris biofilm and biocorrosion pitting. Materials and Corrosion. 2014;**65**:837-845. DOI: 10.1002/maco.201206894.

[62] Xu D, Wen J, Fu W, Gu T, Raad I. D-amino acids for the enhancement of a binary biocide cocktail consisting of THPS and EDDS against an SRB biofilm. World Journal of Microbiology and Biotechnology. 2012;**28**:1641-1646. DOI: 10.1007/s11274-011-0970-5

[63] Hamilton WA. Sulphate-reducing bacteria and anaerobic corrosion. Corrosion. 1985;195-217. DOI: 10.1146/annurev.mi.39.100185.001211

Chapter 13

Effect of Alloying Element on the Integrity and Functionality of Aluminium-Based Alloy

Ojo Sunday Isaac Fayomi,
Abimbola Patricia Idowu Popoola and
Nduka Ekene Udoye

Additional information is available at the end of the chapter

http://dx.doi.org/10.5772/intechopen.71399

Abstract

Aluminum alloy are gaining huge industrial significance because of their outstanding combination of mechanical, physical and tribological properties over the base metal. Alloying elements are selected based on their individual properties as they impact on the structure and performance characteristics. The choice of this modifier affects the materials integrity in service resulting to improved corrosion, tribological and mechanical behavior. Hence, the need to understand typically the exact inoculants that could relatively impact on the low strength, unstable mechanical properties is envisage with the help of liquid stir casting technique. In this contribution, sufficient knowledge on Al alloy produced by stir casting will be reviewed with close attention on how the structural properties impact on the mechanical performance.

Keywords: aluminum alloy, alloying element, liquid stir casting technique, reinforcement

1. Introduction

The chemical composition consisting of an aluminum alloy is added to pure aluminum in order to improve its properties for the primary purpose of increasing the strength. The other elements such as iron, magnesium, manganese, zinc and silicon is added to build up 15% alloy by weight. If the aluminum is in molten form, the other elements is mixed with aluminum to produce the required alloy. [1]. Pure aluminum is not usually used for structural applications and that in order to produce aluminum that is of adequate strength for the manufacture of structural components, it is necessary to add other elements to the aluminum [1, 2]. The strength characteristic of aluminum (1xxx series) makes it a useful product for structural fabrication. The pure aluminum contains levels of impurities such as iron and silicon that enables it to respond to strain hardening even though 1xxx series is the same as pure aluminum [3]. To compare these alloys with other series aluminum alloys we observe it have a very low strength. The major properties considered when chosen these alloys for structural application is their superior corrosion resistance and their high electrical conductivity [4, 5].

The principal method of producing a selection of different materials that can be used in different structural application is the mixture of alloying elements with the aluminum itself. One can deduce that the different alloy element used to produce each of the alloy series from the seven selected aluminum alloy series. We studied the effects of these elements on aluminum [6]. **Table 1** shows the test result of different mechanical properties of alloys.

Alloy	**Temper**	**Proof stress 0.20% (MPa)**	**Tensile strength (MPa)**	**Shear strength (MPa)**	**Elongation A5 (%)**	**Elongation A50 (%)**	**Hardness Brinell HB**	**Hardness Vickers HV**	**Fatigue endurance limit (MPa)**
	H2	85	100	60	12		30	30	
AA1050A	H4	105	115	70	10	9	35	36	70
	H6	120	130	80	7		39		
	H8	140	150	85	6	5	43	44	100
	H9	170	180			3	48	51	
	0	35	80	50	42	38	21	20	50
	T3	290	365	220	15	15	95	100	250
	T4	270	350	210	18	18	90	95	250
AA2011	T6	300	395	235	12	12	110	115	250
	T8	315	420	250	13	12	115	120	250
	H2	115	135	80	11	11	40	40	
	H4	140	155	90	9	9	45	46	130
	H6	160	175	100	8	6	50	50	
	H8	180	200	110	6	6	55	55	150
	H9	210	240	125	4	3	65	70	
	0	45	105	70	29	25	29	29	100
	H2	240	330	185	17	16	90	95	280
	H4	275	360	200	16	14	100	105	280
AA3103	H6	305	380	210	10	9	105	110	
	H8	335	400	220	9	8	110	115	
	H9	370	420	230	5	5	115	120	
	0	145	300	175	23	22	70	75	250
	H2	165	210	125	14	14	60	65	
	H4	190	230	135	13	12	65	70	230
AA5083	H6	215	255	145	9	8	70	75	
	H8	240	280	155	8	7	80	80	250

Alloy	Temper	Proof stress 0.20% (MPa)	Tensile strength (MPa)	Shear strength (MPa)	Elongation A5 (%)	Elongation A50 (%)	Hardness Brinell HB	Hardness Vickers HV	Fatigue endurance limit (MPa)
	H9	270	310	165	5	4	90	90	
	0	80	180	115	26	25	45	46	200
	H2	185	245	150	15	14	70	75	
	H4	215	270	160	14	12	75	80	250
AA6063	H6	245	290	170	10	9	80	85	
	H8	270	315	180	9	8	90	90	280
	H9	300	340	190	5	4	95	100	
	0	100	215	140	25	24	55	55	220
	0	50	100	70	27	26	25	85	110
	T1	90	150	95	26	24	45	45	150
	T4	90	160	110	21	21	50	50	150
	T5	175	215	135	14	13	60	65	150
AA5251	T6	210	245	150	14	12	75	80	150
	T8	240	260	155		9	80	85	
	0	60	130	85	27	26	35	35	120
	T1	170	260	155	24	24	70	75	200
AA6082	T4	170	260	170	19	19	70	75	200
	T5	275	325	195	11	11	90	95	210
	T6	310	340	210	11	11	95	100	210
	T6	240	290		8				
AA6262	T9	330	360		3				
	0	105	225	150		17	60	65	230
AA7075	T6	505	570	350	10	10	150	160	300
	T7	435	505	305	13	12	140	150	300

Table 1. Mechanical properties of different aluminum alloys [7].

1.1. Principal effects of alloying elements in aluminum

1.1.1. Unalloyed aluminium 1xxx series

This alloy consists of 99% of aluminum in high purity.

Properties of aluminum 1xxx series

- Perfect corrosion resistance.
- Effective workability.
- High thermal and electrical conductivity.

Uses of Aluminum 1xxx series

- Transmission of electricity or power grid.
- To connect natural grid across the country [8, 9].

1350 alloy designation is for electrical applications while 1100 alloy designation is for food packaging trays. The most common applications for the 1xxx series alloys are aluminum foil, electrical buss bars, metallizing wire and chemical tanks and piping systems [10].

1.1.2. Copper (Cu) 2xxx

The alloy of aluminum and copper consist of 2–10% of copper with other traces of elements. **Figures 1–3** are the SEM micrograph of the different quantity of Fe 0.53 in Cu sample.

Uses of copper

- It provides strength and facilitates precipitation hardening.
- It reduces ductility and corrosion resistance.
- It contains highest strength heat treatable aluminum alloys.
- It is applied in aerospace, military vehicles and rocket fins [11, 13].

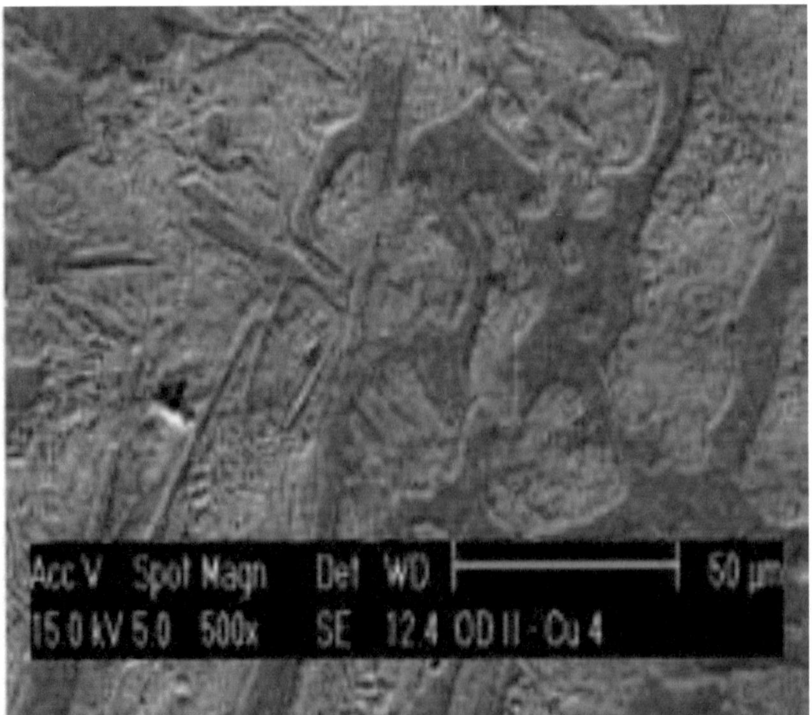

Figure 1. SEM micrograph of the Fe 0.53 Cu sample (×500).

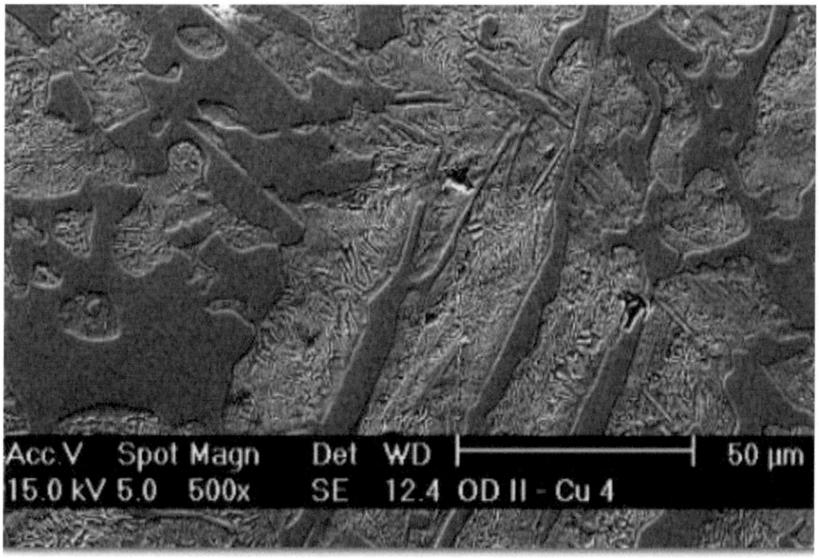

Figure 2. SEM micrograph of the Fe 0.21 Cu sample (×500) [11].

Figure 3. SEM micrograph of the control sample of Fe in Cu (×500) [12].

1.1.3. Manganese (Mn) 3xxx

The alloy of manganese and aluminum results to improvement in strain hardening and strengthening but does not reduce ductility or corrosion resistance. It retains strength when used on non-heat treatable materials. The uses of 3xxx series alloys are cooking utensils, radiators, evaporators, heat exchangers and associated piping systems [14]. **Figure 4** is the SEM micrograph of manganese in iron and EDX of manganese ferrite (Mn-Fe_2O_4) powder prepared by hydrothermal route.

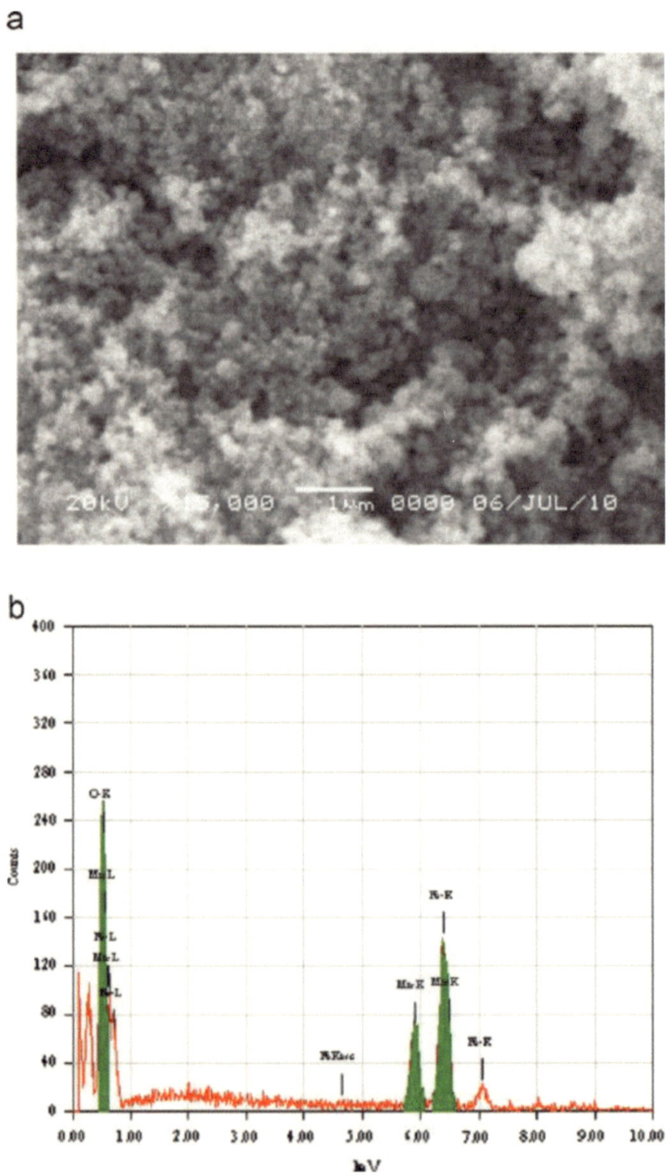

Figure 4. (a) SEM micrograph and (b) EDX of manganese ferrite (Mn-Fe_2O_4) powder prepared by hydrothermal route [15].

1.1.4. Silicon (Si) 4xxx

The alloy of silicon and aluminum reduces the melting point of temperature and enhances fluidity. **Figure 5** is the SEM micrograph of cast aluminum-silicon alloys.

Uses of silicon

- It produces a non-heat-treatable alloy.
- Silicon with magnesium produces a precipitation hardening heat-treatable alloy.
- Alloy of silicon and aluminum are used for the manufacturing of castings [17].
- They are also used as filler wires for fusion welding.

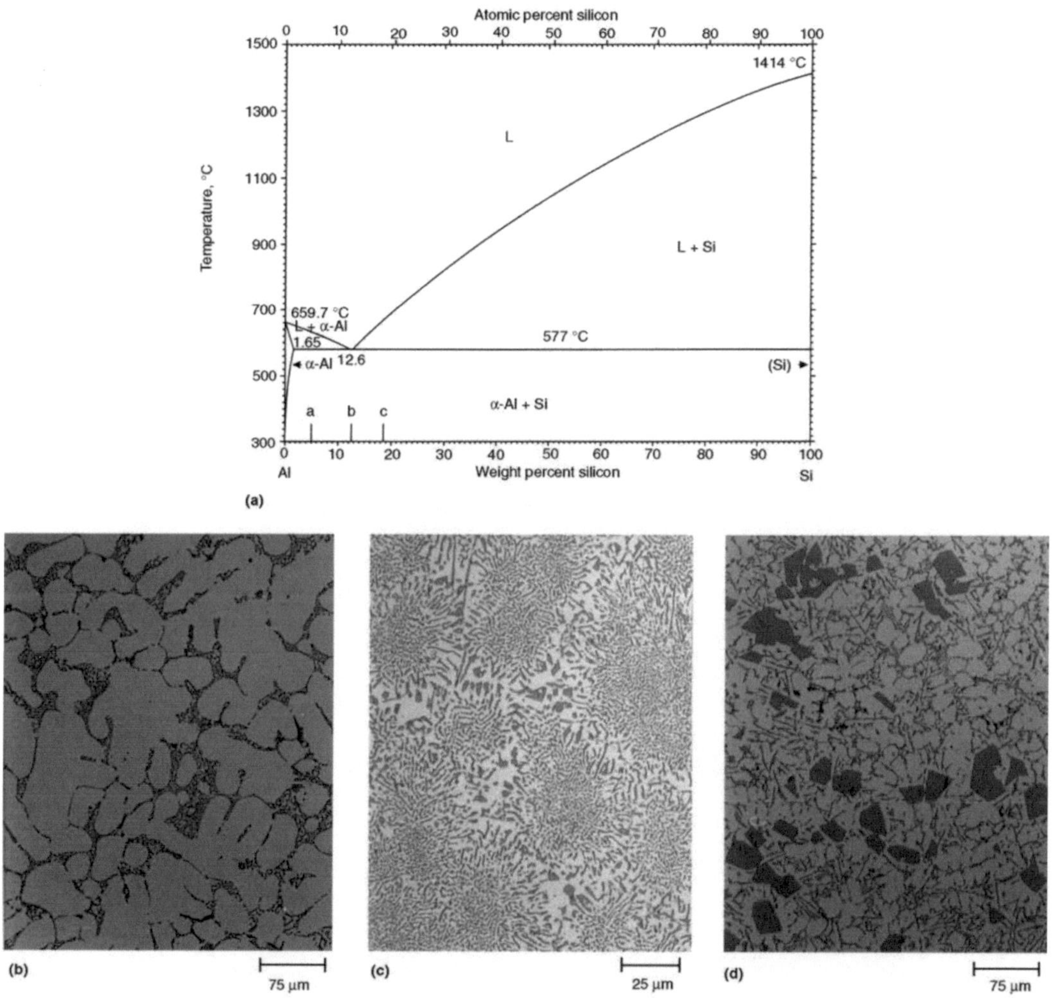

Figure 5. Commercial cast aluminum-silicon alloys. (a) Al-Si equilibrium diagram; (b) microstructure of hypoeutectic alloy (1.65–12.6 wt.% Si) 150; (c) microstructure of eutectic alloys (12.6% Si) 400; and (d) microstructure of hypereutectic alloy (>12.6% Si) 150 [16].

1.1.5. Magnesium (Mg) 5xxx

We used solid solution strengthening to improve strain hardening of metal by the alloy of magnesium with aluminum. **Figure 6** SEM micrograph of a magnesium material with porous microstructure produced using space-holding particles.

Uses of magnesium

- They are used extensively for structural applications.
- The 5xxx series alloys are produced mainly as sheet and plate [19].
- They are also used in truck and train bodies, armored vehicles, ship and boat building, chemical tankers, pressure vessels and cryogenic tanks.

1.1.6. Magnesium and silicon (Mg_2Si) 6xxx

The alloy of magnesium and silicon to aluminum forms the compound magnesium-silicide (Mg_2Si). The 6xxx series provides heat treatability through the formation of this compound. They are easily and economically separated and often found in an extensive selection of extruded shapes. Uses of this alloy: bicycle frames, scaffolding, drive shafts, automotive frame sections,, tubular lawn furniture, stiffeners and braces used on trucks, boats and many other structural fabrication. **Figures 7–9** are SEM image of magnesium and silicon taken at different temperatures.

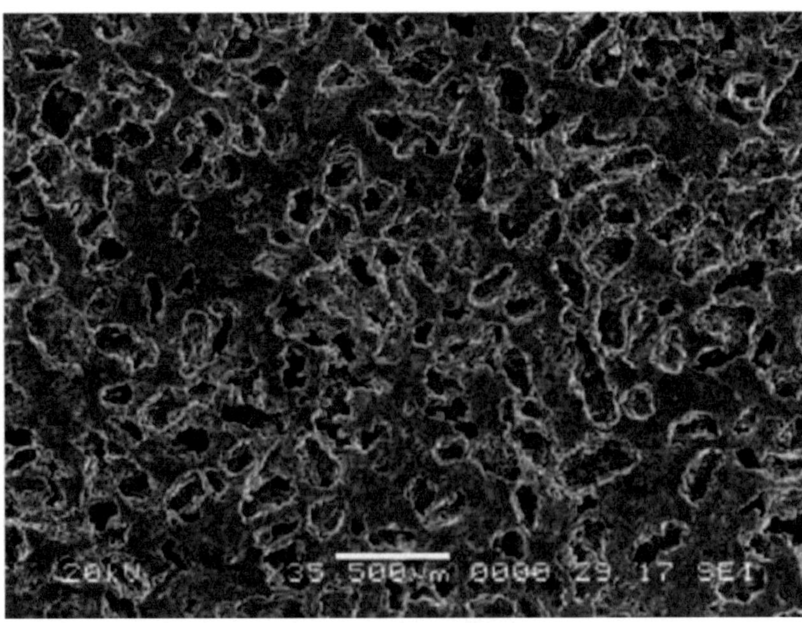

Figure 6. SEM micrograph of a magnesium material with porous microstructure produced using space-holding particles [18].

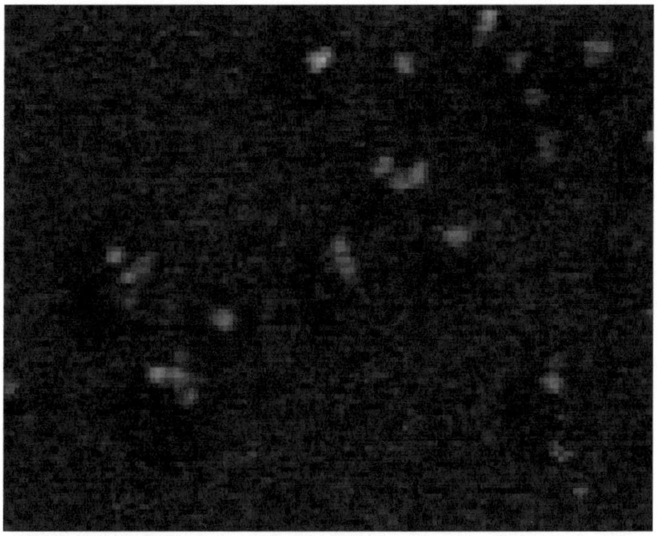

Figure 7. SEM image taken after heating to 400°C and exposure to air.

Figure 8. SEM image after heating to 490°C.

1.1.7. Zinc (Zn) 7xxx

The mixture of zinc and aluminum with other trace element such as magnesium and copper produces heat-treatable strong aluminum alloys. The zinc add to the strength and allow precipitation hardening. Some of these alloys are subject to stress corrosion cracking and for this reason are not usually fusion welded. There is a decrease in the 7xxx series Al alloys with these over-aging treatments. The 7xxx series Al alloys and re-aging (RRA) treatment possess

Figure 9. SEM after heating to 500°C [15].

high strength and good SCC resistance [12]. However, the RRA treatment shows short retrogression time and cannot be used for large-section Al alloys. Novel heat treatment have been developed to keep the high strength of the 7xxx series Al alloys and improve their corrosion resistance simultaneously [15, 20] advanced a novel aging treatment, called high temperature pre-precipitation(HTPP) aging treatment. In the present work, the comparative study of the effects of the various heat treatments, especially the secondary aging and HTPP aging omits tensile properties, corrosion behaviors and microstructures was carried out on different alloy group. Other alloys within this series are often fusion welded with excellent results. Some of the common applications of the 7xxx series alloys are aerospace, armored vehicles, baseball bats and bicycle frame [16]. **Figure 10** is the SEM micrograph for an aluminum alloy that is (a) cast-solutionized, (b) cast-solutionized-aged at 20°C—60 min, (c) hot-rolled and (d) hot-rolled-aged at 200°C—60 min.

1.1.8. Iron (Fe)

Iron is added to some pure alloys to provide the increase in strength which is the most common impurity found in aluminum.

1.1.9. Chromium (Cr)

The essence of chromium to aluminum is to control grain structure, to protect grain growth in aluminum-magnesium alloys, and to prevent recrystallization in aluminum-magnesium-silicon or aluminum-magnesium-zinc alloys during heat treatment. Chromium will also reduce stress corrosion y and improves toughness.

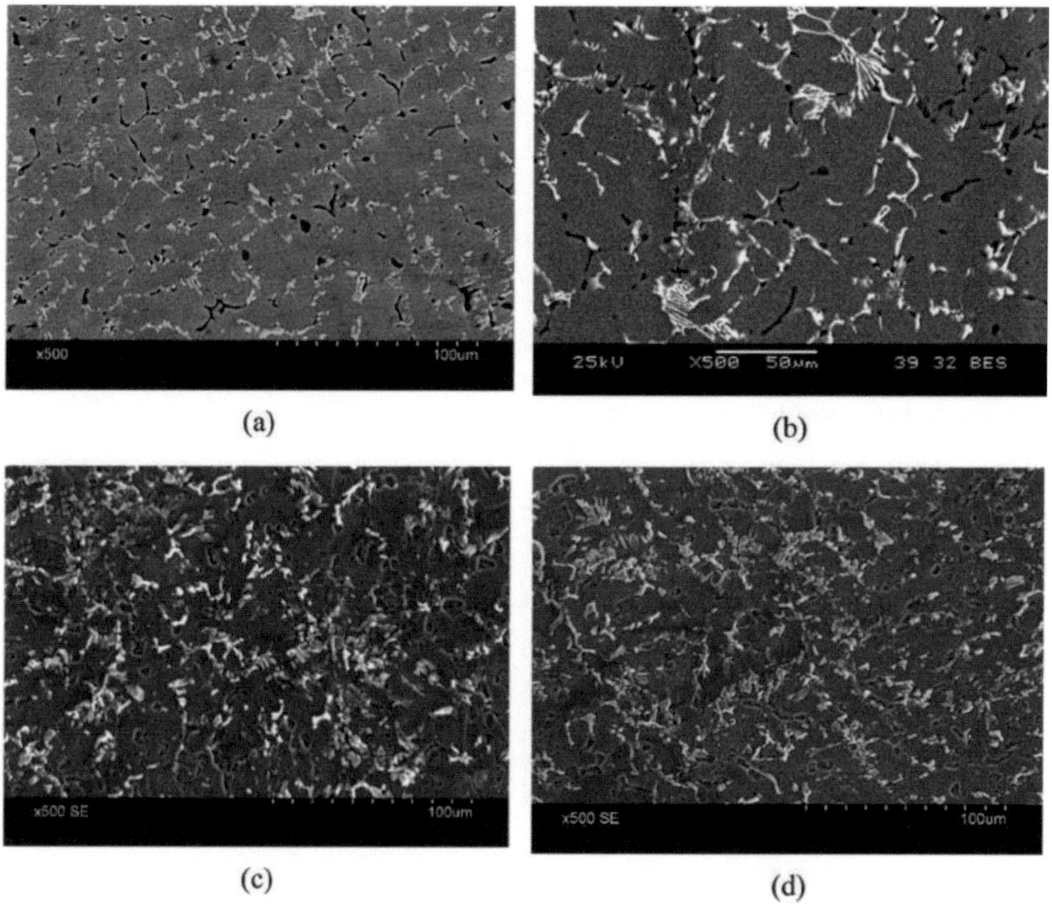

Figure 10. SEM micrograph for an aluminum alloy that is (a) cast-solutionized, (b) cast-solutionized-aged at 20°C—60 min, (c) hot-rolled and (d) hot-rolled-aged at 200°C—60 min [20].

1.1.10. Nickel (Ni)

Alloy of Nickel and aluminum-copper improve hardness and strength at elevated temperatures and reduce the coefficient of expansion.

1.1.11. Titanium (Ti)

Titanium is added to aluminum to serves as a grain refiner. The grain refining effect of titanium is enhanced if boron is present in the melt or if it is added as a master alloy containing boron largely combined as TiB_2. Titanium is a common addition to aluminum weld filler wire as it refines the weld structure and helps to prevent weld cracking.

1.1.12. Zirconium (Zr)

The fine precipitate of intermetallic particles that inhibit recrystallization is produced when Zirconium is added to aluminum.

1.1.13. *Lithium (Li)*

The addition of lithium to aluminum increases strength and Young's modulus, It also provide precipitation hardening and decreases density.

1.1.14. *Lead (Pb) and Bismuth (Bi)*

These are added to aluminum to assist in chip formation and improve machinability. These free machining alloys are not weld able because the lead and bismuth produce low melting constituents and can produce poor mechanical properties and high crack sensitivity on solidification [6]. **Figure 11** is the microstructure of alloy Al-6.16Zn-3.02 Mg-1.98Cu, aged at 172°C for 4 h: (a) SEM micrograph, showing S phase and Al_7Cu_2Fe particle; (b) TEM micrograph, showing η phase and Al_7Cu_2Fe particles [21].

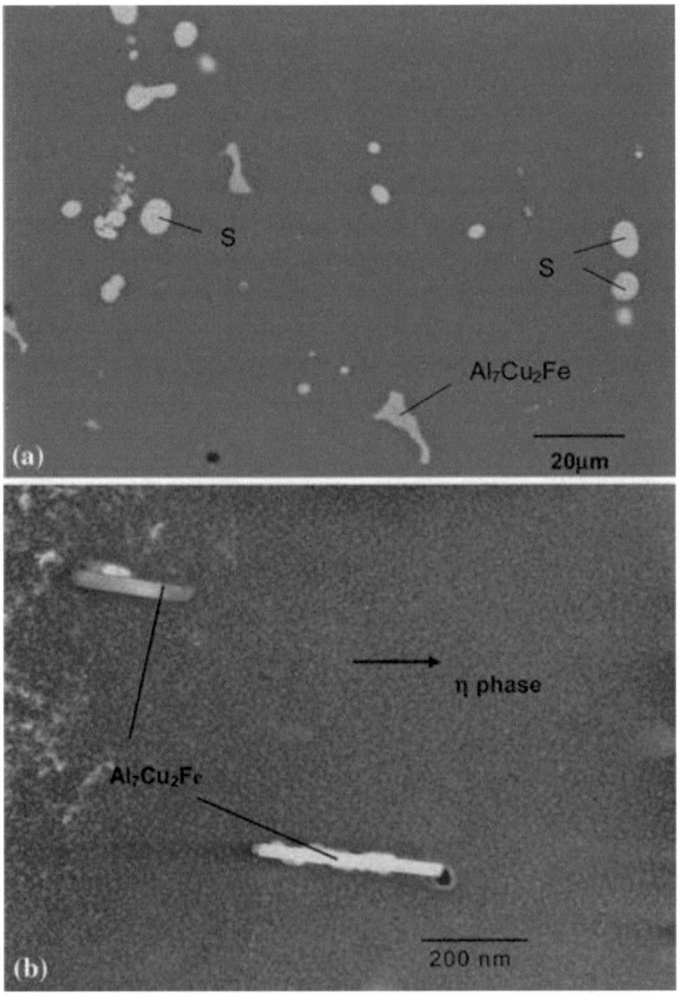

Figure 11. Microstructures of alloy Al-6.16Zn-3.02 Mg-1.98Cu, aged at 172°C for 4 h: (a) SEM micrograph, showing S phase and Al_7Cu_2Fe particle; (b) TEM micrograph, showing η phase and Al_7Cu_2Fe particles [21].

1.2. Other applications of aluminum alloy

- In the chemistry

The properties of aluminum such as strength, density, workability, electrical conductivity and corrosion resistance are affected by adding other elements such as magnesium, silicon or zinc.

- Bradley fighting vehicle

7xxx series and 5xxx series aluminum alloys is used to produce the military Bradley Fighting Vehicle.. It is trusted to keep soldiers safe and mobile, aluminum is also used in many other military vehicles.

- Our favorite beverage container

Aluminum alloys is used to produce America's favorite beverage container, the aluminum can, is made from multiple. The shell of the can is composed of 3004 and the lid is made from 5182. Sometimes it takes more than one alloy to make one, everyday item [7].

- Hot and cold

Application involving the use of aluminum alloys is made stronger through heat-treatment or cold working. The attributes of a particular alloy are different because of their additives and treatment. **Table 2** is the percentage composition of aluminum alloys of different metal.

1.3. Aluminum alloy designation

Aluminum alloys for sheet products are identified by a four-digit numerical system which is administered by the Aluminum Association. The alloys are conveniently divided into eight groups based on their principal alloying element [9]. **Table 3** shows the different alloy group and alloying element.

Element	Si	Fe	Cu	Mn	Mg	Cr	Zn
wt.%	0.4	0–5	0.1	0.2	3.5	0.05	0.25

Table 2. Composition of aluminum alloys.

Alloy group	Principal alloying element	% Aluminum
1xxx	Unalloyed aluminum	Purity of 99.0%
2xxx	Copper	Heat treatable alloys
3xxx	Manganese	Non heat treatable alloys
4xxx	Silicon	Low melting point alloys
5xxx	Magnesium	Non heat treatable alloys
6xxx	Magnesium and silicon	Heat treatable alloys
7xxx	Zinc	Heat treatable alloys
8xxx	Other elements	None

Table 3. Different alloy group and alloying element.

2. Methodology

2.1. Stir casting

Stir casting is generally accepted as a particularly promising route, currently practiced commercially. Its advantages lie in its simplicity, flexibility and applicability to large quantity production. Stir casting is a liquid state method of composite materials fabrication, in which a dispersed phase (ceramic particles, short fibers) is mixed with a molten matrix metal by means of mechanical stirring. It is also attractive because, in principle, it allows a conventional metal processing route to be used, and hence minimize the final cost of the product. The liquid composite material is then cast by conventional casting methods and may also be processed by conventional Metal forming technologies [9]. Factors considered in preparing metal matrix composites by stir casting method are [10],

- To ensure uniform distribution of the reinforcement material
- It is to achieve wettability between the two main substances
- To control porosity in the cast metal matrix composite

The material properties and process parameters are used to determine the final distribution of the particles in the solid such as the wetting condition of the particles with the, melt, relative density strength of mixing,, and rate of solidification. The distribution of the particles in the molten matrix depends on the geometry of the mechanical stirrer, stirring parameters, placement of the mechanical stirrer in the melt, melting temperature, and the characteristics of the particles added [5].

2.2. Process parameters

2.2.1. Stirrer design

Stirrer design is used in stir casting process to form vortex. The blade angle and number of blades give the flow pattern of the liquid metal. The stirrer is immersed to two third the depth in molten metal. The essence is for uniform distribution of reinforcement in liquid metal and perfect interface bonding.

2.2.2. Stirring speed

Stirring speed is an important parameter to promote binding between matrix and reinforcement i.e. wettability. Stirring speed decides formation of vortex which is responsible for dispersion of particulates in liquid metal. In our project stirring speed is 300 rpm.

2.2.3. Stirring temperature

Aluminum melts around 650°C, at this temperature semisolid stage of melt is present. Particle distribution depends on change in viscosity. The viscosity of matrix is mainly influenced by the processing temperature. The viscosity of liquid is decreased by increasing processing temperature with increasing holding time for stirring which also promote binding between matrix and reinforcement. Good wettability is obtained beekeeping temperature at 800°C.

2.2.4. Stirring time

As stirring promote uniform distribution of reinforcement partials and interface bond between matrix and reinforcement, stirring time plays a vital role in stir casting method. Less stirring leads to non-uniform distribution of particles and excess stirring forms clustering of particles at some places. Stirring time is 5 minutes in our case.

2.2.5. Preheat temperature of reinforcement

Casting process of AMC's is difficult due to very low wettability of alumina particles and agglomeration phenomenon which results in non-uniform distribution and poor mechanical properties [1]. Reinforcement is heated to 500°C for 40 minutes. It removes moisture as well as gases present in reinforcement.

2.2.6. Preheat temperature of mold

This is used to remove the entrapped gases from the slurry to go into the mold. It also improves the mechanical properties of the cast AMC. The mold is heated to500°C for 1 h.

2.2.7. Magnesium

Addition of magnesium enhances the wettability. However increase the content above 1 wt.% increases viscosity of slurry and hence uniform particle distribution becomes difficult [7].

2.2.8. Reinforcement feed rate

Non-uniform feed rate promotes clustering of particles at some places which causes the porosity defect and inclusion defect, so to have a good quality of casting the feed rate of powder particles must be uniform. The flow rate of reinforcements measured is 0.5 gram per second [5].

2.2.9. Pouring of melt

Pouring rate and pouring temperature plays significant role in quality of casting. Pouring rate of slurry must be uniform to avoid entrapping of gases. At this stage the temperature of melt is 800°C. The distance between mold and crucible also plays vital role in quality of casting. Apart from this size of reinforcement plays significant role in quality of casting.

2.2.10. Speed of rotation

Speed of rotation issued to influence the structure; increase of speed promotes refinement and very low speed results in instability of the liquid mass. It is logical to use the highest speed to avoid tearing.

2.3. Experimental setup and procedure

The process of stir casting starts with placing empty crucible in the furnace. The heated temperature is then gradually increased up to 800°C. Aluminum alloy is cleaned to remove dust particles, weighed and charged in the crucible for melting. Required quantities of reinforcement powder and magnesium powder are weighed on the weighing machine.

Reinforcements are heated for 45 minutes at a temperature of 500°C. When matrix was in the semisolid stage condition at 650°C, 1% by weight of pure magnesium powder is used as wetting agent. After 5 min the scum powder is added which forms a scum layer of impurity on liquid surface which to be removed. We increase the heater temperature to 800°C. Stirring is started at this heater temperature and continued for 5 min. Speed controller help Stirring rpm to increase from 0 to 300 RPM. We add preheated reinforcement during 5 min of stirring. Conical hopper is used to pour reinforcements manually with the help of. The flow rates measured in 0.5 g/s. It is then gradually lowered to the zero. The molten composite slurry is poured in the metallic mold without giving time for reinforcement to settle down at crucible bottom. Before pouring the molten slurry in the mold, it is preheated at 500°C temperature for 1 h. The flow of the slurry is kept uniform to avoid trapping of gas. This is necessary to maintain slurry in molten condition throughout the pouring. While pouring the slurry in the mold, also distance between crucible and the mold plays a vital role in quality of casting. **Figure 12** is the Schematic view of stir casting setup.

2.4. Hardness

The Brinell hardness tests were carried out on Brinell hardness tester. Six samples of Al/Sic-MMC's for different sizes and weight fraction of SiC particles were prepared. After test and hardness value on dial, the Brinell hardness values with reference to scale HRB were taken for all samples and shown by graphs. Impact Strength and Impact Test were carried out over Charpy

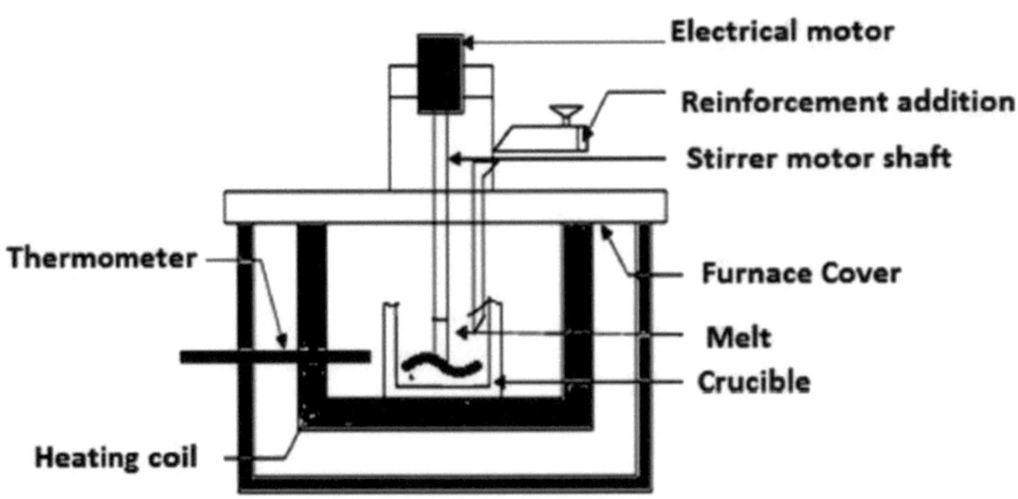

Figure 12. Schematic view of stir casting setup [7].

Impact Testing Machine and results were recorded. According to size and weight fraction of SiC particles, 12 specimens Al/Sic-MMC's of Square cross-section of size (10 × 10 × 55) with single V-notches are planned. The size of V-notches is 45° and 2 mm depth.

3. Results and discussions

Following conclusions are given from present work,

- Stir casing process can successfully be used for manufacturing of AMC's having low density and enhanced mechanical properties.
- Stir casting process is cost effective and conventional route for manufacturing of composite material.
- Material having isotropic nature can be manufactured successfully.
- Preheating of mold reduces porosity and enhances mechanical properties.
- Addition of magnesium is important to increase wettability.
- Design of stirrer decides the flow pattern of melt.
- Stirrer speed, stirring time decides quality of casting.
- Preheat temperature of mold, preheat temperature of reinforcement, reinforcement size, reinforcement feed rate and melt pouring rate are also the important parameters in stir casting method.

4. Conclusion

To conclude majority of authors fabricated the composite by stir casting process with different reinforcements like SiC, Al_2O_3, fly ash, ground nut and rice husk ash. In this study stir casting process is the simplest and cheapest route to fabricate the particulate type metal matrix composites. However, agglomeration of particles added in molten matrix is the difficulty faced by most of the authors during fabrication process. The mechanism to avoid agglomeration of particles is through coating of the reinforcement and inert gas environment during fabrication process. We use two step and electromagnetic stir casting process to improve the homogeneity of particles during fabrication. This method gives high specific strength, greater strength to weight ratio at elevated temperature, greater wear resistance as compare to matrix phase. If you increase the Zn content, modeling results will also show that the Zn contents has been increased, but the electrical conductivity and thermal conductivity reduce slightly with the Zn addition. For Mg variation, the strength property of the alloy improves in the range of 2.7–2.9 wt.% Mg, Increase of Mg contents will reduce the electrical conductivity and thermal conductivity. In general, the experimental evidence for microstructures is in accordance with the predictions in the modeling processes [21].

Author details

Ojo Sunday Isaac Fayomi[1,2]*, Abimbola Patricia Idowu Popoola[2] and Nduka Ekene Udoye[1]

*Address all correspondence to: ojo.fayomi@covenantuniversity.edu.ng

1 Department of Mechanical Engineering, College of Engineering, Covenant University, Ota, Ogun State, Nigeria

2 Department of Chemical, Metallurgical and Materials Engineering, Tshwane University of Technology, Pretoria, South Africa

References

[1] Novateur Publications International Journal of Innovations in Engineering Research and Technology [IJIERT] Issn: 2394-3696. 2016 Jan;**3**(1)

[2] Wahab MN, Daud AR, Ghazali MJ. Preparation and characterization of stir cast-aluminum nitride reinforced aluminum metal matrix composites. International Journal of Mechanical And Materials Engineering (IJMME). 2009;**4**(2):115-117

[3] Sharma P, Chauhan G, Sharma N. Production of amc by stir casting—An overview. International Journal of Contemporary Practises;**2**(1)

[4] Surappa MK. Aluminium matrix composites: Challenges and opportunities. Sadhana. February/April 2003;**28**, Parts 1 & 2:319-334

[5] Estrada-Guel I, Carreňo-Gallardo C, Mendoza-Ruiza DC, Miki-Yoshidaa M. Graphite nanoparticledispersion in 7075 aluminum alloy by means of mechanical alloying. Journal of Alloys and Compounds. 2009;**483**:173-177

[6] Veeresh Kumar GB, Rao CSP, Selvaraj N, Bhagyashekar MS. Studies on Al6061-SiC and Al7075-Al_2O_3 metal matrix composites. Journal of Minerals & Materials Characterization & Engineering. 2010;**9**(1):43-55

[7] Baradeswaran A, ElayaPerumal A. Study on mechanical and wear properties of Al7075/Al_2O_3/graphite hybrid composites. Composites: Part B. 2014;**56**:464-471

[8] Keshavamurthy R, Mageri S. Microstructure and mechanical properties of Al7075-TiB_2 in-situ composite. Research Journal of Material Sciences. November 2013;**1**(10):6-10. ISSN: 2320–6055

[9] Isadarea AD, Aremo B. Effect of heat treatment on some mechanical properties of 7075 aluminium alloy. Materials Research. 2013;**16**(1):190-194

[10] kumar Lakshmipathy J, Kulendran B. Reciprocating wear behavior of 7075Al/SiC in comparison with 6061Al/Al2O3 composites. International Journal of Refractory Metals and Hard Materials. 2014;**46**:137-144

[11] Valsange M, Kulkarni SG, Sonawane SA. Stir casting used in manufacturing of aluminium matrix composite. International Journal for Research in Technological Studies. August 2014;**1**(9)

[12] Dwivedi SP, Sharma S, Mishra RK. A356 "Aluminum alloy and applications—A review". Advanced Materials Manufacturing & Characterization. 2014;**4**(2)

[13] Bharath V, Nagaral M, Auradi V, Kori SA. Preparation of 6061Al-Al_2O_3 MMC's by stir casting and evaluation of mechanical and wear properties. Procedia Materials Science. 2014;**6**:1658-1667

[14] kumar R, Bhandare G, Sonawane PM. Preparation of aluminium matrix composite by using stir casting method. International Journal of Engineering and Advanced Technology. December 2013;**3**(2)

[15] Miller W, Zhuang L, Bottema J, Wittebrood A, Smet PD, Haszler A, Vieregge A. Recent development in aluminium alloys for the automotive industry. Materials Science and Engineering: A. 2000;**280**(1):37-49. DOI: 10.1016/S0921- 5093(99)00653-X

[16] Lu S-Z, Hellawell A. Modification of Al-Si alloys: Microstructure, thermal analysis and mechanics. JOM. 1995;**47**(2):38-40

[17] Jones RM. Mechanics of Composite Materials. 2nd ed. Taylor and Francis Ltd.; 1999

[18] Florian B, Edvin L. Surface Studies on Industrial Aluminium Alloys. Lund, Sweden; 2014. p. 19

[19] Kulkarni SG, Meghnani JV, Lal A. Effect of fly ash hybrid on mechanical property and density of aluminium 356 alloy. Procedia Materials Science. 2014;**5**:746-754

[20] Sepehrband P, Esmaeili S. Application of recently developed approaches to microstructural characterization and yield strength modeling of aluminum alloy AA7030. Materials Science and Engineering A. 2008;**487**:309-315

[21] Yu J, Li X. Journal of Phase Equilibrium and Diffusion. 2011;**32**:350. DOI: 10.1007/s11669-011-9911-0

Chapter 14

Equal Channel Angular Extrusion Characteristics on Mechanical Behavior of Aluminum Alloy

Abiodun Ayodeji Abioye, Ojo Sunday Isaac Fayomi,
Abimbola Patricia Idowu Popoola and
Oluwabunmi Pamilerin Abioye

Additional information is available at the end of the chapter

http://dx.doi.org/10.5772/intechopen.71019

Abstract

Materials strengthened by conventional methods such as strain hardening, solute additions, precipitation and grain size refinement are often adopted in industrial processes. But there is limitation to the amount of deformation that these conventional methods can impact to a material. This study focused on the review of major mechanical properties of aluminum alloys in the presence of an ultrafine grain size into polycrystalline materials by subjecting the metal to an intense plastic straining through simple shear without any corresponding change in the cross-sectional dimensions of the sample. The effect of the heavy strain rate on the microstructure of aluminum alloys was in refinement of the coarse grains into ultrafine grain size by introducing a high density of dislocations and subsequently re-arranging the dislocations to form an array of grain boundaries. Hence, this investigation is aimed at gathering contributions on the influence of equal channel angular extrusion toward improving the mechanical properties of the aluminum alloys through intense plastic strain.

Keywords: equal channel angular extrusion, strain hardening, severe plastic deformation

1. Introduction

1.1. Deformation processes

Solid materials transforms from one shape to another through a process called deformation. Solid materials can be plastically deformed into complex shapes to obtain a material having the desired geometry and properties required [1]. Deformation processes are commonly used with other unit operations such as casting, machining, grinding and heat-treatment. These unit operations occur during the transformation of raw materials to finished parts [2]. Deformation process generally can be classified into two; (1) bulk forming process and (2) sheet forming process. Rolling, extrusion and forging are examples of bulk forming process while stretching, flanging, drawing and contouring are sheet forming process. Input materials for bulk forming processes are usually in form of billets, rods and slabs while the input in sheet forming processes are usually in sheet blank form [2]. Deformation processes work by stressing metal sufficiently to cause plastic flow into desired shape. The processes also alter the grain sizes of the materials and induce plastic strains into the materials. Some deformation processes used mostly are rolling, extrusion, forging, extrusion and wire drawing. Extrusion process is considered for the purpose of this study.

Hall-Petch relation shows that the yield stress increases as refinement of grain size increases. It implies that the mechanical behavior of a metal will not be the same if its grain size changes. In order to improve properties of metal, methods of changing grain size is very important. There has been reported limitation of conventional metallurgy processes (such as rolling, forging, drawing and extrusion) in that they cannot supply metals whose grain sizes are substantially smaller than engineering micro-components in dimension [3]. In conventional metal forming operation the amount of plastic strain produced is often limited. Recently, materials with grain structures in the nanoscale range are of high interest. Nanomaterials' ability to decrease the geometrical dimensions of different mechanical devices makes nano structural materials attractive in engineering world. Successful shaping methods of engineering nanocomponents are required for building nanodevices, this can be achieved through bottom-up approach method where the building blocks are in atomic and molecular levels [3, 4]. Merging the manufacturing process of nanomaterials with material fabrication is highly challenging in terms of cost of making many of the raw components for functional nanomaterials. Time required for performing any engineering work at nano scale is also a considerable challenge. These challenges often reduced mechanical response and electrical properties; likewise caused spatial distortions, suboptimal thermal behavior, which weakens the general system performance [3, 4]. Another way is to use top down approach where bulk materials are restructured to nanoscale level using traditional shaping methods while constructing systems and devices at the micro and nano scales.

Severe plastic deformation (SPD) process has been shown as one of the major ways of fabricating bulk nanostructured samples and billets out of different metals and alloys [5]. The first developments and investigations of nanostructured materials processed using SPD methods were fulfilled by Valiev and his co-workers over than 10 years ago.

1.2. Severe plastic deformation (SPD)

Severe plastic deformation was defined as the intense plastic straining under high imposed pressure [6, 7]. Severe plastic deformation methods are used to convert coarse grain metals and alloys into ultrafine grained (UFG) materials [8, 9]. The ultra-fine grain materials obtained possesses improved mechanical and physical properties which impart on them a wide commercial use. Severe plastic deformation is a new metal forming process capable of generating very large or severe plastic deformation in a material without a major change in the billet geometry [5]. The method covers all metal forming processes which are based on simple shear and/or repetitive reversed straining and tend to preserve the initial shape of the billet. Different SPD processes are shown in **Figure 1**. Where ε is the strain; t is the torsion; n is the number of passes; φ is half of the inner die angle; D is the initial diameter; d is the final diameter; H is the height; W is the width; T is the initial thickness; t is the final thickness and r is the radius.

1.3. Equal channel angular extrusion (ECAE) process

Equal channel angular extrusion (ECAE), also called equal channel angular pressing (ECAP) has been used to produce ultrafine-grained materials [10–12]. ECAP is a convenient forming procedure to extrude material by the use of specially designed channel dies without a substantial change in geometry [13]. Extrusion by ECAP method enables obtaining of a fine-grain structure in larger volumes. There has been a significant progress in the use of ECAP from ordinary metal processing method to well establish procedure for ultrafine grain refinement. This ultrafine grain refinement improved the strength and toughness in metal and alloy. At present, ECAP is the best developed of all severe plastic deformation (SPD) processing techniques [14–16]. Industrial significance is given to production by processing bulk materials through ECAE. Furthermore, useful tools that can be made use of in the development of new SPD techniques and improving on the existing ones can be formed through the basic principles of ECAE, dealing with the mechanics of setal flow and the microstructural evolution. The mechanical and physical properties of all crystalline materials are determined by several factors, the average grain size of the material generally plays a very significant, and often a dominant, role. One of the most promising SPD processes is equal channel angular pressing (ECAP) [10]. Large plastic deformation is involved in ECAP for the deformation of work-piece in a deforming work-piece. This required moving the work piece through two intersecting channels—usually at an angle of 90° or 120°—of identical cross sections in a die as shown in **Figure 2**. ECAE is an effective method of producing a large amount of simple shear deformation in a material by passing it around a corner of two intersecting channels with equal cross-sections as seen in **Figure 2**. The major advantage of ECAE over normal extrusion is that the cross section of the material undergoing ECAE remains the same after the process. The material can be passed through the same die to repeat the process and accumulate higher plastic deformation [17]. Segal, in order to change the texture of material processed metals by method of equal channel angular pressing [10, 18, 19]. The microstructural analysis carried out on materials that passed through the process showed that grain size was refined to nanometer level [20]. Different strain rates during the process of ECAP has been employed to investigate the evolutional characteristics of the material microstructure [21].

The deformation during the ECAP is a mixed form of shearing deformation and bending deformation which affect the orientation of grain crystal of the microstructure [22]. Apart from orientation of grain crystal, low-angular boundary and critical angle of partition grain have significant effect on the flow stress during ECAP [23]. A schematic diagram of ECAP showing its geometry is shown in **Figure 3** where P symbolizes the deformation force.

Process name	Schematic representation	Equivalent plastic strain
High-pressure torsion (HPT), Valiev, 1997		$\varepsilon = \frac{\mathrm{tg}\gamma}{\sqrt{3}}$
Equal channel angular processing (ECAP), Segal, 1977	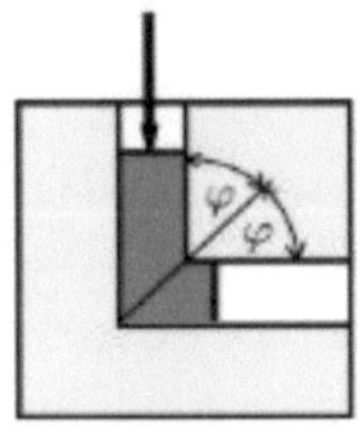	$\varepsilon = n\frac{2}{\sqrt{3}}\cot\varphi$
Cyclic extrusion-compression (CEC), J. and M. Richert, Zasadziński, Korbel, 1979	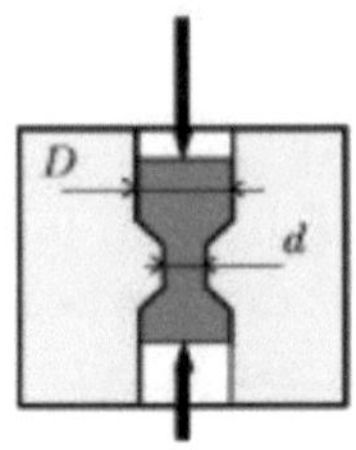	$\varepsilon = n4\ln\left(\frac{D}{d}\right)$
Multiaxial forging (MF), Ghosh, 1988	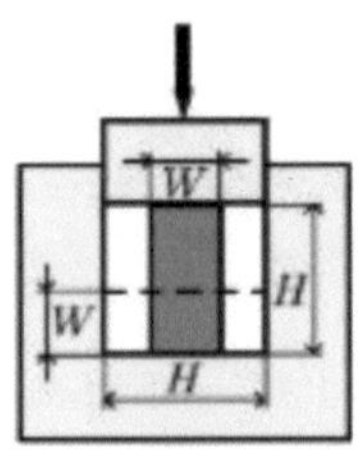	$\varepsilon = n\frac{2}{\sqrt{3}}\ln\left(\frac{H}{W}\right)$
Accumulative roll-bonding (ARB), Saito, Tsuji, Utsunomiya, Sakai, 1998	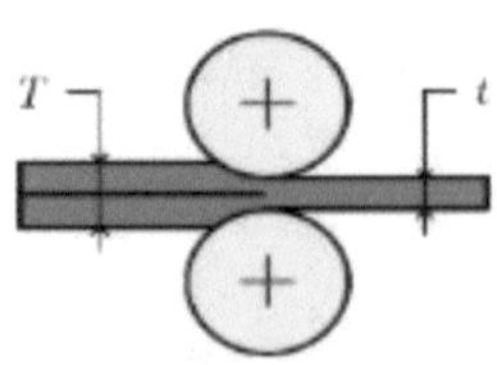	$\varepsilon = n\frac{2}{\sqrt{3}}\ln\left(\frac{T}{t}\right)$
Repetitive corrugation and straightening (RCS), Zhu, Lowe, Jiang, Huang, 2001	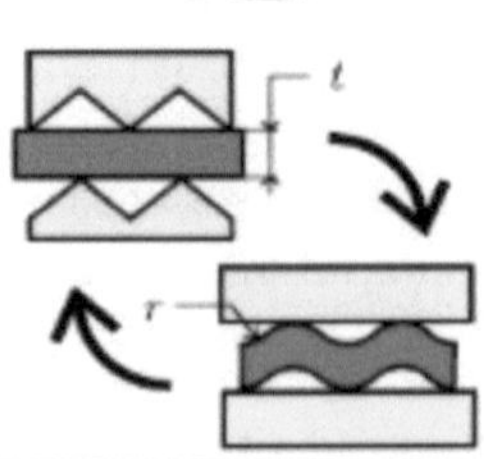	$\varepsilon = n\frac{4}{\sqrt{3}}\ln\left(\frac{r+t}{r+0.5t}\right)$

Figure 1. SPD processes developed for grain refinement [4].

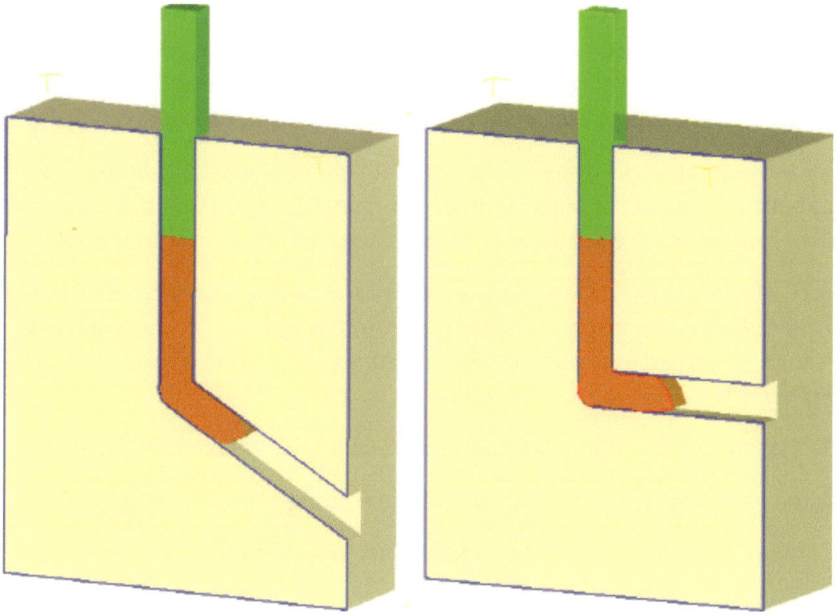

Figure 2. ECAP principles (channel intersecting at 120° and 90°, respectively) [10].

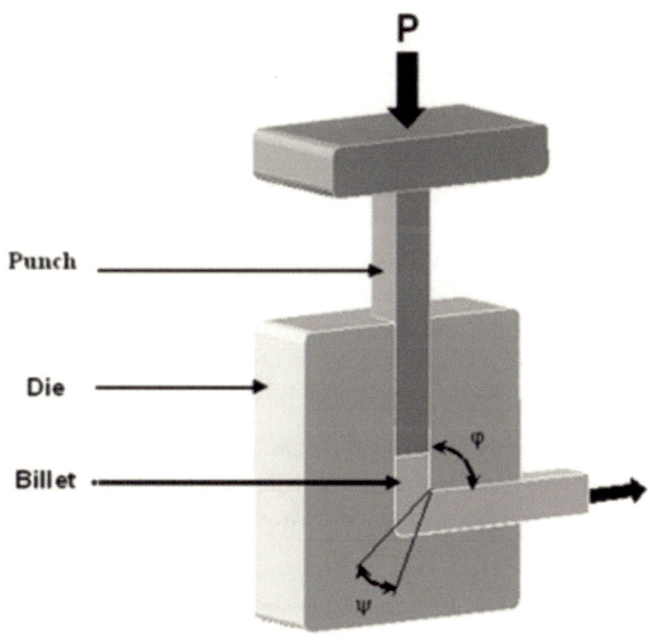

Figure 3. Schematic diagram of an ECAP die showing the inner die angle (∅) and the outer die angle (Ψ) [26].

The properties of materials processed by ECAP are strongly dependent on the plastic deformation behavior during pressing [16, 24]. These properties are governed mainly by the inner die angle Ø and outer die angle y, the material properties like strength and hardening behavior [11], and process variables such as lubrication and deformation speed [25].

As the billet moves as a rigid body in the vertical channel, all deformation is restricted to a small area about the channel's meeting line. The metal is subjected to a simple shear strain under relative low pressure compared to the traditional extrusion process [25].

During ECAP, the work-piece cross-sectional area remains unchanged; this ensures process repeatability until the deformation reached a required grain size level. As a result of cumulative nature, high strain can be achieved with multiple passes. In multiple pass, different paths may be employed, such paths include:

i. Path A: the work-piece orientation remains unchanged in successive passes

ii. Path B: the work-piece is rotated by 90° about its longitudinal axis

iii. Path C: the work-piece is rotated by 180° about its longitudinal axis.

2. Aluminum 6063 alloy

Pure aluminum is a ductile and weak material; however the presence of a relatively small percentage of impurities in aluminum considerably increases its tensile strength and hardness properties [27]. The mechanical properties of aluminum and its alloys also depend on the amount of work it has been subjected to and not only its purity. The major purpose of working is to fragment the grains in the aluminum alloy resulting in an increase in tensile strength and hardness but decrease in ductility [27].

The 6000 series alloys have recently found increased application in automotive and construction industry. Therefore, several research works have been undertaken to strengthen the alloys either by small addition of copper, magnesium, zinc and/or silicon or by a pre deformation treatment [35]. There is a great increase in market for extrusion intricate shape, medium strength and good toughness as a result of development of Al-Mg-Si alloys for light structures. These developed alloys are required to meet precise tensile properties and fatigue strength, welding characteristics and formability [28].

Aluminum alloy 6063 commonly referred to as an architectural alloy because is a medium strength alloy. It is normally used in intricate extrusions. It has a better surface finish, better corrosion resistance, and better formability than iron, readily suitable to welding and can be easily anodised.

The principal alloying elements in Al 6xxx series alloys are Mg and Si, both having low solid solubility in Al at room temperature [29]. The presence of Fe impurities together with Mg and Si, influences the material microstructure through the formation of intermetallic particles, such as Al (Fe,Mn)Si, Mg_2Si, and Al_3Mg_2. It is known that the Fe-rich particles cannot be re-dissolved during homogenization owing to their high melting point (>700°C) this behavior has helped in improve on the mechanical properties of Al alloy. A range of intermetallic particle phases, with their crystallographic structures, have been identified [30]. Distinguishing between the phases is of importance for aluminum alloy metallurgy, but may be difficult based purely on cross-sectional microscopy [31].

2.1. Properties of aluminum 6063 alloy

Al 6063 alloy is an extrusion alloy that is heat treatable for strengthening; machinability is considered to be average for this alloy; likewise, its forming ability, either hot or cold, is good; the alloy is readily welded by all of the conventional methods; hot working (as with forging) can be done on this alloy; Cold working characteristics 6063 are good for all conventional forming methods; it hardens due to aging heat treatment and cold working; its electrical conductivity is 50% of copper. Al 6063 alloy is called architectural aluminum for two reasons—firstly, it has good surface finish smoother than other available alloys, and secondly, its strength is approximately half the strength of 6061, making it suitable in applications where ultimate strength is not a requirement. This class of Al alloy is classed as "good" for forming and cold working requirements, "excellent" for anodizing, and "fair" for machining. It has a good resistance to general corrosion, including resistance to stress-corrosion cracking in the heat treated condition. The mechanical properties of 6063 aluminum alloy depend greatly on the temper, or heat treatment, of the material. Some of the physical properties and temper designations of Al 6063 are stated in **Tables 1** and **2** respectively.

2.2. Uses of aluminum 6063 alloy

6063 is mostly used in extruded shapes for architecture as discussed above, particularly for window frames, door frames, and roofs making aluminum 6063 an important extrusion alloy. It possesses moderate strength and has excellent finishing characteristics. Al 6063 has become the prime architectural alloy and its response to a wide variety to surface finishes further demonstrates its versatility therefore finding usage in decorative applications. The corrosion resistance is very good and the grade is easily welded and brazed, and is heat-treatable as well. Moreover, Al 6063 is commonly used in decorative applications, pipes and tubings for

Properties	Value
Density	2.71 g/cm^3
Melting point	600°C
Modulus of elasticity	67G Pa
Electrical resistivity	0.035E−6Ωm
Thermal conductivity	180 W/m K
Thermal expansion	23E−6/K

Table 1. Physical properties of Al 6063 alloy [32].

Standard tempers	Temper definition
F	As fabricated. There is no special control over thermal conditions, and there are no mechanical property limits.
O	Annealed. Applies to products that are annealed to obtain the lowest strength temper.
T1	Cooled from an elevated-temperature shaping process and naturally aged.
T4	Solution heat-treated and naturally aged.
T5, T52, T53	Cooled from an elevated temperature shaping process and artificially
T54, T55	Aged.
T6	Solution heat treated and artificially aged.

Table 2. Temper designations and definitions [33].

irrigation systems, lawn furniture, esthetic applications and trim. Alloy 6063 is often used for electrical applications in the T5, T52, and T6 conditions due to its good electrical conductivity (the definition has been shown in **Table 2**).

3. Equal channel angular extrusion of aluminum 6063 alloy

Up till date, 6063 is widely used in the production of extrusions—long constant-cross-section structural shapes produced by pushing metal through a shaped die. These include "L" and "U" shaped channels and angles. The influence of magnitude of plastic deformation on properties of metallic materials is connected with increase of internal energy. Due to the result of non-homogeneity of deformation at ECAE technique the internal energy gain differs at different places of formed alloy [34].

ECAE as earlier defined is a technique using severe plastic deformation to produce ultra-fine grain sizes in the range of hundreds of nanometers to bulk course grained materials [10, 35, 36]. ECAE is performed by pressing Al 6063 alloy billet of material through a die that has two channels which intersect at an angle. The billet experiences simple shear deformation, at the intersection, without any precipitous change in the cross section area because the die does not allow for lateral expansion. This means the billet can be pressed more than once and can be rotated about the pressing axis during subsequent pressings. A single pass with channels 90° to each other, induces approximately 1.15 equivalent strains in the billet.

Different deformation routes can be applied depending on the billet rotation; these routes are shown in **Table 3**. ECAE technique can be applied to commercial pure metals and metal alloys [37]. The schematic diagrams of these routes are shown in **Figures 4** and **5**.

3.1. Superplasticity of aluminum alloys via ECAE

The ability of a material to pull out to a high tensile elongation without the development of necking is termed superplasticity. Superplasticity is usually in the range of 1–10 μm fine grain

Routes	Description
Route A	No rotation of the billet
Route B_A	Rotated counter clockwise 90° on even number of passes and clockwise 90° on odd number of passes [39]
Route B_c	Rotated counterclockwise 90° after every pass
Route C	Rotated 180° after ever pass

Table 3. ECAE routes [38].

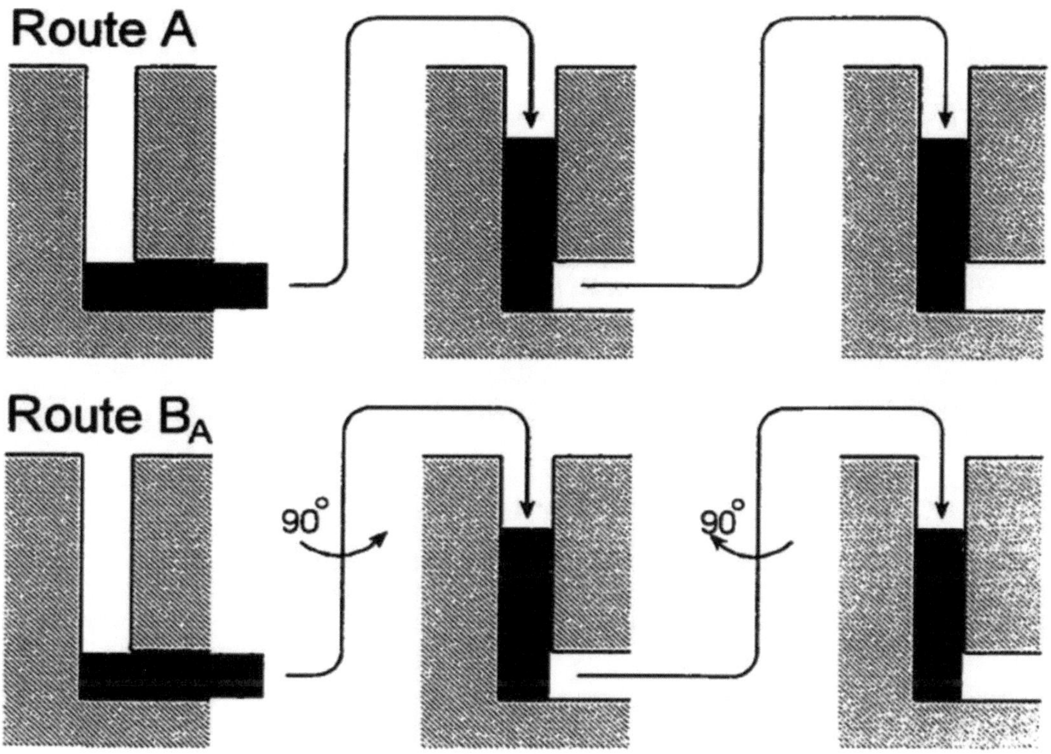

Figure 4. Schematic illustration of route A and route B_A [6].

size. Much finer grain sizes in the near-nanometer range has been achieved in Al-based alloys in experiments by using an intense plastic straining technique such as equal-channel angular pressing (ECAP) [40]. Superplastic properties in Aluminum alloys cannot be achieved under conventional processing conditions. Superplasticity is the ability of a material to undergo very large uniform neckless tensile deformation normally over 500% elongation prior to failure at a temperature well below its melting point (*Tm*) because the deformation mechanisms fall into the grain boundary sliding (GBS) regime [41, 42] fine grain size of 10 μm [43], high operation temperature of 0.9*Tm* and slow strain rate are required as for some Al alloy material. The aerospace industry first developed the AA2004 (Al-6Cu-0.5Zr) alloy, also known as Supral 100, which is a good example for superplastic applications [44]. The alloy composition corresponds to a relatively large addition of zirconium which provides a dispersion of very fine Al_3Zr particles

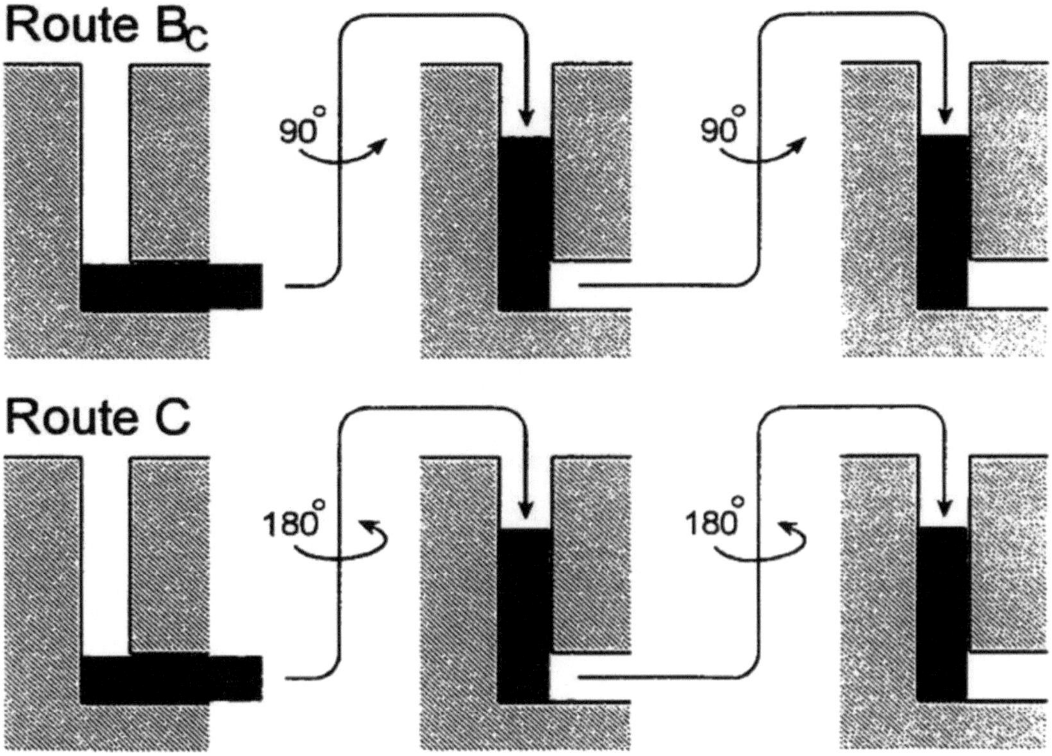

Figure 5. Schematic illustration of route B_C and route C [6].

that stabilize the wrought structure developed during hot/cold rolling and prevent recrystallization until the onset of superplastic forming [45, 46]. This alloy is a medium strength alloy with mechanical properties similar to AA6061 and 2219. It is usually used in lightly loaded or non-structural applications. Automotive industry demands an increasing use of aluminum alloys to reduce weight so as to improve performance and fuel consumption. However, in comparison to steel aluminum alloy sheet materials have lower formability in cold stamping processes. Some aluminum alloy offers an alternative approach which can be deformed to quite a high percentage at elevated temperature by the so-called superplastic forming (SPF) process [45, 46]. Although fine grain size can be achieved in Al alloy through significant thermomechanical process, it is very costly and also the low deformation strain rate results in long forming times [6, 47]. Therefore reduction of the grain size in the sub-micrometer or nanometer scale is usually encouraged by applying Equal-Channel Angular Pressing, or ECAP [48].

Superplasticity has been utilized for the fabrication of complex parts from sheet metal because of the large elongation [49, 50] There is considerable current interest in developing materials with ultrafine grain sizes, since experimental evidences have shown that reducing grain size increases the superplastic strain rate and/or decreases the superplastic temperature [51]. Materials processed by ECAE have the potential for superplasticity, especially at very high strain rates and low temperatures. The validity of this proposal has been demonstrated by several recent reports of high strain rate superplasticity above 10^{-2} s^{-1} and low temperature superplasticity in aluminum and magnesium alloys processed by ECAE [52–55].

4. Conclusions

From the current review, it can be concluded that severe plastic deformation via ECAP is a very useful process on increasing mechanical properties with only partial and acceptable decrease in ductility. Strengthening of material was caused by grains refinement and strain hardening of solid solution. Because the cold-worked material was a normal material that has already been extended through part of its allowed plastic deformation by ECAE, dislocation motion and plastic deformation have been hindered enough by dislocation accumulation, and stretching of electronic bonds. Elastic deformation then reached their limit, a third mode of deformation then occurs faster (i.e. fracture). Strain hardening thus reduces ductility and increases brittleness.

Processing by ECAE led to grain refinement and arrays of ultrafine grains that are significantly smaller than those generally produced using conventional thermo mechanical processing. ECAE is a simple process that can be readily applied to a wide range of materials without the requirement of developing specific and different treatments for each alloy composition. The presence of these exceptionally small grain sizes provides an opportunity for achieving superplastic ductility, and thus a superplastic forming capability at a very fast strain rates (extremely higher than what was used in the experiment).

Author details

Abiodun Ayodeji Abioye[1], Ojo Sunday Isaac Fayomi[1,2]*, Abimbola Patricia Idowu Popoola[2] and Oluwabunmi Pamilerin Abioye[1]

*Address all correspondence to: ojosundayfayomi3@gmail.com

1 Department of Mechanical Engineering, Covenant University, Ota, Ogun State, Nigeria

2 Department of Chemical, Metallurgical and Materials Engineering, Tshwane University of Technology, Pretoria, South Africa

References

[1] Altan T. Metal Forming: Fundamentals and Applications – American Society of Metals. ASM Series in Metal Processing. Asm Intl, Metals Park. 1983;**1**:20-57

[2] BMED. Unit Manufacturing Process. Issues and Opportunities in Research. The National Academy Press; Washington, DC: 1995. 79-92. http://www.nap.edu/openbook.php?record_id=4827&page=79

[3] Olejnik L, Rosochowski A. Methods of fabricating metals for nano-technology. Bulletin of the Polish Academy of Sciences, Technical Sciences. 2005;**53**(4):413-423

[4] Furukawa M, Nemoto M, Horita Z. Processing of metals by equal-channel angular pressing. Journal of Materials Science. 2001;**36**:2835-2843

[5] Valiev RZ, Islamgaliev RK, Alexandrov IV. Bulk nanostructured materials from severe plastic deformation. Progress in Materials Science. 2000;**45**:103-189

[6] Valiev RZ. Investigations and Applications of SPD. Norwell: Kluwer Academic Publishers; 2000. p. 211-229

[7] Estrin Y, Vinogradov A. Extreme grain refinement by severe plastic deformation: A wealth of challenging science. Acta Materialia. 2013;**61**(3):782-817

[8] Zrnik J, Dobatkin SV, Mamuzi I. Processing of metals by severe plastic deformation (SPD)—Structure and mechanical properties respond. METABK. 2008;**47**(3):211-216

[9] Zhilyaev AP, Shakhova I, Belyakov A, Kaibyshev R, Langdon TG. Wear resistance and electroconductivity in copper processed by severe plastic deformation. Wear. 2013; **305**:89-99

[10] Segal VM. Materials processed by simple shear. Materials Science and Engineering: A. 1995;**197**:157-164

[11] Kumar SR, Gudimetla K, Venkatachalam P, Ravisankar B, Jayasankar K. Microstructural and mechanical properties of Al 7075 alloy processed by Equal Channel Angular Pressing. Materials Science and Engineering: A. 2012;**533**:50-54

[12] Chegini M, Fallahi A, Shaeri MH. Effect of equal channel angular pressing (ECAP) on wear behavior of Al-7075 alloy. Procedia Materials Science. 2015;**11**:95-100

[13] Kim HS, Seo MH, Hong SI. Plastic deformation analysis of metals during equal channel angular pressing. Journal of Materials Processing Technology. 2001;**113**:305-764

[14] Han BQ, Matejczyk D, Zhou F, Zhang Z, Bampton C, Lavernia EJ. Mechanical Behavior of a Cryomilled Nanostructured Al-7.5 pct Mg Alloy. Metallurgical and Materials Transactions A. 2004;**35A**:947-950

[15] Zhilyaev AP, Lee S, Nurislamova GV, Valiev RZ, Langdon TG. Microhardness and microstructural evolution in pure nickel during high-pressure torsion. ScriptaMater. 2001;**44**:2753-2758

[16] Jin H, Loyd DJ. The tensile behaviour of fine grains 5754 alloy produced by asymmetric rolling and annealing. Metallurgical Transactions A. 2004;**35A**:997-1006

[17] Segal VM. Equal channel angular extrusion: From macromechanics to structure formation. Materials Science and Engineering: A. 1999;**271**:322-333

[18] Zhu YT, Lowe TC. Observations and issues on mechanisms of grain refinement during ECAP process. Materials Science and Engineering: A. 2000;**291**(1-2):46-53

[19] Segal VM. Engineering and commercialization of equal channel angular extrusion (ECAE). Materials Science and Engineering: A. 2004;**386**:269-276

[20] Valiev RZ, Islamgaliev RK, Semenova IP. Superplasticity in nanostructured materials: New challenges. Materials Science and Engineering: A. 2007;**463**:2-7

[21] Mazurina I, Sakai T, Miura H, Sitdikov O, Kaibyshev R. Effect of deformation temperature on microstructure evolution in aluminum alloy 2219 during hot ECAP. Materials Science and Engineering: A. 2008;**486**:662-671

[22] Han WZ, Zhang ZF, Wu SD, Li SX. Influences of crystallographic orientations on deformation mechanism and grain refinement of Al single crystals subjected to one-pass equal-channel angular pressing. Acta Materialia. 2007;**55**:5889-5900

[23] Reihanian M, Ebrahimi R, Moshksar MM, Terada D, Tsuji N. Microstructure quantification and correlation with flow stress of ultrafine grained commercially pure Al fabricated by equal channel angular pressing. Materials Characterization. 2007;**25**:1312-1323

[24] Iwahashi Y, Wang J, Horita Z, Nemoto M, Langdon TG. Principle of equal-channel angular pressing for the processing of ultra-fine grained materials. Scripta Materialia. 1996;**35**:143-146. http://dx.doi.org/10.1016/1359-6462(96)00107-8

[25] Wu Y, Baker I. An experimental study of equal channel angular extrusion. Scripta Materialia. 1997;**37**:437

[26] Saravanan M, Pillai RM, Pai BC, Brahmakumar M, Ravi KR. Equal channel angular pressing of pure aluminium—An analysis. Bulletin of Materials Science. 2006;**29**(7):679-684. © Indian Academy of Sciences

[27] Omotoyinbo JA, Oladele IO. The effect of plastic deformation and magnesium content on the mechanical properties of 6063 aluminium alloys. Journal of Minerals and Materials Characterization and Engineering. 2010;**9**(6):539-546

[28] Korsunsky AM, Hofmann F, Abbey B, Gaucherin G. Synchrotron X-ray analysis of microstructure and microdeformation in a recast AA6063 aluminium alloy. 2009;**45**(5): 351-364

[29] Attallah MM. Microstructure Property Development in Friction Sir Welds of Aluminum based Alloys. [Ph.D. thesis]. Birmingham: University of Birmingham; 2008

[30] Hatch JE. Aluminum: Properties and Physical Metallurgy. ASM International, Metals Park. 2004; 1-24

[31] Beyerlein I, Lebensohn RA, Tome CN. Ultrafine Grained Materials II. Seattle: TMS; 2002. p. 585

[32] Segal VM, Reznikov VI, Drobyshevskiy AE, Kopylov VI. Plastic working of metals by simple shear. Russian Metallurgy (English Translation). 1981;**1**:99-105

[33] Valiev RZ, Korznikov AV, Mulyukov RR. Structure and properties of ultrafine-grained materials produced by severe plastic deformation. Materials Science and Engineering: A. 1993;**168**:141-148

[34] Stolyarov VV, Zhu YT, Lowe TC, Valiev RZ. Microstructure and properties of pure Ti processed by ECAP and cold extrusion. Materials Science and Engineering: A. 2001;**303**:2-89

[35] Furukawa M, Iwahashi Y, Horita Z, Nemoto M, Langdon TG. The shearing characteristics associated with equal-channel angular pressing. Materials Science and Engineering: A. 1998;**257**:328-332

[36] Valiev RZ, Jslamgaliev RK, Tsenev NK. Superplasticity of Ultra-fine Grained Aluminium Alloys. London, England: United States Army, European Research Office of the U.S. Army; 1997

[37] Sherby OD, Wadsworth J. Superplasticity—Recent advances and future-directions. Progress in Materials Science. 1989;**33**:169

[38] National Research Council (U.S.). Committee on New Materials for Advanced Civil Aircraft, New Materials for Next Generation Transports. Washington, D.C.: National Academy Press; 1996. p. 84

[39] Shaeri MH, Salehi MT, Seyyedein SH, Abutalebi MR, Park JK. Characterization of microstructure and deformation texture during equal channel angular pressing of Al–Zn–Mg–Cu alloy. Journal of Alloys and Compounds. 2013;**576**:350-357

[40] Flower HM, Boyle GJ, Dashwood RJ, Grimes R. The influence of Al_3Zr on the superplastic behaviour of aluminum alloys. In: 8th International Conference on Superplasticity in Advanced Materials; 2003. pp. 28-30

[41] Shaeri MH, Salehi MT, Seyyedein SH, Abutalebi MR, Park JK. Characterization of microstructure and deformation texture during equal channel angular pressing of Al–Zn–Mg–Cu alloy. Journal of Alloys and Compounds. 2013;**576**:350-357

[42] Al-Mufadi F, Djavanroodi F. Equal-channel angular pressing of thin-walled copper tube. Arabian Journal for Science and Engineering. 2015, September 2015;**40**(9):2785-2794

[43] Shaeri MH, Shaeri M, Salehi MT, Seyyedein SH, Djavanroodi F. Microstructure and texture evolution of Al-7075 alloy processed by equal channel angular pressing. Transactions of Nonferrous Metals Society of China. 2015;**25**:1367-1375

[44] Grimes R, Dashwood RJ, Flower HM, Jackson M, Katsas S, Todd G. Progress towards high superplastic strain rate aluminum alloys. In: ICSAM, 8th International Conference on Superplasticity in Advanced Materials; 28-30 July 2003; Oxford, UK; 2004. pp. 213-220

[45] Hojas M, Kuhlein W, Siegert K, Werle T. Superplastic aluminum sheet, metallurgical requirements, production and properties. Metall. 1991;**45**:130

[46] Vetrano JS, Lavender CA, Hamilton CH, Smith MT, Bruemmer SM. Superplastic behaviour in a commercial 5083-aluminum-alloy. Scripta Metallurgica Et Materialia. 1994;**30**:565

[47] Figueiredo RB, Kawasaki M, Xu C, Langdon TG. Achieving superplastic behaviour in fcc and hcp metals processed by equal-channel angular pressing. Materials Science and Engineering: A. 2008;**493**:104

[48] Herling DR, Smith MT. Improvements in superplastic performance of commercial AA5083 aluminium processed by equal channel angular extrusion. Superplasticity in Advanced Materials. 2001;**3**:357-465

[49] Barnes AJ. Superplastic forming of aluminum alloy. Materials Science Forum. 1994; **170-172**:701-714

[50] Medvedev, Ng HP, Lapovok R, Estrin Y, Lowe TC, Anumalasetty VN. Comparison of laboratory-scale and industrial-scale equal channel angular pressing of commercial purity titanium. Materials Letters. 2015;**145**(0):308-311

[51] Tsutomu T, Kenji H. Superplasticity at room temperature in Zn-22Al alloy processed by equal channel angular extrusion. Materials Transactions. 2004;**45**(4):1261-1265

[52] Chung SW, Somekawa H, Kinoshita T, Kim WJ, Higashi K. The non-uniform behaviour during ECAE process by 3-D FVW simulation. Scripta Materialia. 2004;**50**:1079-1083

[53] Berbon PB, Komura S, Utsunomiya A, Horita Z, Furukawa M, Nemoto M, Langdon TG. Nanomaterials by severe plastic deformation. Materials Transactions, JIM. 1999; **40**:772-778

[54] Watanabe H, Mukai T, Ishikawa K, Higashi K. Scripta Materialia. 2002;**46**:851-856

[55] Watanabe H, Mukai T, Mubuchi M, Higashi K. Scripta Materialia. 1999;**41**:209-213

Chapter 15

Microstructural Characteristics and Mechanical Behaviors of New Type SIMA Processed Aluminum Alloy

Chia-Wei Lin, Fei-Yi Hung and Truan-Sheng Lui

Additional information is available at the end of the chapter

http://dx.doi.org/10.5772/intechopen.70462

Abstract

In this chapter, a new type strain-induced melt activation (SIMA) process for Al-Mg-Si alloys was used. The microstructural characteristics, formability at elevated temperature and mechanical properties were estimated. The high-hardness globular grain boundaries are formed by eutectic phases. This new type SIMA process has proved that it can decrease high temperature compression resistance and improve ability of metal flowing at high temperature. After SIMA forming process, the mechanical properties of materials can be improved via artificial aging and can be competed with general artificial aged materials. All results show that this SIMA process is a potential process.

Keywords: strain-induced melt activation, microstructural evolution, high temperature formability, mechanical properties, erosion resistance

1. Introduction

1.1. Strain-induced melt activation (SIMA) process

Semi-solid metal forming process is a hybrid net shape process that combines the advantages of forging and casting [1, 2]. Materials are manufactured at the temperature of solid-liquid coexistence and perform good high temperature formability [3–5]. Magneto-hydrodynamic (MHD) stirring casting, spray casting and strain-induced melting activation (SIMA) are three common semi-solid process for getting fine and globular grains that can enhance formability at elevated temperatures [6, 7].

The SIMA process is useful because of its low cost, simple equipment and high stability. **Figure 1(a)** reveals steps of the traditional SIMA process [8]. Casting, hot work, cold work and heat treatment for partial melting and grain spheroidization are four major steps of traditional SIMA process. A promoted and new type of SIMA process is proposed here. The two main dissimilarities between two SIMA processes are: (1) this new process uses severe hot extrusion instead of cold work to introduce sufficient amount of strain energy and (2) the new process uses a salt bath for heating to enhance heating uniformity and to reduce heating duration. **Figure 1(b)** shows the steps of this process.

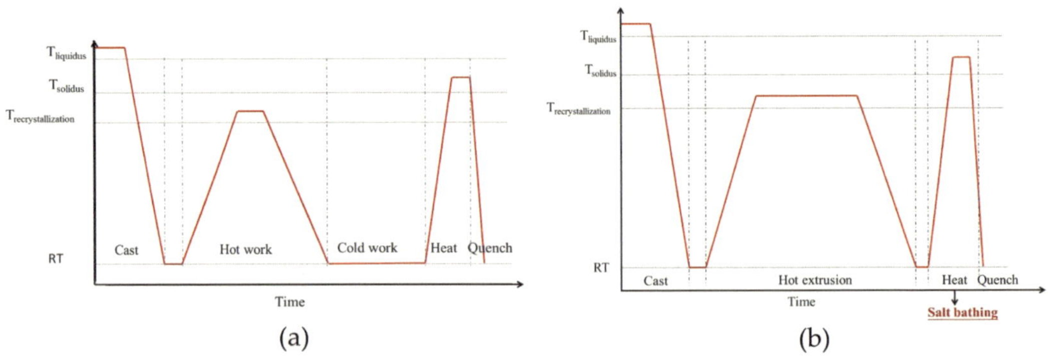

Figure 1. Procedures of (a) traditional strain-induced melt activation (SIMA) process and (b) new type SIMA process. RT, room temperature.

The mechanisms of globular grain formation for the traditional SIMA [8, 9] process are shown in **Figure 2**. The procedures are: (1) casting; (2) hot work for disintegrating the casting structure; (3) cold work for introducing sufficient strain energy; and (4) heat treatment for making alloys recrystallize and partially melt at the temperature of muzzy zone. In the step of heat treatment, grains recrystallize and phases with low melting point and parts of eutectic composition start to melt partially and penetrate to the grain boundaries of recrystallized grains due to their high free energy. It results in liquid phases surrounding the fine recrystallized grains. Finally, grain growth occurs and generates spheroidized grains as heating carry on because the surface energy of global grains is the smallest.

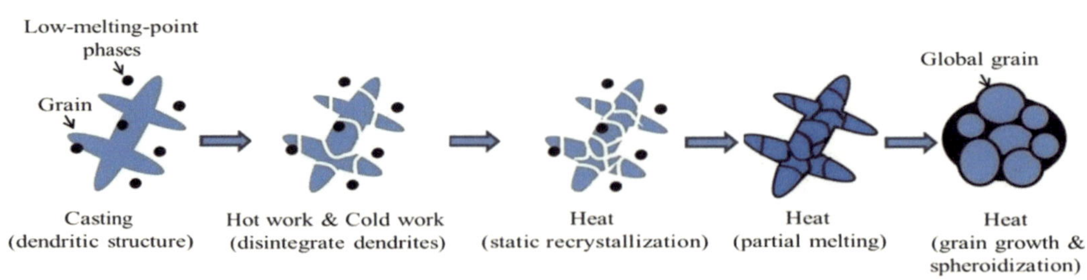

Figure 2. Evolution of globular grain formation in traditional SIMA process.

For the growth mechanism of globular grains, when alloys are heated to the temperature of solid-liquid coexistence, two coarsening mechanisms are considered [7, 10–13]: grain coalescence and Ostwald ripening. Ostwald ripening dominates for higher liquid fractions and long-duration heating, and grain coalescence dominates for lower liquid fractions and short-duration heating. In general, Oswald ripening should be the major growth mechanism of globular grains. The globular grain growth is based on Lifshitz-Slyozov-Wagner (LSW) theory [1, 14–18]. The formula is $d^n - d_0{}^n = Kt$, where d is the average globule size that depends on salt bath duration, d_0 is the initial globule size, t is the duration of salt bath, K (units: $\mu m^3\ min^{-1}$) is the coarsening rate constant and exponent n is determined by the diffusion mechanism of grain growth. Ostwald ripening is the major mechanism of globular grain growth of SIMA alloys in the case of liquid fraction being sufficient [10–13]. As $n = 3$, the mechanism of grain growth is Ostwald ripening. The theoretical formula can be rewritten as $d^3 = Kt + d_0{}^3$. Ostwald ripening is a dissolution-precipitation diffusion-controlled mechanism [10–13]. Large grains become larger and small grains become smaller and even disappear.

1.2. 6xxx aluminum alloys

6xxx Al alloys, a series of precipitation-hardened Al alloys [19], are widely used. More than 70% of extruded Al alloys are 6xxx Al alloys, therefore, this series alloy is chosen for this chapter of new type SIMA process, as hot extrusion is a major step in this process. Similar results and discoveries can be seen in other series of aluminum alloys.

The major components of 6xxx series aluminum alloys are Mg and Si. The Mg_2Si phase is generated first since its forming free energy is lower than those of other precipitating phases [20–22]. 6xxx aluminum alloys with more Si that are required to form stoichiometric Mg_2Si commonly have relatively high strength and low ductility due to the hard and brittle Si particles in their matrix. In contrast, 6xxx aluminum alloys with excess Mg have relatively low strength and high ductility. Besides Mg and Si, Cu, Mn, Cr, Zr, Sc and V are added to these alloys. Adding Mn inhibits grain growth, enhances strength, and increases recrystallization temperature [20, 23, 24]. Adding Cu improves strength and hardness due to the refining of the precipitated phase during artificial aging [13–16]. Adding Cr promotes corrosion resistance [23]. Adding Zr enhances strength since Al_3Zr is generated and grain growth is inhibited, and adding Sc or V retards grain growth [24, 25].

1.3. Topics of this study

In short, 6xxx Al-Mg-Si alloys are the major materials for the new type SIMA process. There are three essential topics to be exhibited: (1) microstructural characteristics and microstructure evolution of SIMA processed Al-Mg-Si alloys and their relationship with parameters of SIMA process; (2) formability and deformation mechanism at elevated temperatures of SIMA-processed Al-Mg-Si alloys and (3) the mechanical properties improvement of SIMA forming alloys and its strengthening mechanism. These above topics are discussed in following sections in sequence.

2. Microstructure evolution and microstructural characteristics of new type SIMA-processed alloys

2.1. Brief introduction

The degree of spheroidization and globule size affect high temperature formability obviously in SIMA process [3–5]. Highly spheroidized, fine and uniform globules lead to high formability at elevated temperatures. Hence, how to control hot extrusion conditions, how to design proper material composition and how to determine parameters of SIMA process to get fine, uniform and highly spheroidized globule is the major theme of discussion in this section.

Seven kinds of hot-extruded Al-Mg-Si alloys with different composition or fabricated by distinct extrusive ratio were used for discussing the effects of composition and extrusive ratio. The target is to build a principle for designing suitable materials for new type SIMA process and to choose most proper alloy for further researches [26].

2.2. Experimental methods

2.2.1. Designing and choosing alloys

The compositions and codes of seven kinds of Al-Mg-Si alloys are shown in **Table 1**. All the materials are extruded sheet except for 6069-rod is an extruded rod. They were used for determining the influences of extrusion conditions and understanding how to design composition of alloys for SIMA process by observing their microstructure evolution.

2.2.2. New type strain-induced melt activation (SIMA) process

The flow of new type SIMA process is shown in **Figure 1(b)**. The considerable three parameters of new type SIMA process are extrusive ratio (it also means the introduced strain energy), temperature of salt bath and duration of salt bath in sequence of its flowchart. The temperatures of salt bath are set as 570–630°C due to partial melting starting at about 570°C and severe deformation occurring above 630°C in Al-Mg-Si alloys. The durations of salt bath are set from 1 to 60 min for understanding the effect of salt bath duration on globule size and degree of spheroidization.

Material	Mg	Si	Cu	Mn	Fe	Cr	V	Zr	Al
6061	0.88	0.67	0.16	0.03	0.19	0.04	0	0	Bal.
6061-Mn	0.83	0.75	0.20	0.31	0.19	0.07	0	0	Bal.
6066_3 mm	1.02	1.29	0.95	1.02	0.19	0.18	0	0	Bal.
6066_9 mm	1.02	1.29	0.95	1.02	0.19	0.18	0	0	Bal.
6069-sheet	1.25	0.75	0.73	0.05	0.13	0.16	0	0.11	Bal.
6069-rod	1.30	0.85	0.72	0.09	0.13	0.14	0.12	0	Bal.
6082	0.71	1.10	0.06	0.71	0.19	0.23	0	0	Bal.

Unit: wt.%.

Table 1. Compositions of several Al-Mg-Si alloys used in this research.

The codes of salt bath specimens are interpreted by the following example: SB620-10 means a material which was heated by salt bath at 620°C for 10 min.

2.2.3. Microstructure analysis and globule size measuring

The microstructure evolution and globule size were investigated using optical microscopy (OM). The concept of an equivalent circle is used in this study. When the shape of a grain is not circular in a two-dimensional plane, grain size is represented by the diameter of the equivalent circle, as shown in **Figure 3**. Two shape parameters, x and z, were defined for the degree of spheroidization [4]. In **Figure 3**, a, b, c and A represent the major axis, minor axis, perimeter, and area of a grain, respectively. According to the definitions $x = (b/a)$ and $z = (4\pi A)/c^2$, x is the ratio of the minor axis to the major axis and z becomes closer to 1 as the shape becomes more circular. As x and z become closer to 1, the grains become more equiaxial and the degree of spheroidization increases.

In salt bath step of new type SIMA process, low melting point second phases and part of matrix with high dislocation density were partially melted at the temperature of solid-liquid coexistence, liquid generated and penetrated into recrystallized grain boundaries. It is defined as liquid phase. The fraction of liquid phases was measured using Image J software. High liquid fraction results in improvement of high temperature formability according to Refs. [3, 6, 7].

2.2.4. Elements distribution analysis and phases identification

In order to estimate the distribution of elements and phases, energy dispersive spectroscopy (EDS) integrated in scanning electron microscope (SEM) and electron probe X-ray micro-analyzer (EPMA) were used. The evolution of elements distribution and diffusion can be semi-quantified by these two instruments.

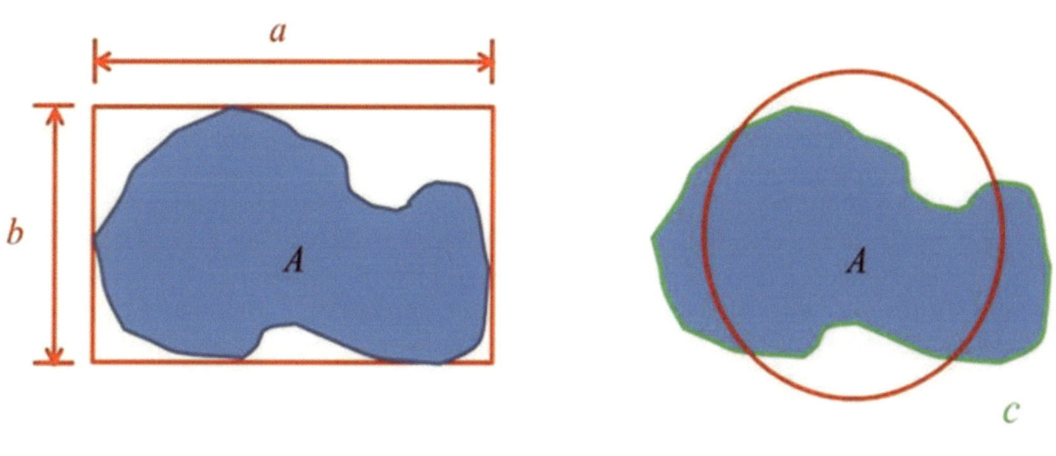

Figure 3. Parameter of spheroidization degree definition. a, b, c and A represent the major axis, minor axis, perimeter and area of a grain, respectively.

2.3. Effects of extrusive condition and chemical composition on microstructure evolution of SIMA-processed Al-Mg-Si alloys

The microstructures of the initial extruded alloys are shown in **Figure 4**. Their initial grain sizes are shown in **Table 2**. Even though the extrusion ratios of 6061 and 6061-Mn are identical, the initial grain size of 6061 is larger than that of 6061-Mn due to Mn retarding grain growth.

As the extrusion ratio increased, the level of dynamic recrystallization also increased. This can be shown by comparing the microstructures of 6066_3 mm and 6066_9 mm. The higher level of recrystallization of 6066_3 mm led to a more uniform and smaller initial grain size. Dynamic recrystallization in 6061 alloys is easier than that in 6066 alloys due to Mn addition, which increases recrystallization temperature, in the latter.

Both coarse and fine grains can be seen in 3 mm-thick 6069-sheet alloys. According to the grain coalescence mechanism, if the crystal orientations of neighboring grains are similar, the grains will easily merge, forming coarse grains. This led to the nonuniform grain distribution in 6069-sheet. In the 28 mm-diameter 6069-rod alloy, fine grains appear in the center and coarse and fine grains appear at the edge. This is due to the lower strain in the center of the rod creating a smaller driving force for grain growth, and the higher strain at the edge of the rod making some crystals with similar orientations grow and then merge. Although 6069-sheet and 6069-rod did not contain Mn, the level of grain growth of 6069 alloys was lower than that of 6061 alloys due to the addition of V and Zr.

Equiaxial grains resulted from dynamic recrystallization in the 3 mm-thick 6082 alloy due to Mn content. This confirms that Mn inhibits grain growth.

The microstructural evolution in the salt bath step for all alloys is shown in **Figure 5**. The codes of specimens subjected to a salt bath are marked with the prefix "SB." After the salt bath, grains were spheroidized and grain boundaries were broadened due to the low melting point phases melting and penetrating the grain boundaries. With grain growth, melting phases aggregated to form liquid pools [3, 4]. The alloys with spheroidized grains are defined as SIMA alloys. **Table 2** shows the liquid fraction of all alloys after a salt bath at 620°C for 10 min. Fine grains, high spheroidization degree and high liquid fraction are the major factors for promoting high temperature formability [7–9].

It is assumed that all grain growth mechanism of all materials is Oswald ripening for comparing the grain growth rate. The grain growth rate of all materials is shown in **Table 2**. This table shows that the grain growth rate of 6061-Mn is lower than 6061, and the grain growth of 6066 alloys and 6082 alloys are relatively lower than other alloys. It proves that adding Mn can decrease the rate of globular grain growth. Besides, after the salt bath, the grain size of 6066-3 mm is smaller than that of 6066-9 mm even though their grain growth rates K are almost identical. It indicates finer and more uniform dynamic recrystallized grains are beneficial for getting smaller and more uniform spheroidized grains.

Table 2 also shows the liquid fractions of the four kinds of aluminum alloy during the 10-min salt bath. The order of liquid fractions is those for 6066, 6069, 6082 and 6061 (from high to low). The liquid fractions of a given alloy are similar, except for 6061. In general, the liquid fractions

http://dx.doi.org/10.5772/intechopen.70462

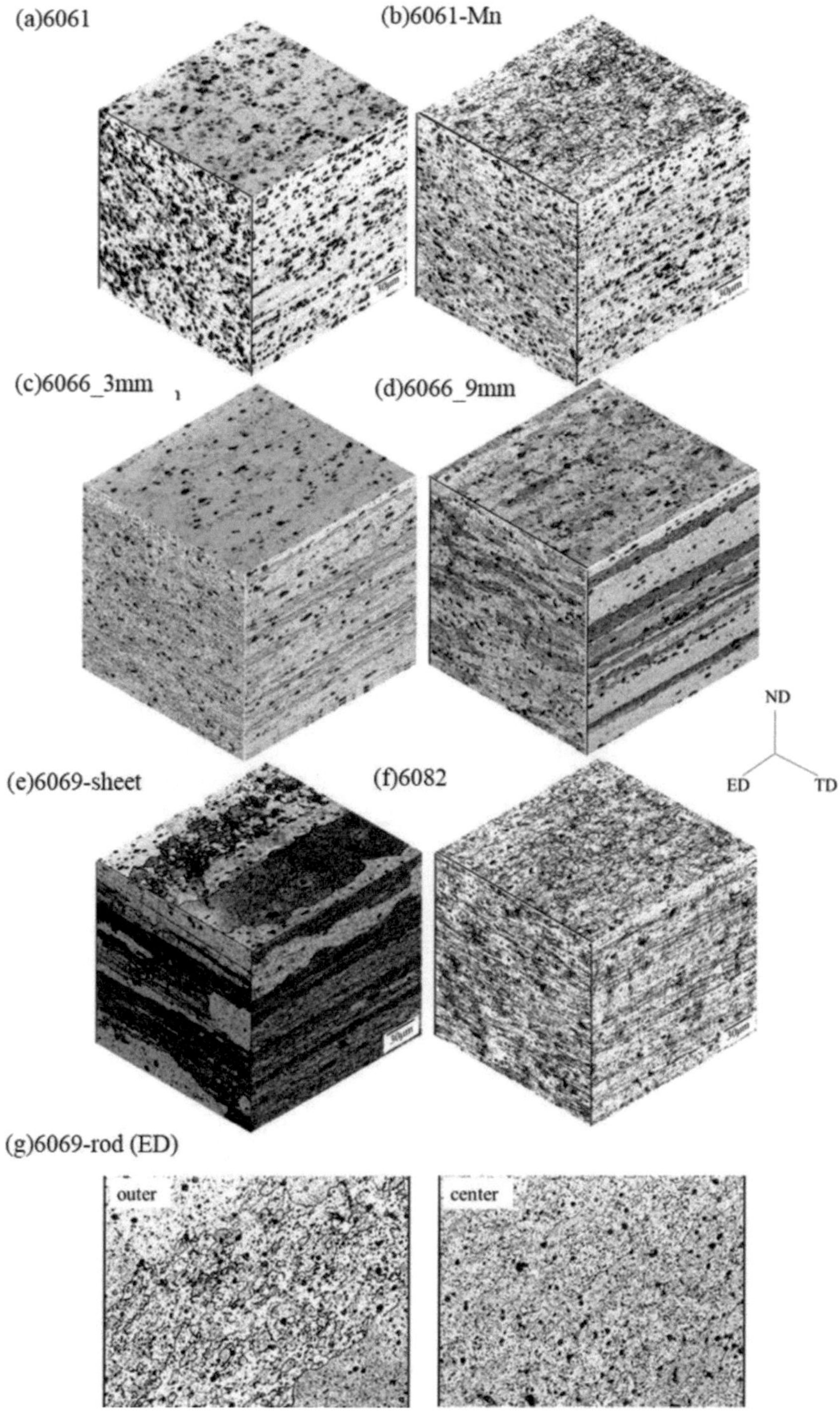

Figure 4. Microstructures of extruded Al-Mg-Si alloys: (a) 6061, (b) 6061-Mn, (c) 6066_3 mm, (d) 6066_9 mm, (e) 6069-sheet, (f) 6082 and (g) 6069-rod.

Material	Initial grain size (μm)	Liquid fraction (%) at 10 min salt bath	K (μm^3 min^{-1})
6061	40	3.2	42,329
6061-Mn	7	7.2	30,054
6066_3 mm	3–5	16	16,806
6066_9 mm	5–7	15.5	17,024
6069-sheet	7; 300	10.4	31,579
6069-rod (center)	15	11.2	32,678
6082	8	7.5	24,357

Table 2. Grain size, liquid fraction and growth rate of Al-Si-Mg alloys.

of a given alloy should be similar. However, the high grain growth and low liquid fraction of 6061 alloys led to position affecting the results.

The degrees of spheroidization of all alloys are shown **Figure 5**. The degrees of spheroidization of 6066 alloys and 6069-rod are higher than those of other samples. The degrees of spheroidization of 6066-3 mm, 6066-9 mm, and 6069-rod are higher due to their fine and uniform initial recrystallized grains and high liquid fraction.

The elemental distributions obtained using EPMA are shown in **Figure 6**. Mg, Si and Cu are the major elements located at grain boundaries, and Fe, Mn, Cr and V exist in the particle phases. It indicates that the melting phases in the salt bath step are mostly composed of Mg, Si, Cu and Al. The three phase diagrams in **Figure 7** show that the eutectic points of Al and Si, Al and Mg_2Si and Al and Al_2Cu are 577, 595 and 548°C, respectively. Moreover, **Table 3** shows the weight percentages of Mg_2Si, Al_2Cu, excess Si and the sum of these three phases. The order of the sums of these phases is those for 6066, 6069, 6082 and 6061 (from high to low). This order is consistent with that for the liquid fraction shown in **Table 3**. These results show that the major phases distributed at globule boundaries are these three phases.

Three main factors affect the generation of fine, uniform, high-liquid fraction and highly spheroidized globular grains in this new type of SIMA process: (1) a proper composition with sufficient elements (e.g., Mg, Si and Cu), which can form low melting point phases for enhancing the liquid fraction, is necessary; (2) the initial extrusion microstructure should be fine and uniform grains for generating fine and uniform spheroidized grains; and (3) particular elements (e.g., Mn, V and Zr) should be added to inhibit grain growth not only to create fine recrystallized grains in the hot extrusion step but also to create fine spheroidized grains in the salt bath step. In this part, 6066 aluminum alloys were the most suitable material for obtaining finest and most spheroidized globules. Therefore, 6066 alloy is used in further researches.

2.4. The parameters of new type SIMA process on microstructure evolution of 6066 aluminum alloy

The influences of salt bath temperatures on microstructural evolution are revealed in **Figure 8**. After salt bath for 30 min at different temperatures, the grains were not spheroidized uniformly as temperature lower than 610°C, and the alloys deformed severely or partially melted as temperature higher than 630°C.

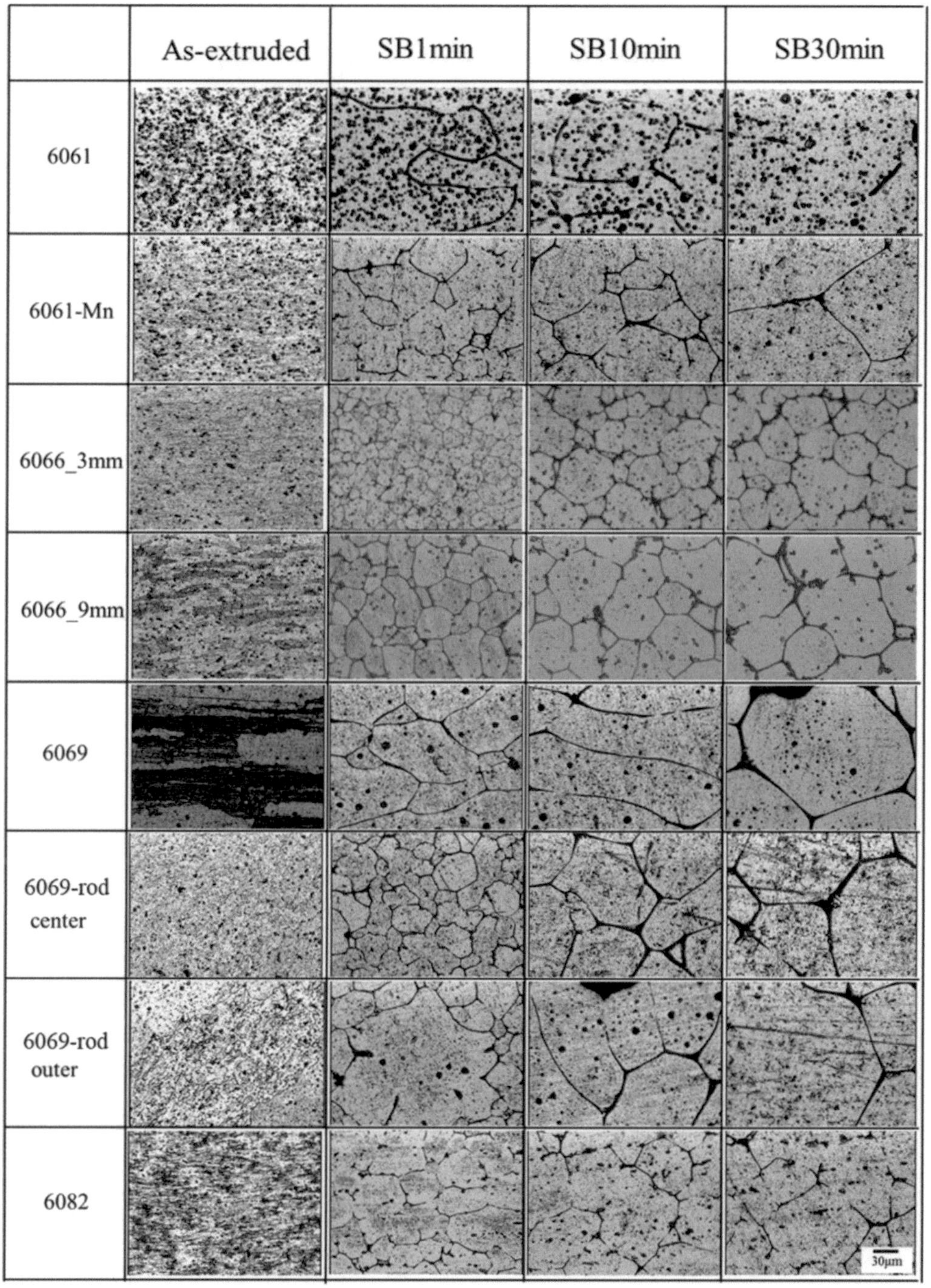

Figure 5. Microstructure evolution of Al-Mg-Si alloys via salt bath.

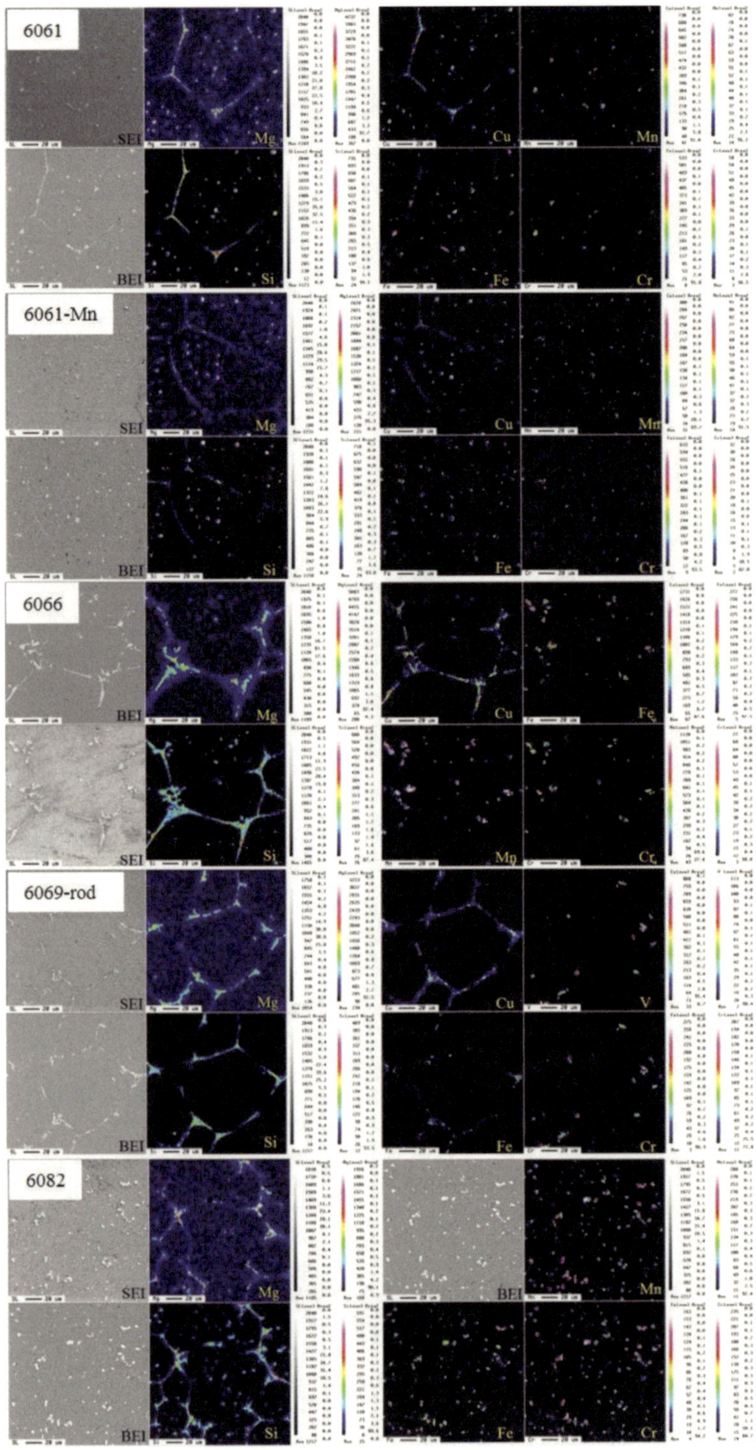

Figure 6. Elementary distribution of Al-Mg-Si alloys.

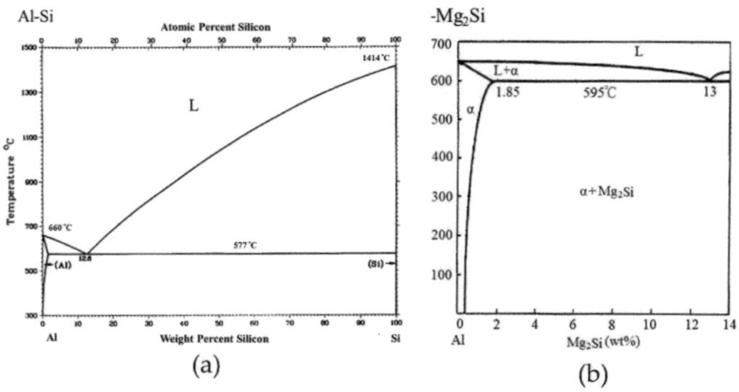

Figure 7. Phase diagrams of (a) Al-Si, (b) Al-Mg_2Si and (c) Al-Cu.

Material	Mg_2Si	Al_2Cu	Excess Si	Sum
6061	1.11	0.33	0.28	1.72
6066	1.61	1.76	0.7	3.97
6069	2.01	1.35	0.07	3.43
6082	1.12	0.11	0.7	1.93

Table 3. The weight ratio of Mg_2Si and Al_2Cu in Al-Mg-Si alloys (wt.%).

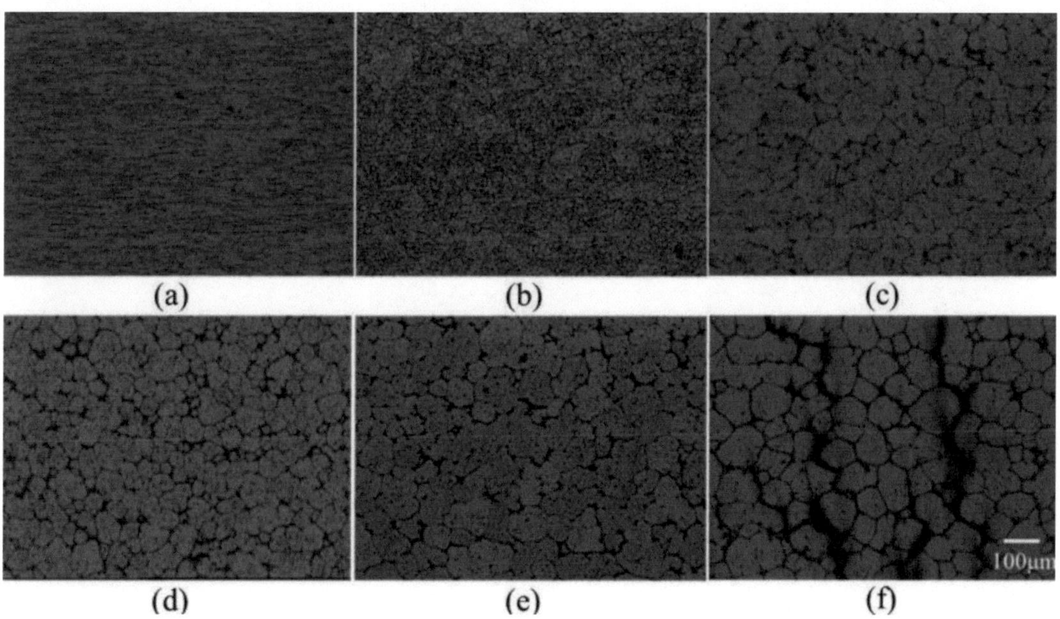

Figure 8. Microstructure of 3 mm-thick 6066 alloy after immersion in salt bath at various temperature and duration: (a) 550°C for 30 min, (b) 570°C for 30 min, (c) 590°C for 30 min, (d) 610°C for 30 min, (e) 620°C for 30 min and (f) 630°C for 7 min.

Figure 9 exhibits the microstructural evolution of 6066 alloys in the salt bath at 610, 620 and 630°C with various duration of salt bath. When salt bath duration increases, the grain growth becomes obvious. A higher salt bath temperature led to higher grain growth for a given period time of salt bath. The grain growth rates for various salt bath temperatures were calculated based on the LSW theory. The linear fitting results are shown in **Table 4**. The results reveal grain growth was very severe at 630°C.

Broadening of globule boundaries was occurred for all temperatures of salt bath, as shown in **Figure 9**. It was resulted from the low melting point phases melting [5–7] and the liquid

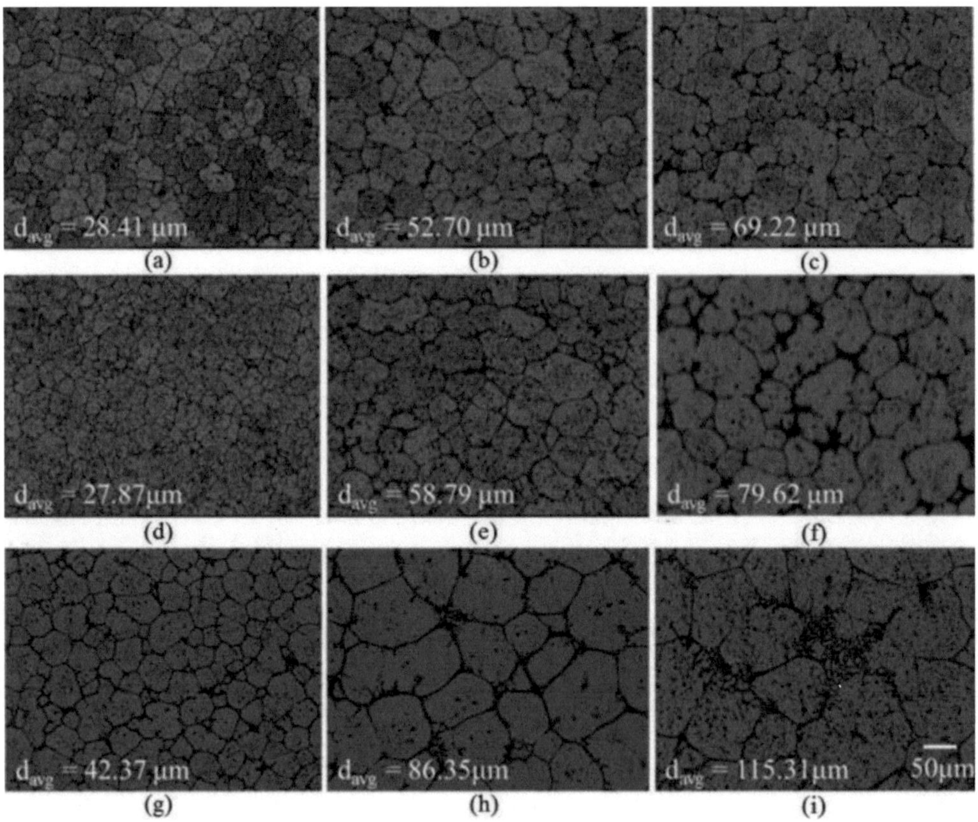

Figure 9. Microstructure evolution with different salt bath conditions of 3 mm-thick 6066 alloy: (a) 610°C for 1 min, (b) 610°C for 10 min, (c) 610°C for 30min, (d) 620°C for 1 min, (e) 620°C for 10 min, (f) 620°C for 30 min, (g) 630°C for 1 min, (h) 630°C for 10 min and (i) 630°C for 30 min.

	6066_3 mm/610°C	6066_3 mm/620°C	6066_3 mm/630°C
K (μm^3 min^{-1})	10,751	16,806	50,820
R^2	0.9852	0.9926	0.9818

Table 4. *K* values and coefficients of determination obtained via LSW theory for 6066_3 mm with various salt bath temperature.

penetrating the grain boundaries. With the increasing salt bath duration, liquid pools formed due to large amounts of liquid phases aggregating. **Figure 10** shows the liquid fraction. With the increasing salt bath temperature and duration, the liquid fraction increased. Moreover, cracks and voids generated in the materials at 630°C, as shown in **Figure 8(f)**, at which the partial melting occurred and the volume shrank too rapidly during quenching. Even though liquid phase formation at 630°C was the fastest, defects are easily generated. Therefore, the most suitable salt bath temperature was 620°C because liquid formation is sufficiently fast and stable.

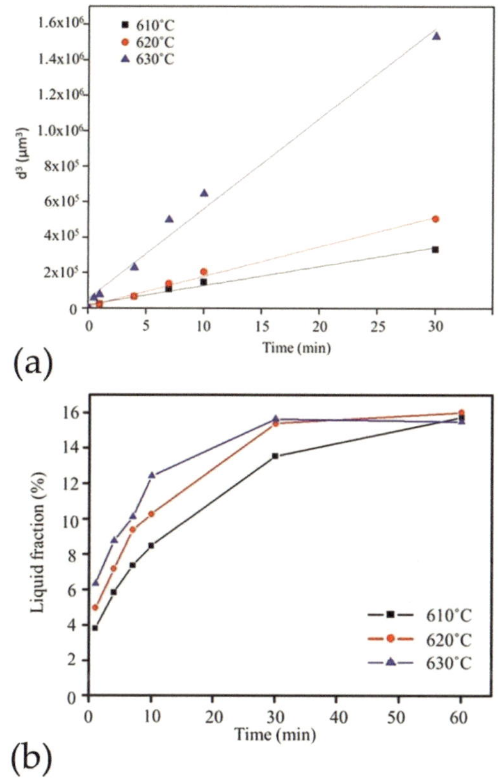

Figure 10. (a) Plot of grain size versus salt bath duration from Lifshitz-Slyozov-Wagner (LSW) theory and (b) liquid fraction for various salt bath temperature (for 3 mm-thick 6066 alloy).

The effect of extrusive ratio on microstructure evolution was discussed by 6066 aluminum alloys with different thickness (6066_3 mm and 6066_9 mm). The microstructure evolution of these two alloys with various salt bath duration at 620°C is exhibited in **Figure 11**. It shows that the average globule size of 6066_9 mm is larger than that of 6066_3 mm at same salt bath duration. High-angle with high free energy grain boundary of recrystallized grains is necessary for SIMA process because they can provide paths for liquid penetrating. However, the amounts of strain introduced by hot extrusion are different in these two alloys with diverse thickness. The recrystallization degree of these two alloys is also different. In 6066_3 mm,

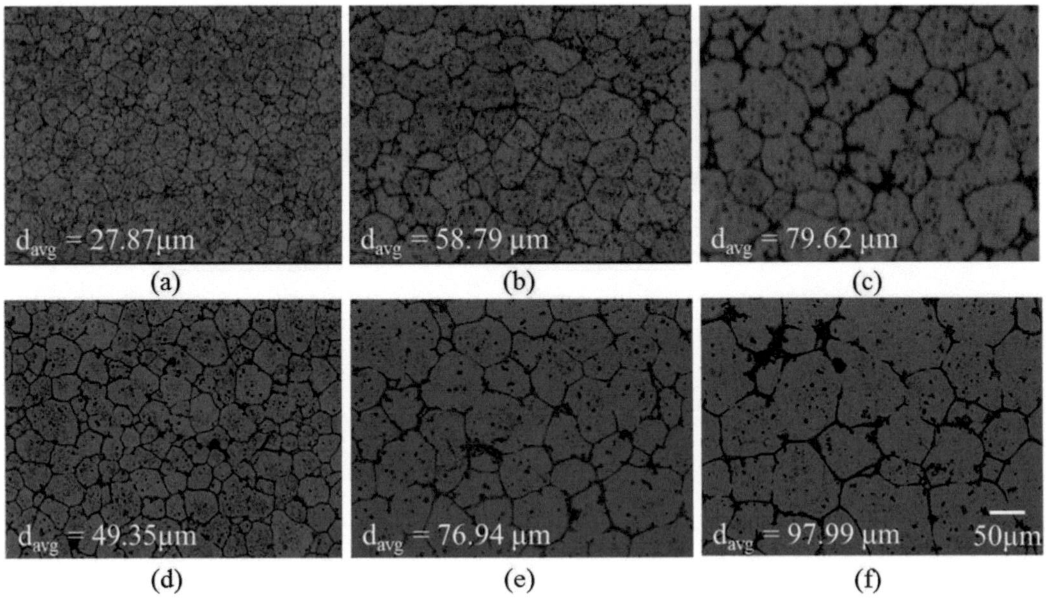

Figure 11. Microstructure of 6066 alloy with two thickness after salt bath at 620°C for various duration: (a) 3 mm-thick and 1 min salt bath, (b) 3 mm-thick and 10 min salt bath, (c) 3 mm-thick and 30 min salt bath, (d) 9 mm-thick and 1 min salt bath, (e) 9 mm-thick and 10 min salt bath and (f) 9 mm-thick and 30 min salt bath.

severe dynamic recrystallization occurred at the hot extrusion step overall the alloys. On the other hand, in 6066_9 mm alloy, dynamic recrystallization occurred only in part of alloy, and static recrystallization also occurred at the step of salt bath. It leads to the different initial recrystallized grain sizes of two as-extruded alloys and also results in different globule size after salt bath. Even though the recrystallization states of these two alloys are different, the grains of those both can be spheroidized.

The relationship between degree of spheroidization and parameters of proposed SIMA process is shown in **Figure 12**. It demonstrates that two shape factors (x and z) are closer to 1 with increasing the duration of salt bath, lower z results from lower temperature of salt bath, and the extrusive ratio do not affect the value of z. These results show that the degree of spheroidization is positive relative to temperature and duration of salt bath but it is not influenced by extrusive ratio. In other words, higher liquid fraction of an alloy leads to higher degree of spheroidization of that.

2.5. Formation mechanism for the two-step strain-induced melt activation (SIMA) process

The mechanisms of globular grain formation for the new type SIMA process are shown in **Figure 13**. The steps are: (1) a suitable composition alloy is cast with uniformly distributed low melting points second phases; (2) disintegrate initial casting microstructure, introduce sufficient strain energy and dynamically grain recrystallize over the alloy through severe hot extrusion; (3) second phases with lower melting points melt and the liquid penetrates grain

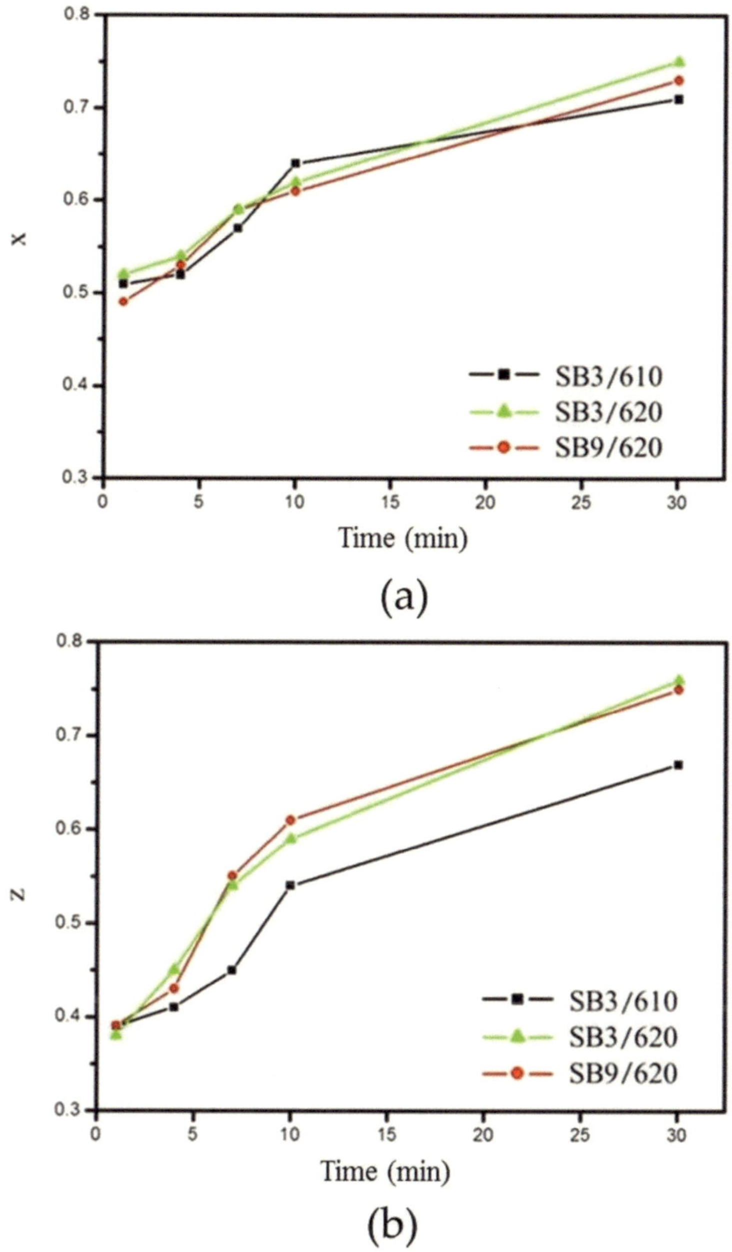

Figure 12. Variation of degree of spheroidization with salt bath conditions: (a) *x* and (b) *z*. SB3: salt bath material with 3 mm thickness; SB9: salt bath material with 9 mm thickness.

boundaries and then surrounds the grains at the step of salt bath and (4) grains grow and become globular with adequate duration of salt bath. This new type SIMA process can produce globular grains faster, and the grains spheroidize uniformly and are finer compared to the traditional SIMA process.

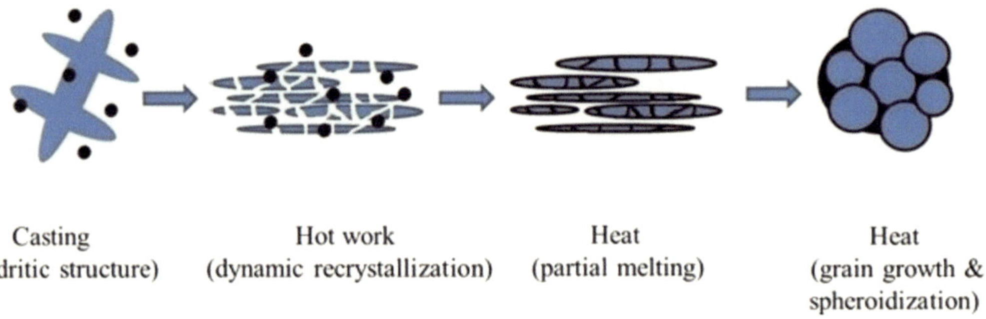

Figure 13. Evolution of globular grain formation in new type SIMA process.

The grains of the 6066_9 mm material can also be spheroidized after the salt bath. It indicates that the static recrystallized grains in the salt bath also provided high-energy grain boundaries for liquid penetration. Both dynamic and static recrystallization can generate grain spheroidization.

2.6. Summary

1. Three major factors to generate fine, uniform, high-liquid fraction and highly spheroidized globular grains in this new type SIMA process: (a) sufficient elements to form low melting point phases, (b) adding some elements to inhibit grain growth and (c) proper extrusive parameters to get fine and uniform initial extrusive microstructure.

2. In 6xxx series alloys, the 6066 aluminum alloy is the most suitable alloys for new type SIMA process. Mg, Si and Cu can generate low melting point phases and Mn can inhibit grain growth.

3. The eutectic phases of Al and Mg_2Si, Al and Al_2Cu and Al and Si are the major phases located on the globule boundaries.

3. High temperature formability and mechanism of deformation of SIMA-processed alloys

3.1. Brief introduction

In this section, mechanical properties and formability under tensile and compressive stress at 500–600°C of the SIMA-processed 6066 alloys were evaluated. And the high temperature deformation mechanism and forming performance were investigated. Besides, the ability of metal flowing and compressibility in a mold of screw at elevated temperatures of SIMA-processed alloys were also estimated.

3.2. Experimental methods

3.2.1. The preparation and codes of specimens

The 6066 alloy is originally a casting column of 6-inch diameter. It was hot extruded in the first step of the new type SIMA process to a 9 mm-thick sheet and to a 3 mm-thick sheet.

Hereafter, "F" will be adopted to abbreviate the hot-extruded sheet material. A 3 mm-thick sheet is used in high temperature tensile test and 9 mm-thick sheet is used in high temperature compressive test, respectively.

In the second step of the new type SIMA process, F was held at 620°C in salt bath (620°C is the best temperature confirmed in Section 2). To vary the size of globule, for different duration, namely 1, 4, 10 and 30 min, were set in this treatment. The specimens of these four different durations will be denoted in the following as S1, S4, S10 and S30, respectively.

Aluminum alloys always manufactured in fully annealed status. The fully annealed alloys will be designated as "O."

3.2.2. Micro-hardness distribution analysis

In order to investigate the difference of hardness between interior globule and globule boundaries, nanoindentation test was performed to evaluate the micro distribution of hardness, using triangular diamond probe with 0.25 nm/s drift velocity and 800 nm depth. The space between adjacent measurement points was 5 μm.

3.2.3. Mechanical test at elevated temperatures and deformation mechanism discussion

A universal material tester was used for the tensile test and compression test. The tensile test was conducted at 500, 550, 600°C and room temperature with the initial strain set at $1.67 \times 10^{-3}\ s^{-1}$. The dimensions of the tensile plate-specimens were shown in **Figure 14(a)**. The broken specimens were used to analyze the fracture mechanism by SEM.

The compression test was performed at 550 and 600°C. Cylindrical specimens with a height of 9 mm and a diameter of 6 mm, as depicted in **Figure 14(b)**, were used. In order to explore the influence of compressive strain rate, two crosshead speeds, namely 3.6 and 36 mm/min, were chosen for the test, which correspond to the initial strain rats of $6.67 \times 10^{-3}\ s^{-1}$ and $6.67 \times 10^{-2}\ s^{-1}$, respectively. All specimens were compressed to 70% strain and the flow curves were analyzed to determine flow resistance. The compressed specimens were sectioned longitudinally and subjected to metallographic analysis.

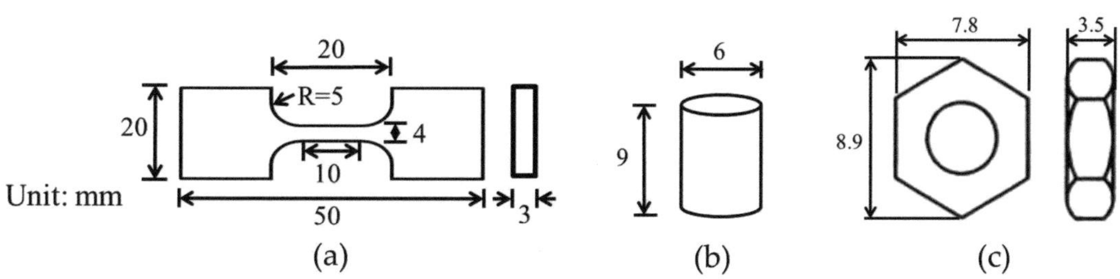

Figure 14. Dimensions of (a) tensile specimen, (b) compressive and forming specimen and (c) screw nut for forming test at high temperature.

3.2.4. *High temperature formability evaluation*

3.2.4.1. *Compression with a given compression rate*

In the high temperature compression test with a given compression rate, as-extruded alloys, fully annealed alloys, and SIMA alloys were tested to compare their high temperature formability. The alloys were machined into compression specimens with dimensions of 40 mm (length) × 20 mm (width) × 9 mm (thickness). The compression loading of the dissimilar alloys were measured and compared as 50% compression ratio. The compression ratio is defined as $R\% = (t_0 - t_f)/t_0$, where t_0 is the thickness of the initial sheet (9 mm) and t_f is the thickness after compression. The compression was at 600°C and with 20 mm/min compression rate. The compression resistance is lower as the compression ratio is higher, which indicates that high temperature formability is higher. The compressed specimens were used in further researches of mechanical properties improvement.

3.2.4.2. *Forming test at elevated temperature*

The high temperature forming test was conducted to 550°C. In the test, cylindrical specimen as those in the compression test was compressed and shaped into a small screw with 0.8 mm pitch. The size of the screw is shown in **Figure 14(c)**. The compressive displacement was set as 6 mm to compare the degree of filling the fringe of different specimens. The crosshead speed was 36 mm/min because the forming efficiency is a great concern in practice. The curves of compressive load versus compressive displacement and the metallographic data of the specimens compressed into screw nut were obtained to determine the resistance of forming, ability of metal flowing and high temperature formability.

3.3. Micro-hardness distribution

The hardness of globule boundaries and liquid pool, where large amount of second phase formed by partial melting aggregated, is twofold to threefold higher than that of the α-Al matrix in grain interior as shown in **Figure 15**.

3.4. Tensile properties of SIMA alloy at different temperatures

Temperatures of tensile test were set at room temperature, 500, 550 and 600°C. F, S10 and S30 were chosen to use in tensile test. **Figure 16** exhibits the breaking features of three materials at room temperature, 500, 550 and 600°C. These figures reveal dimple fracture for extruded alloys, and intergranular fracture for SIMA-processed alloys. Spheroidized grains can be identified on the breaking surface. Quasi-continuous distribution of second phases with low melting point on globule boundaries results in intergranular fracture.

The yield stress (YS) and ultimate tensile stress (UTS) of four materials are below 20 MPa as shown in **Figure 17**. It shows the deformation resistance of 6066 Al alloy is very low at elevated temperatures.

Figure 17 also reveals that the uniform elongation (UE) and total elongation (TE) of F are obviously higher than those of S10 and S30 at all temperatures. The results are due to occurrence of intergranular fracture, which results from quasi-continuous distribution of second

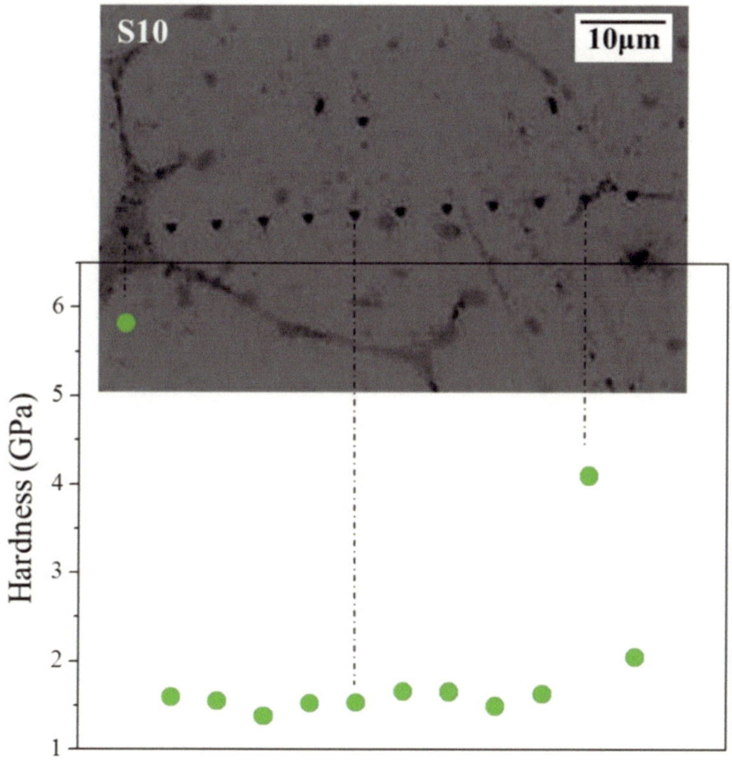

Figure 15. The distribution of hardness evaluated by nanoindentation.

phases with low melting point on globule boundaries. These phases are hard and brittle at room temperature, and they can be softened or partial melted at 500 and 550°C. Therefore, intergranular fracture is the main tensile breaking mechanism of S10 and S30 at room temperature and elevated temperatures. The phases with low melting point on globule boundaries are the weakness when SIMA-processed 6066 Al alloys are subjected to tensile loading at all temperatures. SIMA-processed alloys are not proper for forming with tensile loading.

3.5. Compressive properties and deformation mechanism of SIMA-processed alloys at elevated temperatures

Six materials (F, O, S1, S4, S10 and S30) were compressed at 500, 550 and 600°C to 70% compression ratio with 3.6 mm/min (slow rate, abbreviated as "s") and 36 mm/min (fast rate, abbreviated as "f") in the compression test. No cracks or fractures took place for all materials tested under all conditions as shown in **Figure 18**.

The general tendency of decreasing flow resistance with decreasing compression velocity and increasing temperature as shown in **Figure 19**, which reveals typical flow curves of six materials. The SIMA-processed alloys all had higher flow resistance when they were compressed with slow compression velocity at 500°C. In view of high temperature formability, this result implies poor performance of the SIMA-processed alloy at 500°C.

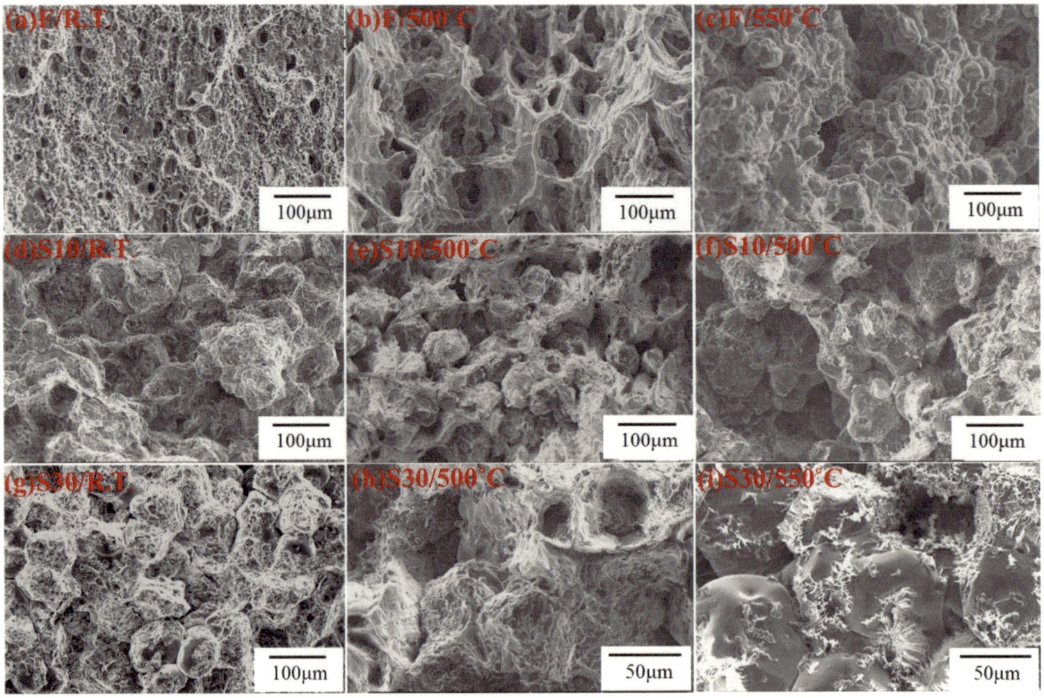

Figure 16. Fracture surfaces of different specimens stretched at various temperatures.

Given in **Figure 19(a)** and **(b)** is the flow curves compressed at 550 and 600°C with 3.6 mm/min. It reveals compression resistance of the SIMA-processed materials are lower than those of F, and they are similar or slightly lower than those of O. Compared to O and F compressed with fast compression rate, **Figure 19(d)** exhibited lower flow resistance of S10 and S30, and **Figure 19(b)** indicates lower flow resistance of S4, S10 and S30. Productivity can be enhanced by a higher compression velocity in practical forming applications. The SIMA-processed alloy compressed at 550 and 600°C with 36 mm/min (fast compression rate) performs low deformation resistance. Therefore, SIMA process is potential to substitute annealing in the pretreatment process of manufacturing at elevated temperatures involving high compression rate.

Figure 20 reveal the metallographic characteristics of the extruded and the SIMA-processed materials of various salt bath durations compressed at 600°C with slow compression velocity. The compressed microstructure of extruded alloy is very uniform, but the compression of the SIMA-processed alloy results in three zones of different microstructure: (1) large deformation zone in center of specimen, in which the spheroidized SIMA grains are severely flattened; (2) free deformation zone near the top and bottom edges of the specimen, in which the spheroidized grains still undeformed; and (3) a transition zone in-between the above two zones. The laminated structure is more distinctive and the globules in free deformation zone become larger when duration of salt bath increased.

Three deformation zones are proposed here and they are formed by the mechanism shown schematically in **Figure 21**. Free deformation zone is generated by globule boundary sliding and flow of liquid, which is formed by low melting point second phases melting. Large

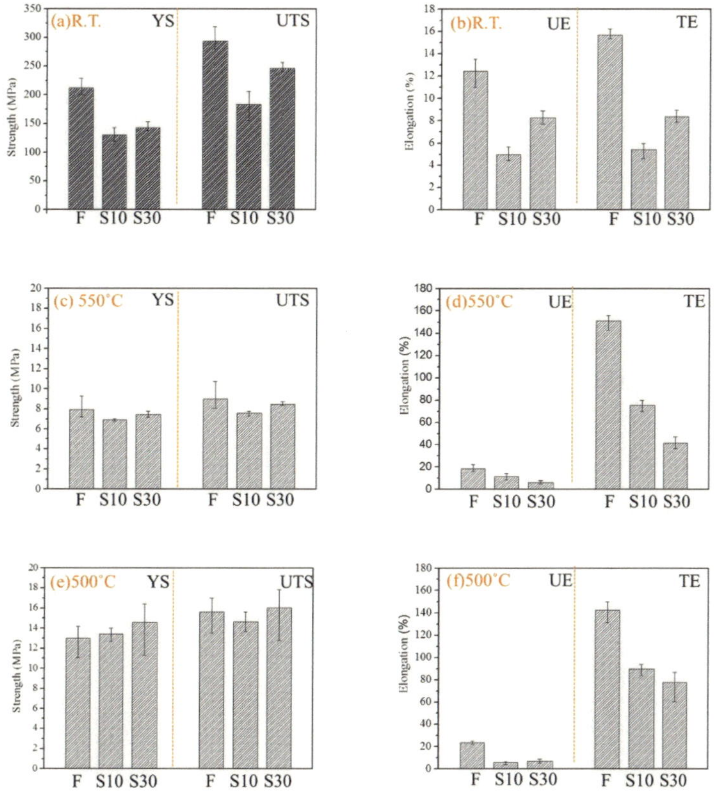

Figure 17. Mechanical properties of F, S10 and S30 at various temperatures (a) strength at room temperature, (b) elongation at room temperature, (c) strength at 550°C, (d) elongation at 550°C, (e) strength at 500°C and (f) elongation at 500°C.

deformation zone is compressed majorly by plastic deformation of the globules, infrequently by flow of liquid and globule boundary sliding. In transition zone, the mechanisms of above two zones co-dominate and compete with each other, hence the microstructure alters with position.

Figure 22 reveals the thickness dissimilarity of the deformation zone of the SIMA alloy for different durations of salt bath. **Figure 22(a)** describes the compression zones and **Figure 22(b)**

Figure 18. Appearances of several representative specimens.

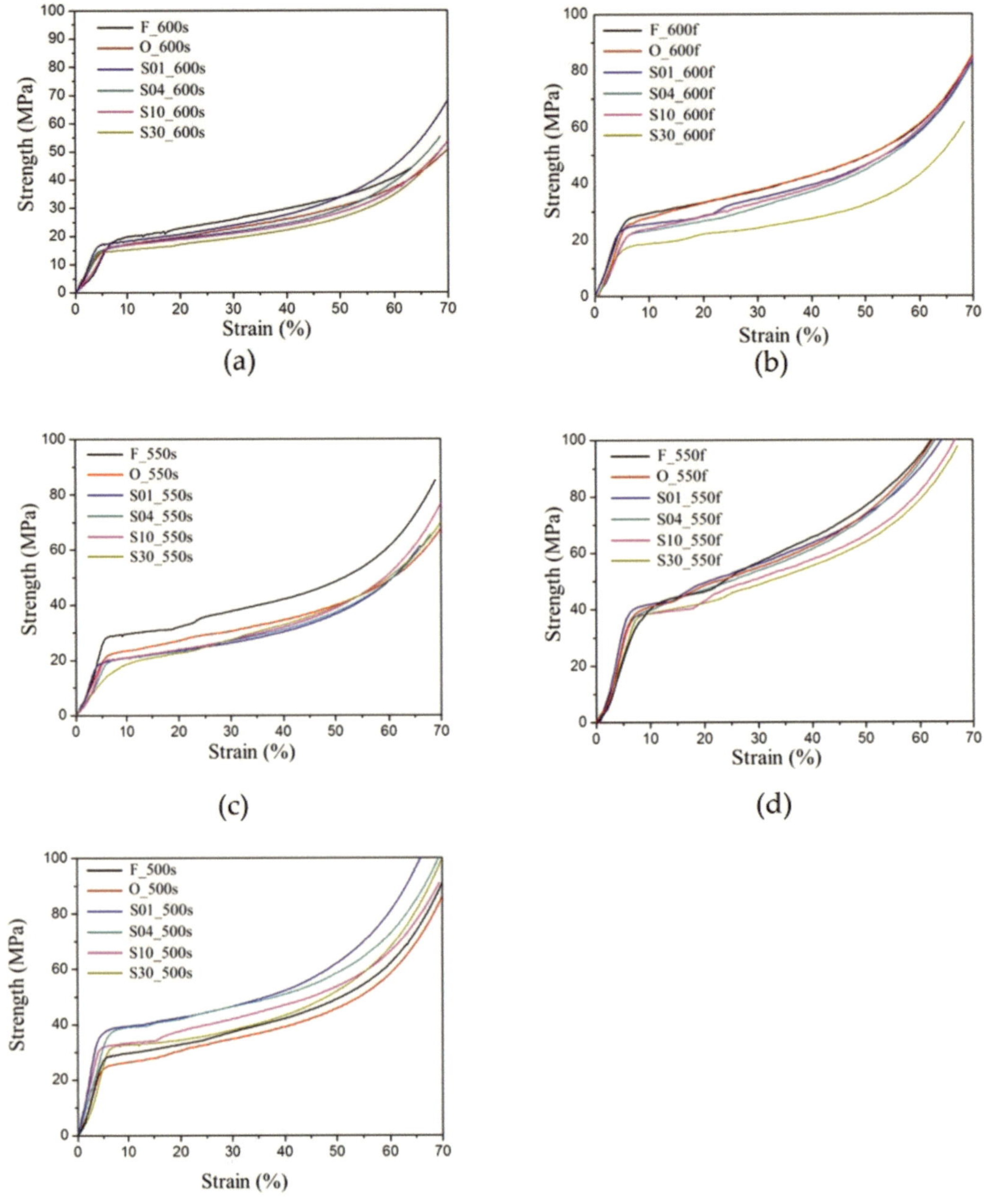

Figure 19. Curves of specimens compressed with different strain rate and at various temperatures: (a) with slow strain rate at 600°C, (b) with fast strain rate at 600°C, (c) with slow strain rate at 550°C, (d) with fast strain rate at 550°C and (e) with slow strain rate at 500°C.

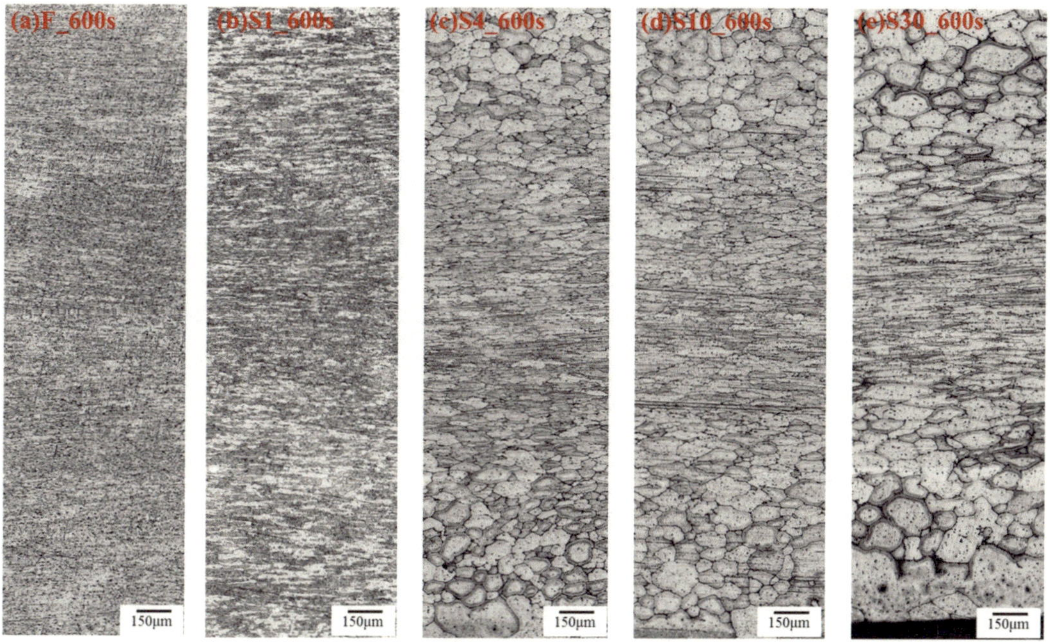

Figure 20. Overall microstructures of various compressing specimens at 600°C.

reveals the thickness of large deformation layer of SIMA alloys. The higher liquid fraction, higher fluidity and more deformed grains at 600°C lead to the thickness was obviously lower than that of 500°C. The thickness of the large deformation layer of the three SIMA-processed materials is close to each other with the slow compression velocity. It is caused by compress stress being decreased by dynamic recrystallization. The thickness of the large deformation layer of S4 is the largest with the fast compression velocity, followed by those of S10 and S30. The reasons are: (1) grain in the large deformation zone (in the center of the samples) limits flowing and sliding of each other, and thus grains deform during compression in this zone. The deformation of small grains (e.g., S4) is more difficult than that of large grains (e.g., S30) [27–29]. (2) The thickness of the large deformation layer decreases with increasing fluidity during compression. The fluidity of the SIMA-processed alloys enhances because of increasing in the fraction of the liquid phases with the salt bath duration increasing.

3.6. High temperature forming behavior of SIMA-processed alloy

3.6.1. Compression test with a given loading

The compression loading at a 50% compression ratio for various materials for the compression at 600°C is shown in **Figure 23**. It shows the compression loading of S10 is the lowest. The new type SIMA process reduced compression loading by about 35% compared with that of the as-extruded alloys, and full annealing reduced that by only about 9%. It indicates that the new

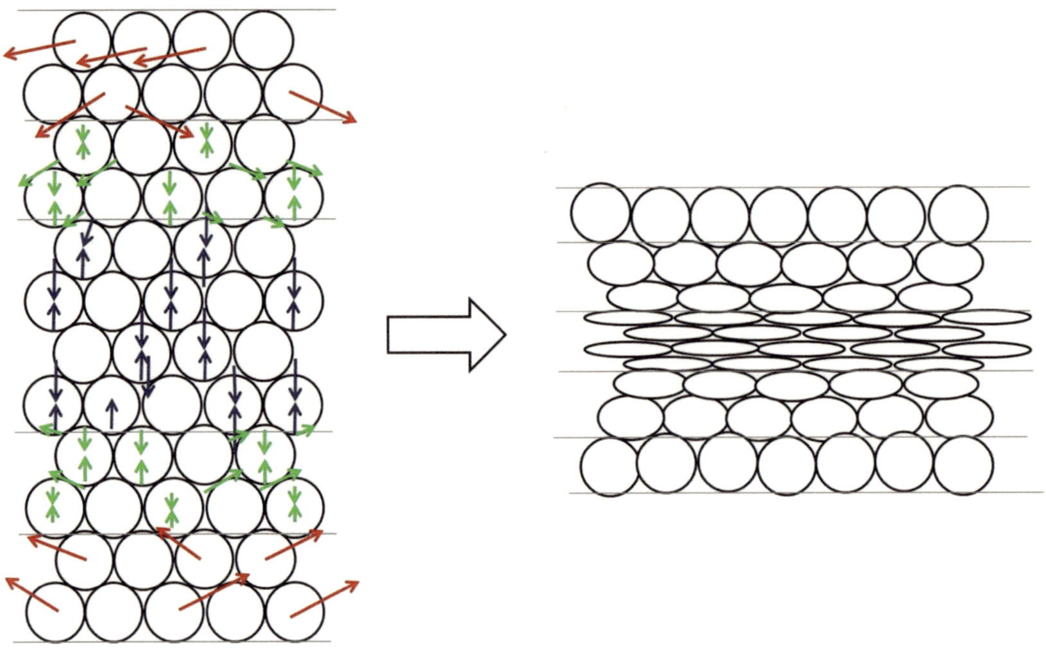

Figure 21. Deformation mechanism of SIMA alloy as compressing.

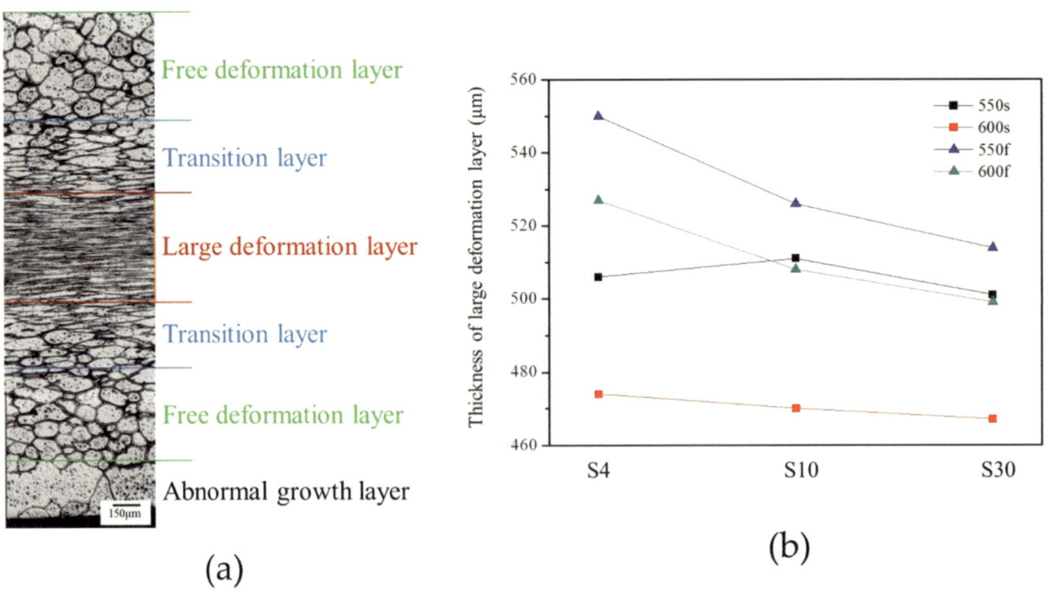

Figure 22. (a) Diagram of compressing microstructure distribution of SIMA alloy and (b) varying trend of large deformation layer with different specimens.

type SIMA process can promote the compressibility at elevated temperatures. The SIMA-processed alloys have the lowest compression resistance.

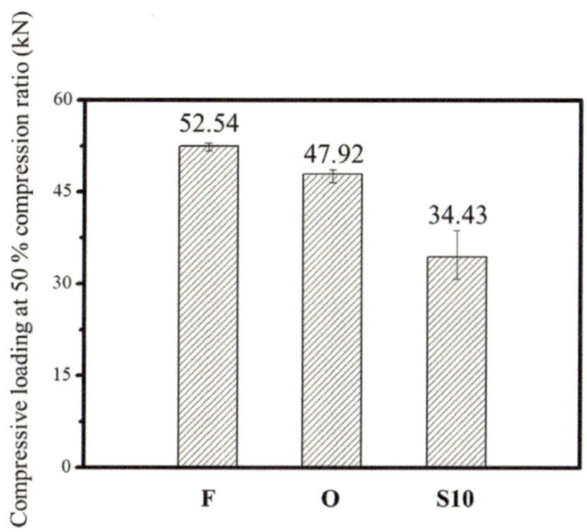

Figure 23. Deformation resistance of several materials.

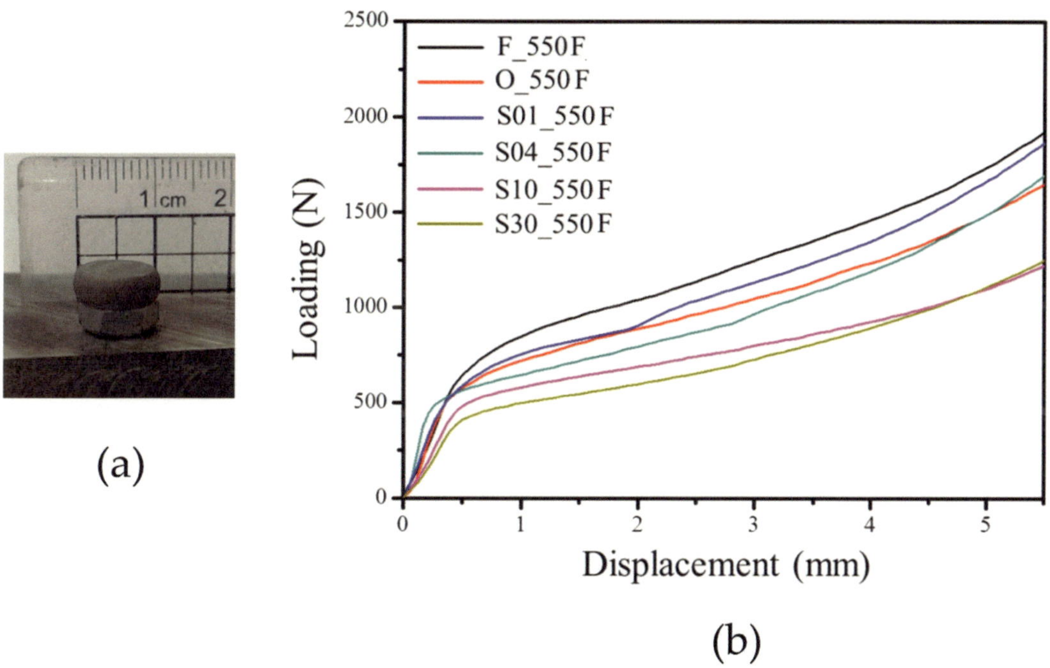

Figure 24. (a) A photograph of real compressed specimen and (b) curves of specimens compressed with slow strain at 550°C.

3.6.2. Forming test at elevated temperature

F, O, S1, S4, S10 and S30 were compressed and shaped into a small screw nut as shown in **Figure 24(a)**. **Figure 24(b)** shows the order of loading (from small to large) is S30, S10, S4, O, S1

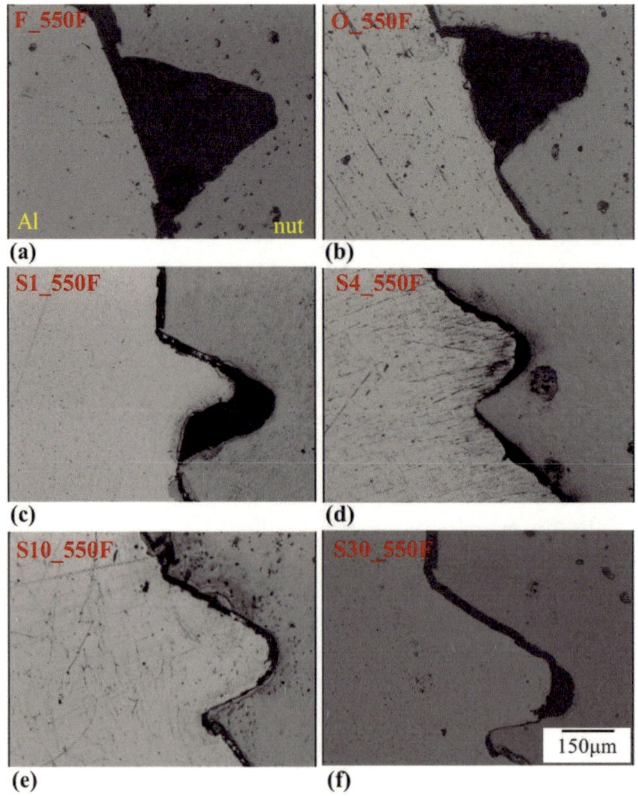

Figure 25. The cross-section microstructures of forming specimens (a) F_550F, (b) O_550F, (c) S1_550F, (d) S4_550F, (e) S10_550F and (f) S30_550F.

and F. The compression loading of S30 is the smallest at any compressive displacement. After salt bath for 4 min, the forming resistance of the SIMA-processed alloy is lower than that of F and O. When salt bath duration is adequate, the forming resistance of the SIMA-processed alloy is relative lower.

Figure 25 shows the microstructure of sectioned-shaped specimens. The position of observation is the first fringe of the screw nut. The degree of filling the fringe (from high to low) is S10, S4, S30, S1, O and F. It indicates S10 has the highest ability of metal flowing. It is due to its suitable globule size and sufficient liquid fraction.

SIMA-processed alloys are suitable for forming with compressive stress. The deformation resistance of the SIMA-processed alloy is low and ability of metal flowing of that is high during compression at elevated temperatures with high compression velocity. The compression formability of SIMA-processed alloys is even higher than that of O. These results prove that the new type SIMA process is applicable for manufacturing with compressive stress. The compression loading can reduce by about 30% and production capability can be enhanced by this SIMA process.

3.7. Summary

1. High temperature tensile data show that even though high temperature tensile yield stress of SIMA alloys is small, the elongation and formability cannot compete with F. It is due to the multi-elements and hard grain boundaries melting or softening at high temperatures and generating cracks. Cracks are easy to proceed along grain boundaries and result in intergranular fracture.

2. As compressing temperature is higher than 550°C, the deformation resistance of SIMA alloys descends obviously. With duration of salt bath increasing, deformation resistance decreases more. This advantage is more manifest when compressing rate increases due to dynamic recrystallization appearing as compressing rate being slower. And the results of high temperature forming test prove that sufficient fraction of liquid phase and suitable size of globular grains promotes the fluidity of material and improve the high temperature formability.

3. At high temperatures, the compressive deforming mechanism of SIMA alloy includes two ways: (1) flowing, flow of liquid incorporating solid grains and grain boundary sliding of outer layer of material, which is named free deformation zone; (2) plastic deformation occurring in the center of material due to grains restrict each other to slide and flow, which is named large deformation zone. Between above two zones, transition zone exists. Two mechanisms compete with each other in this zone and result in the microstructure of this zone has gradually variation.

4. S10 is the best one for high temperature forming. It performs low deformation resistance and good fluidity. It proves that suitable size of spheroidized grains is necessary. Notably, intergranular fracture still should be avoided, that is, tensile stress cannot be introduced too much in application.

4. Mechanical properties improvement of SIMA-processed alloys

4.1. Brief introduction

In this section, the improvement of the mechanical properties of SIMA alloys was investigated. The SIMA-processed alloy with 10-min salt bath duration (S10) was used and compared its mechanical properties with that of the as-extruded alloy (F). And the mechanical properties of the SIMA alloy and the as-extruded alloy both strengthened via artificial aging (T6) heat treatment and compared their tensile mechanical properties. Besides, the tensile fracture mechanism of these alloys will be demonstrated.

4.2. Experimental methods

The specimens of tensile test are the compressed specimens with a given compression rate in Section 3.6. It defined as SIMA forming material and marked then with a prefix "C-."

4.2.1. *T6 heat treatment and its influences*

T6 heat treatment is used for enhancing the mechanical properties. T6 heat treatment includes solution heat treatment and artificial aging. The parameters of solution treatment are set as 530 and 550°C for 2 h, and the parameters of artificial aging is set as 175°C for 8 h (peak aging). Specimens subjected to T6 heat treatment are marked the suffix "/$T6_{530}$" or "/$T6_{550}$." Different solution temperature is used for investigating the effect of amount of solid solution.

4.2.2. *Tensile test and tensile fracture mechanism analysis*

A universal tester is used in tensile test. Six specimens, namely C-F, C-S10, C-F/$T6_{530}$, C-F/$T6_{550}$, C-S10/$T6_{530}$ and C-S10/$T6_{550}$, were selected to conduct tensile test at room temperature with the initial strain set as 1.67×10^{-3} s^{-1}. The dimension of tensile specimen is shown in **Figure 14(a)**. The tensile direction is parallel to extruded direction.

The tensile fracture mechanism of different specimens was analyzed by tensile sub-surface observing with optical microscope (OM) and fracture characteristic investigating with scanning electron microscope (SEM).

4.3. Mechanical properties improvement of SIMA forming alloys

The tensile properties data of compressed and heat-treated materials are shown in **Figure 26**. **Figure 26(a)** shows the strength data. T6 heat treatment promote the strength of specimens, and the strength of compressed SIMA alloys was slightly lower (by about 10–20 MPa) than that of as-extruded alloys. The ultimate tensile strength (UTS) of SIMA-processed alloys can reach about 430–440 MPa. It proves the strength of SIMA forming alloys after T6 heat treatment is high enough for general applications.

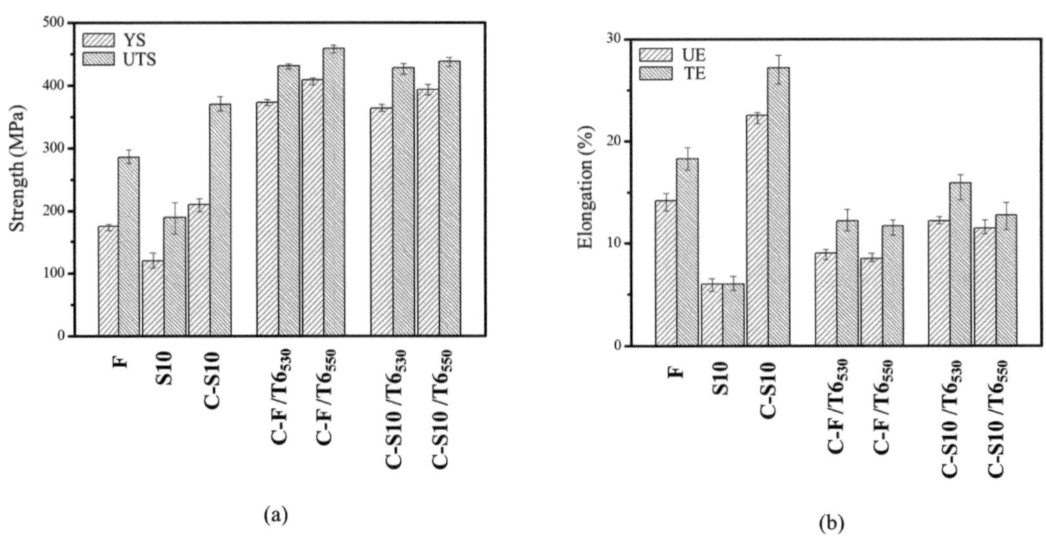

Figure 26. Mechanical properties of specimens: (a) tensile strength and (b) tensile elongation.

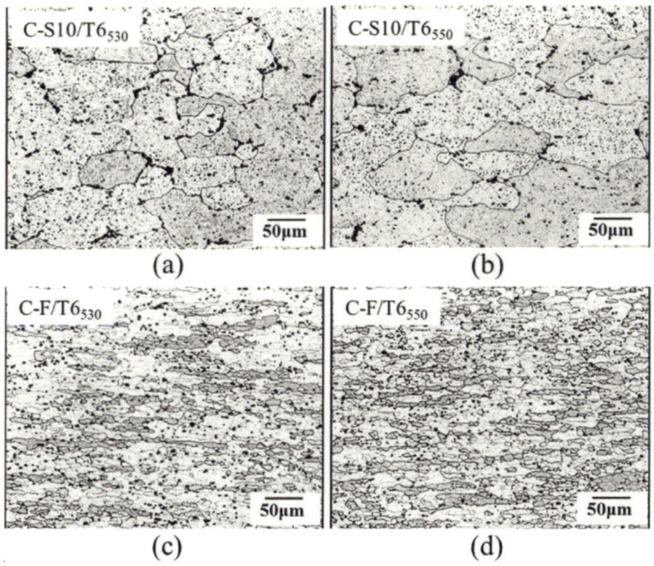

Figure 27. Morphologies of (a) C-S10/$T6_{530}$; (b) C-S10/$T6_{550}$; (c) C-F/$T6_{530}$ and (d) C-F/$T6_{550}$.

The tensile elongation of SIMA-processed alloy is much lower than that of as-extruded alloys as shown in **Figure 26(b)**. Compression at 600°C can enhance the uniform elongation (UE) to about 23% and total elongation (TE) to about 27% because the hard and brittle phases at globule boundaries composed of Al, Mg, Si and Cu diffused into the matrix. Elongation is increased after T6 heat treatment, and it increases with decreasing solution heat treatment temperature. Uniform elongation is about 12% and total elongation is 16% when the solution heat treatment temperature was 530°C. In short, high temperature compression at elevated temperature and T6 heat treatment can improve the mechanical properties of SIMA-processed alloys. An appropriate heat treatment can promote the strength higher than 400 MPa and elongation higher than 10%.

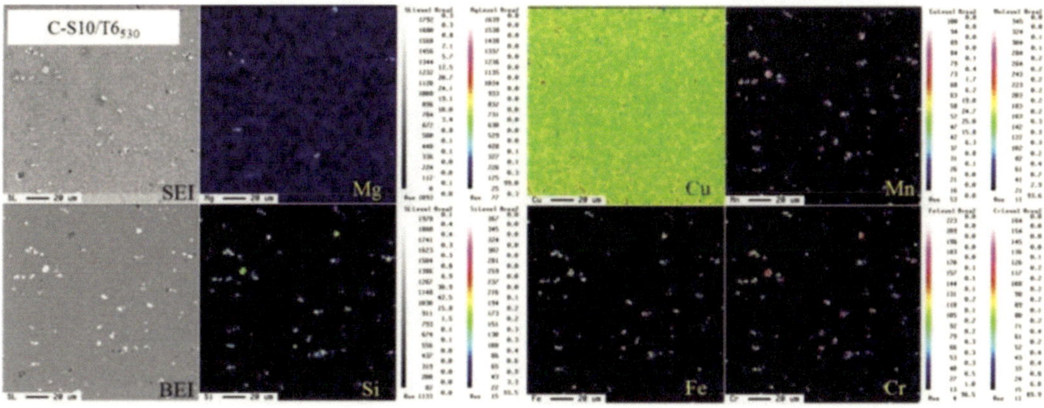

Figure 28. Elemental distribution of compressed TS-SIMA alloy after T6 heat treatment (C-S10/$T6_{530}$) obtained using EPMA.

4.4. Mechanism of tensile fracture of SIMA forming alloys

Figure 27 shows the microstructures of compressed SIMA alloys after T6 heat treatment. It shows that the original compressed globular grains of compressed SIMA alloys grew during solution treatment, as shown in **Figure 27(a)** and **(b)**. The microstructure of T6-heat-treated as-extruded alloys remained as fine recrystallized grains, as show in **Figure 27(c)** and **(d)**. The degree of precipitation strengthening of these two kinds of 6066 Al alloys should be identical

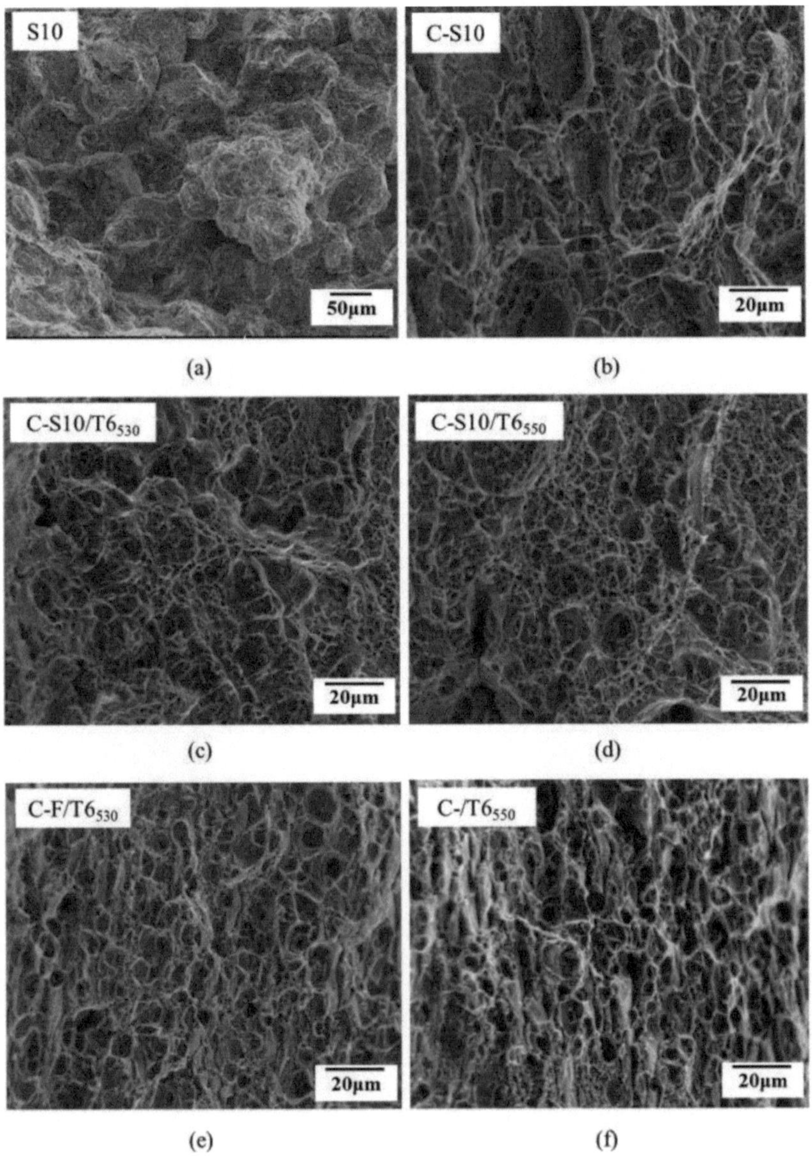

Figure 29. Fracture surfaces of (a) S10; (b) C-S10; (c) C-S10/T6_{530}; (d) C-S10/T6_{550}; (e) C-F/T6_{530} and (f) C-F/T6_{550}.

because their compositions and T6 heat treatment conditions are the same. The slight difference of mechanical properties is due to grain size [27–29]. The strength of as-extruded alloys was higher than that of SIMA forming alloys because of the former's fine grains.

Figure 28 reveals the elemental distribution of compressed SIMA alloys after T6 heat treatment. It reveals all elements distribute uniformly. It indicates T6 heat treatment makes Cu, Mg and Si solid-solute and precipitate completely.

Figures 29 and **30** interpret the fracture mechanism of all materials. The intergranular fracture characteristic of SIMA-processed materials is shown in **Figures 29(a)** and **30(a)**. The phases with low melting point melted, penetrated and solidified at globule boundaries, resulting in the globule boundary being more brittle and harder than the matrix. It leads to stress concentration and the generation of cracks. These cracks initiated at globule boundaries and connected with each other, leading to intergranular fracture. Otherwise, **Figure 29(c)–(f)** show the dimple fractures on the fracture surfaces of compressed SIMA alloys and T6-heat-treated compressed SIMA

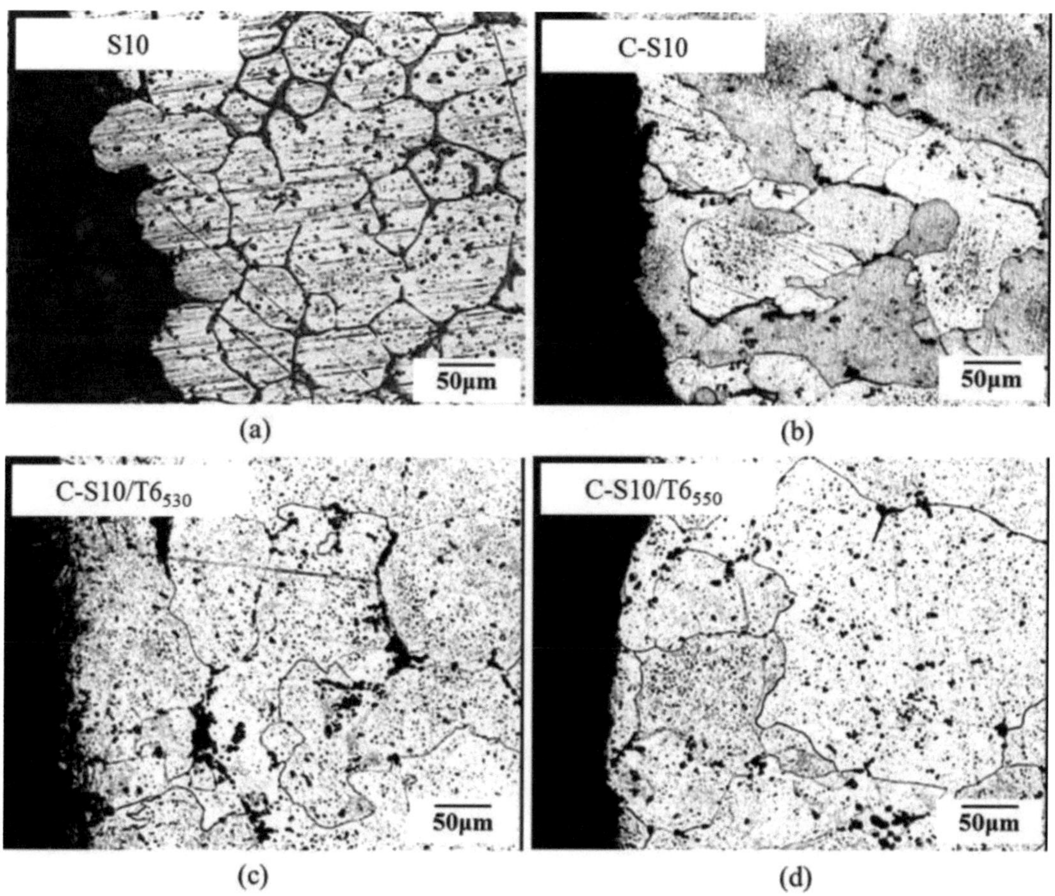

Figure 30. Sub-surfaces of (a) S10; (b) C-S10; (c) C-S10/$T6_{530}$ and (d) C-S10/$T6_{550}$.

alloys. Dimple fracture is a ductile fracture characteristic caused by micro-voids coalescence. **Figure 30(b)-(d)** reveals the sub-surface morphologies of compressed SIMA alloys and T6-heat-treated compressed SIMA alloys. The brittle phases with low melting point disappeared after compression at elevated temperatures, leading to improvement of mechanical properties. After T6 heat treatment, the mechanical properties of SIMA forming alloys are high enough for general applications.

4.5. Summary

1. The mechanical properties of SIMA-processed alloys can be improved obviously (especially elongation) even though that of original SIMA-processed alloys are very low, and T6 heat treatment enhances the mechanical properties of SIMA alloys more to be sufficient for common applications.

2. The improvement of mechanical properties is due to the elimination of brittle and hard phases located on globule boundaries via high temperature compression and solution treatment of T6.

5. Conclusions

1. Three major factors to generate fine, uniform, high-liquid fraction and highly spheroidized globular globules in this new type SIMA process are the following: (a) sufficient elements to form low melting point phases, (b) adding elements for inhibiting grain growth and (c) proper extrusive parameters to get fine and uniform initial extrusive microstructure. For 6xxx Al-Mg-Si alloys, low melting point phases can be composited of Mg, Si, Cu and Al, and globule growth can be inhibited by adding Mn (or V and Zr). Heating method of salt bath results in Ostwald ripening being the major globule growth mechanism. Degree of spheroidization is positive relative to liquid fraction but it is independent on extrusive ratio.

2. At the temperature of solid-liquid coexistence, compressive stress is suitable for SIMA materials forming. Tensile stress will lead to intergranular fracture because brittle and hard phases locate on globule boundaries. Compression of the SIMA-processed alloy results in three zones of different microstructure, namely large deformation zone, free deformation zone and transition zone. SIMA materials perform high compressibility at high compression rate. High spheroidization, high liquid fraction and small globule size enhance the ability of metal flowing and improve the compressibility at elevated temperatures.

3. High temperature compression can improve the elongation of SIMA alloys. The mechanical properties of SIMA alloys can be enhanced by T6 heat treatment. The improvement of mechanical properties is due to the elimination of brittle and hard phases located on globule boundaries via high temperature compression and solution treatment of T6.

Author details

Chia-Wei Lin, Fei-Yi Hung* and Truan-Sheng Lui

*Address all correspondence to: fyhung@mail.ncku.edu.tw

Department of Materials Science and Engineering, National Cheng Kung University, Tainan, Taiwan

References

[1] Wang N, Zhou Z, Lu G. Microstructural evolution of 6061 alloy during isothermal heat treatment. Journal of Materials Science and Technology. 2011;**27**(1):8-14

[2] Fan Z. Semisolid metal processing. International Materials Reviews. 2002;**47**:49-85

[3] Tzimas E, Zavaliangos A. A comparative characterization of near-equiaxed microstructures as produced by spray casting, magnetohydronamic casting and the stress induced, melt activated process. Materials Science and Engineering A. 2000;**289**:217-227

[4] Tzimas E, Zavaliangos A. Evolution of near-equizxed microstructure in the semisolid state. Materials Science and Engineering A. 2000;**289**:228-240

[5] Song YB, Park KT, Hong CP. Recrystallization behavior of 7175 Al alloy during modified strain-induced melt-activation (SIMA) process. Materials Transactions. 2006;**47**:1250-1256

[6] Parshizfard E, Shabestari SG. An investigation on the microstructural evolution and mechanical properties of A380 aluminum alloy during SIMA process. Journal of Alloys and Compounds. 2011;**509**:9654-9658

[7] Paes M, Zoqu EJ. Semi-solid behavior of new Al-Si-Mg alloy for thixoforming. Materials Science and Engineering A. 2005;**406**:63-73

[8] Young KP, Kyonka CP, Courtois JA. United States Patent 4415374. Nov.15 1983

[9] Glickman EE, Nathan M. On the kinetic mechanism of grain boundary wetting in metals. Journal of Applied Physics. 1999;**85**:3185-3191

[10] Zhangm L, Liu YB, Cao ZY, Zhang YF, Zhang QQ. Effects of isothermal process parameters on the microstructure of semisolid AZ91D alloy produced by SIMA. Journal of Materials Processing Technology. 2009;**209**:792-797

[11] Qin QD, Zhao YG, Xiu K, Zhou W, Liang YH. Microstructure evolution of in situ Mg2Si/Al-Si-Cu composite in semisolid remelting process. Materials Science and Engineering A. 2005;**407**:196-200

[12] Ji Z, Hu M, Suguyama S, Yanagimoto J. Formatioin process of AZ31B semi-solid microstructures through strain-induced melt activation method. Materials Characterization. 2008;**59**:905-911

[13] Wang JG, Lu P, Wang HY, Liu JF, Jiang QC. Semisolid microstructure evolution of the predeformed AZ91D alloy during heat treatment. Journal of Alloys and Compounds. 2005;**395**:108-112

[14] Wang Z, Ji Z, Hu M, Xu H. Evolution of the semi-solid microstructure of ADC12 alloy in a modified SIMA process. Materials Characterization. 2011;**62**:925-930

[15] Yan G, Zhao S, Ma S, Shou H. Microstructural evolution of A356.2 alloy prepared by the SIMA process. Materials Characterization. 2012;**69**:45-51

[16] Bolouri A, Shahmiri M, Kang CG. Coarsening of equiaxed microstructure in the semi-solid state of aluminum 7075 alloy through SIMA processing. Journal of Materials Science. 2012;**47**:3544-3553

[17] Hardy SC, Voorhees PW. Ostwald ripening in a system with a high volume fraction of coarsening phase. Metallurgical and Materials Transactions A. 1988;**19**:2713-2721

[18] Atkinson HV, Liu D. Coarsening rate of microstructure in semi-solid aluminum alloys. Transaction of Nonferrous Metals Society of China. 2010;**20**:1672-1676

[19] ASM International Handbook Committee. Properties and Selection: Nonferrous Alloys and Special-Purpose Materials. 1990. p. 51

[20] Hatch JE. Aluminum: Properties and Physical Metallurgy. Ohio: American Society for Metals; 1984. pp. 224-240

[21] Totten GE, Mackenzie DS. Handbook of Aluminum-Physical Metallurgy and Process. New York: Metal Dekker Inc.; 2003. pp. 168-185

[22] Zhen L, Fei WD, Kang SB, Kim HW. Precipitation behavior of Al-Mg-Si alloys with high silicon content. Journal of Materials Science. 1997;**32**:1895

[23] Jeniski Jr RA. Effects of Cr addition on the microstructure and mechanical behavior of 6061-T6 continuously cast and rolled redraw rod. Materials Science and Engineering A. 1997;**237**:52-64

[24] Claves SR, Elias DL, Misiolek WZ. Analysis of the intermetallic phase transformation occurring during homogenization of 6xxx aluminum alloys. Materials Science Forum. 2002;**396**:667-674

[25] Cabibbo M, Evangelista E, Scalabroni C, Bonetti E. A transmission electron microscopy study of the role of Sc+Zr addition to a 6082-T8 alloy subjected to equal channel angular pressing. Materials Science Forum. 2006;**503-504**:841-846

[26] Huang HJ, Cai YH, Cui H, Huang JF, He JP, Zhabg JS. Influence of Mn addition on microstructure and phase formation of spray-deposited Al-25Si-xFe-yMn alloy. Materials Science and Engineering A. 2009;**502**:118-125

[27] Petch NJ. The cleavage strength of polycrystals. Journal of Iron Steel Institute. 1953;**174**: 8-25

[28] Armstrong R, Codd I, Douthwaite RM, Petch NJ. The plastic deformation of polycrystalline aggregates. Philosophical Magazine. 1962;7:45-58

[29] Tilly GP. In: Scott D, editor. Treatise on Materials Science and Technology. 13th ed. New York: Academic Press; 1979. pp. 287-319